Transgenerational Toxicology in Caenorhabditis elegans

Dayong Wang

Transgenerational Toxicology in Caenorhabditis elegans

Springer

Dayong Wang
Southeast University
Medical School
Nanjing, Jiangsu, China

ISBN 978-981-92-1388-7   ISBN 978-981-92-1389-4 (eBook)
https://doi.org/10.1007/978-981-92-1389-4

This Springer imprint is published by the registered company Springer Nature Singapore Pte Ltd.
The registered company address is: 152 Beach Road, #21-01/04 Gateway East, Singapore 189721, Singapore

Preface

How is the enduring vitality of organisms across multiple generations impaired by stresses and pollutants?

So far, most of the performed toxicological studies have focused on what happened in the parental generation (P0-G) after exposure to stresses and pollutants. However, at the most cases, the toxicity of stresses and pollutants may not be limited to the P0-G, and will last for multiple generations. The long-term and continuous exposure to stresses and pollutants in the environment further exacerbates the possible induction of transgenerational toxicity.

In nematode *Caenorhabditis elegans*, the knowledge systems of exposure toxicology, molecular toxicology, target organ toxicology, and toxicology at environmentally relevant concentrations have been established. Moreover, *C. elegans* has very short life cycle, high sensitivity to environmental exposures, high reproductive capacity and large population of progeny, and well-described background of molecular toxicology. These backgrounds make it possible to provide a powerful platform in *C. elegans* for systematically performing the study on transgenerational toxicology. The exposure scenarios discussed in this book is that the parental hermaphrodite nematodes are exposed to stresses and pollutants, while their offspring (F1-Fn) will be cultured under the normal condition.

In this book, we have raised three important concerns:

1. What will happen during the transgenerational process after exposure to stresses and pollutants?
2. What are the key regulation mechanisms for induction of transgenerational toxicity of stresses and pollutants?
3. What are the "drivers" and the "effectors" for the induction of transgenerational toxicity after exposure to stresses and pollutants?

Based on these concerns, in Chap. 1, we first introduced the value of *C. elegans* for the study of transgenerational toxicology. In Chaps. 2, 3, and 4, we introduced and discussed related information relevant to the first raised concern. In Chaps. 5, 6,

7, 8, 9, 10, 11, 12, 13, and 14, we introduced and discussed related information surrounding the second raised concern. In Chaps. 15 and 16, we introduced and discussed related information for the third raised concern. In Chap. 17, we raised several basic conclusions for transgenerational toxicology based on obtained data in *C. elegans*.

Nanjing, Jiangsu, China Dayong Wang

Contents

Chapter 1
Research Values of *Caenorhabditis elegans* for the Study of Transgenerational Toxicology

Abstract *Caenorhabditis elegans* is a classic model animal. Because of high sensitivity to environmental exposures and short life cycle, *C. elegans* has been widely used in assessment of transgenerational toxicity and study of transgenerational toxicology after exposure to pollutants at parental generation. In this chapter, we discussed the research values of *C. elegans* at various aspects for the study of transgenerational toxicology after exposure to stresses and toxicants.

Keywords Transgenerational toxicology · Research value · Model animal · *Caenorhabditis elegans*

1.1 Introduction

Largely due to systematic work in *Caenorhabditis elegans*, the research paradigm of model organisms has been raised in life science. This means that we can perform related researches using certain model organisms (such as model animals) with defined and clearly described genetic and molecular backgrounds. Besides acting as a classic model animal for the study of life science, *C. elegans* shows high sensitivity to different pollutants including both soluble pollutants and particulate pollutants [1–8]. Additionally, *C. elegans* exhibits susceptible property to stresses, such as simulated microgravity [9–12]. Using *C. elegans* as an experimental model, molecular toxicology, target organ toxicology, exposure toxicology, and toxicology at environmentally relevant concentrations have been systematically performed after exposure to stresses and toxicants [13–16].

At the most cases, in nematodes, the toxicity of stresses and toxicants was assessed at parental generation (P0-G). However, usually the toxicity of stresses and toxicants is not only limited at P0-G and can be transferred to the next generation or even across multiple generations [17–20]. To assess toxicity induction for multiple generations, there are mainly two forms of exposure [21]. One is the exposure to stresses and toxicants only at P0-G, which is defined as the transgenerational toxicity from P0-G to their offspring. Another is the exposure to stresses and toxicants at

D. Wang, *Transgenerational Toxicology in Caenorhabditis elegans*,
https://doi.org/10.1007/978-981-92-1389-4_1

continuous generations, which can be defined as the toxicity of multigenerational exposure. The transgenerational toxicology focuses on examination of underlying mechanisms of toxicity of stresses and toxicants on germline of *C. elegans* at P0-G and the consequences of this germline damage on the offspring.

To successfully carry out the study of transgenerational toxicology, the selected animal models should satisfy these requirements: (1) relatively short life cycle, (2) high sensitivity to environmental exposures, (3) high reproductive capacity and large population of progeny, and (4) well-described background of molecular toxicology. Based on these needs, *C. elegans* has been selected as the important animal model for the study of transgenerational toxicology. In this chapter, we introduced and discussed the following several aspects of usefulness and advantages of *C. elegans* for assessing transgenerational toxicity of stresses and toxicants and elucidating underlying mechanisms.

1.2 Very Short Duration to Assess Transgenerational Toxicity of Toxicants or Stresses

In life science, *C. elegans* provides a powerful platform for the study of genetics and related studies. This is based on at least two prerequisites. One is that *C. elegans* has a very short life cycle, and one generation usually lasts 4–5 days. Another is that *C. elegans* can produce a large number of progeny population. These properties meanwhile allow *C. elegans* as an important animal model to carry out assessment of transgenerational toxicity of stresses and toxicants. This suggests that, during 1 month duration, we can possibly finish the evaluation of toxicity of stresses and toxicants across six generations. This can also allow us to design and to perform a large scale of screen and comparison of pollutants with transgenerational toxicity with the aid of some endpoints, which can be easily manipulated. Nevertheless, during a relatively short duration, a huge amount of work has to be carefully designed, arranged, and finished.

1.3 Evaluation of Long-Term Effect of Toxicants or Stresses After Exposure at Parental Generation

At P0-G, after exposure to certain toxicants and stresses, the long-term effects can be assessed from two aspects in *C. elegans*. At P0-G, the long-term effects of stresses and toxicants can be evaluated by assessment of lifespan and physiological changes during the aging, which can also be called healthspan [22–23]. After the exposure to stresses and toxicants at P0-G, the longer effects of stresses and toxicants can be further determined by assessment of transgenerational changes of endpoints reflecting different aspects of response, development, and function of different organs, as well as the physiological and metabolic changes [24–27]. This means that after exposure to stresses and toxicants at P0-G, we can ask how many generations their adverse effects can last. Nevertheless, this is an assumed simplified exposure pattern. In the real environments, the exposure frequency and generations may be very complex, which needs to be carefully considered, examined, and judged according to the real exposure conditions for certain stresses and toxicants.

1.4 Assessment of Transgenerational Toxicity of Toxicants at Environmentally Relevant Concentrations or Stresses at Low Doses

In the recent several years, in the fields of both toxicology and environmental science, increasing attention has been received, and increased efforts have been performed for assessing toxicity of pollutants at environmentally relevant concentrations (ERCs). ERCs for toxicants or environmentally relevant doses (ERDs) for stresses refer to the measured or predicted concentrations or doses in actual contaminated sites or regions. *C. elegans* has been widely used for evaluating toxicity of pollutants (such as nanomaterials including the nanoplastics) at ERCs at the P0-G [16, 28–30]. Moreover, in *C. elegans*, toxicity of pollutants at ERCs in inducing transgenerational toxicity can be detected [31–34]. The ERCs reflect the external exposure concentrations of pollutants in the real environments. In contrast to the external exposure concentrations of pollutants, many factors affect the internal concentrations of pollutants, including species, developmental structure, property of biological barriers, defecation, and intestinal flora. In *C. elegans*, it is a challenge to mimic the internal exposure of pollutants to environmental organisms and mammals.

1.5 Assessment of Transgenerational Toxicity of Toxicants or Stresses After Exposure at Different Developmental Stages at Parental Generation

C. elegans contains embryo stage, four larval stages (L1, L2, L3, and L4), young adult stage, and adult stage. This developmental property allows us to design the exposure to stresses and toxicants at different developmental stages at P0-G. Usually, the larvae have more sensitivity than young adults or adults to pollutants [35–36]. Therefore, exposure of eggs or larvae (such as L1-larvae) to pollutants can normally induce severe transgenerational toxicity [37–40]. After the post-treatment, this also provides the possibility to examine the pharmacological effects of certain drugs or compounds against transgenerational toxicity of stresses and toxicants. Additionally, the adult stage can be further divided into detailed developmental stages. Approximately after adult day 8, the lifespan of nematodes would be sharply reduced, and healthspan will be severely inhibited. This suggests that, if the beginning of pollutant exposure is L1-larvae or egg stage, when exposure ends needs to be carefully designed.

1.6 Evaluation of Physiological, Molecular, and Developmental Changes During Transgenerational Process After Exposure to Toxicants or Stresses

In *C. elegans*, the physiological changes reflect the corresponding functions of different organs and tissues. For example, alterations in locomotion behaviors act as markers of functional state of motor neurons, and changes of sensory perception to environmental stimuli act as markers of functional state of sensory neurons. The change of reproductive capacity reflects the function of gonads in reproductive system. Endpoints indicating locomotion behavior and reproductive capacity have been more frequently used to detect transgenerational toxicity of pollutants, such as nanoplastics [41–42]. Besides these, some other physiological indicators, such as defection, were also applied to assess transgenerational toxicity of pollutants, such as traffic-related fine particulate matter ($PM_{2.5}$) [29]. Accompanied with the effort to elucidate underlying molecular mechanism of transgenerational toxicity of toxicants and stresses, transgenerational changes of certain molecular signals (especially those in the germline) have also been widely assessed. For example, nanoplastics at ERCs could induce alterations in expressions of germline genes encoding nuclear hormone receptors (NHRs) and G protein-coupled receptor (GPCR) CED-1 across multiple generations [43–44]. Exposure to pollutants could also cause transgenerational developmental changes; however, the degrees for the induced transgenerational developmental changes are usually limited. For example, polystyrene nanoparticle (10 μg/L) could only cause developmental deficits in D-type motor neurons at P0-G and F1-G, which was accompanied with decreased locomotion behavior from P0-G to F3-G [45].

1.7 Assessment of Transgenerational Toxicity of Toxicants or Stresses Under Certain Genetic Backgrounds

In *C. elegans*, certain genetic backgrounds can be generated by mutation or RNAi knockdown of genes. Three aspects of nematodes with certain genetic backgrounds are suggested to be used for assessing transgenerational toxicity of pollutants. Firstly, mutation or RNAi of some genes can provide the susceptible properties to toxicity of toxicants and stresses. For example, mutation or RNAi knockdown of genes encoding transcriptional factors in insulin, p38 MAPK, and Wnt signaling pathways caused susceptibility to toxicity of pollutants, including nanoplastics and graphene oxide [46–51]. Using nematodes with mutation or RNAi knockdown of these genes, it is assumed that more severe transgenerational toxicity of toxicants and stresses would be detected. For *C. elegans*, the primary target organs contain both intestine and epidermis. Secondly, deficits in biological barriers (such as intestinal barrier and epidermal barrier) can be generated by mutation or RNAi knockdown of certain genes, which could also show susceptibility to toxicity of pollutants due to their enhanced accumulation and translocation in the body of nematodes [52–55]. For example, in *acs-22* mutant nematodes, more severe nanoplastic accumulation and translocation in intestinal cells and gonad were observed together with formation of more severe nanoplastic toxicity on animals [56]. Using nematodes with deficits in intestinal barrier, it is assumed that more severe pollutants would be translocated and accumulated in gonad to cause the damage on germline and induce the more severe transgenerational toxicity. Thirdly, nematodes with mutation or RNAi of genes controlling certain aspects of physiological behaviors (such as defecation behavior and pharyngeal pumping) or developments (such as gonad or oocyte developments) are also suggested to be applied for evaluating transgenerational toxicity of toxicants and stresses.

1.8 Assessment of Transgenerational Toxicity of Toxicants or Stresses Under Certain Physiological Conditions

In *C. elegans*, some physiological conditions can be generated by treating the animals with certain stresses. After treatment with heat stress, UV stress, osmotic stress, hypoxic stress, alcohol stress, proteotoxic stress, or high glucose stress, normally, the susceptible property would be formed for the following exposure to pollutants. In *C. elegans*, after high glucose treatment, the following exposure to nanoplastics caused more severe toxicity compared to exposure to nanoplastics alone [57]. Therefore, under these physiological conditions, the more severe transgenerational toxicity of pollutants is supposed to be induced in nematodes. This may be helpful for detecting the possibility of acute or short-term exposure to pollutants in causing transgenerational toxicity in *C. elegans*. To a certain degree, these generated physiological conditions can also be considered as certain forms of pathological conditions in nematodes. Nevertheless, before the following exposure to pollutants, the doses of stress to generate certain physiological conditions should be carefully considered or modulated.

1.9 Assessment of Transgenerational Toxicity of Toxicants or Stresses Under Certain Pathological Conditions

In *C. elegans*, some pathological models have been generated for the study of pharmacology [58]. Meanwhile, these pathological models can also be helpful for the toxicological study [13]. Some pathological models for neurodegeneration diseases (such as Alzheimer's disease (AD), Parkinson's disease (PD), Huntington's disease (HD), amyotrophic lateral sclerosis (ALS), and polyglutamine and glutamate-associated neurodegenerative disease) and muscle dysfunction–related diseases (such as channelopathy model and muscle dystrophy model) are available in nematodes. These pathological models can at least provide two aspects of support for assessment of transgenerational toxicity of pollutants and elucidation of underlying mechanisms. One support is that some of these pathological models (such as PD and AD models) can provide the relatively susceptible backgrounds for the assessment of transgenerational toxicity of pollutants. Another support is that these pathological models can provide certain aspects of altered developmental structures, which may be useful for evaluating possible transgenerational toxicity on development after pollutant exposure at P0-G. Besides these pathological models, pathogen infection model and immunosuppression model can also be generated in *C. elegans* for the study of immunology and pharmacology [59–63]. These bacterial and fungal pathogen–induced infection models and immunosuppression models can be considered as a form of stress induction. Therefore, these pathogen-induced infection models and immunosuppression models can provide the susceptible property to toxicity of pollutants at P0-G [64] and possibly across multiple generations.

1.10 Assessment of Transgenerational Toxicity of Toxicants with Different Physical Properties

The chemical pollutants have different properties, including size, shape, surface charge, etc. During the study of nanotoxicology, it has been recognized that the smaller nanoparticles usually have more severe toxicity than larger nanoparticles [1], which is at least partially associated with the more severe accumulation of smaller nanoparticles across biological barriers. With polystyrene nanoparticle as example of nanoplastics, it was reported that 20 nm nanoplastics caused more severe transgenerational inhibition in locomotion behavior and reproductive capacity compared with 100 nm nanoplastics [24]. Additionally, 20 nm nanoplastics caused transgenerational suppression in Mn-SOD expression, mitochondrial unfolded protein response (mt UPR), and endoplasmic reticulum UPR (ER UPR), whereas 100 nm nanoplastics induced transgenerational activation of these protective responses [24]. Nanomaterials have different shapes, including particle, fiber, needle, and fragment [65–68], and these different shapes may affect the toxicity induction of nanomaterials. On the surface of nanomaterials, different charges can be formed, including positive charge and negative charge, and the toxicity induction of nanomaterials could also be influenced by surface charges [1].

Besides the physical properties discussed above, some other physical properties also affect toxicity induction of pollutants. The impurity and additives in nanoplastics or microplastics affected their toxicity induction [1–2]. Additionally, particulate pollutants may have adsorption property, when they exist together with soluble pollutants in the environment. Nanoplastics could enhance the toxicity of other pollutants [69–71], and one of the important reasons was the adsorption of other pollutants [69–70]. During assessment of transgenerational toxicity of pollutants, these physical properties need to be carefully considered.

1.11 Assessment of Transgenerational Toxicity of Toxicants with Different Chemical Properties

On the surface of chemical pollutants, there are different chemical groups, and the toxicity of pollutants is associated with the existence of these groups to a great degree. In the environment, pollutants will undergo different chemical transformations after certain chemical reactions, and the toxicity usually can be changed accordingly. After transformation of plastic additive dimethyl terephthalate to mono-methyl terephthalate, the toxicity was reduced [72].

Meanwhile, certain forms of artificial chemical modifications can also affect toxicity of pollutants. Some of these chemical modifications (such as surface PEG and carboxylic acid modification) could reduce toxicity of carbon-based nanomaterials in nematodes [28, 73]. In contrast to these, some other forms of chemical modifications (such as amino and sulfonate modifications) enhanced toxicity of nanoplastics at P0-G in nematodes [74, 75]. Moreover, in *C. elegans*, it was observed that both transgenerational reproductive toxicity on reproductive capacity and transgenerational neurotoxicity on locomotion and neuronal development could be enhanced by amino modification on nanoplastics [25, 45].

1.12 Different Forms of Exposure for Evaluating Toxicity of Toxicants or Stresses Across Multiple Generations

For toxicity assessment of toxicants and stresses across multiple generations, several exposure methods can be selected. So far, at the most cases, the transgenerational toxicity of pollutants was assessed after exposure to pollutants at P0-G, and no further pollutant exposure was performed from the F1-G. This exposure method can help us clearly determine how parental germline signals governed the induction of transgenerational toxicity of toxicants and stresses. The second exposure method is to perform the exposure to toxicants and stresses at different generations [76], which can be further divided into continuous exposure and intermittent exposure across multiple generations. While performing this form of exposure to assessing toxicity of toxicants and stresses, the influence of adaptive response should be carefully considered and excluded. Normally, the exposure to toxicants and stresses was performed in hermaphrodites, and the offspring was generated by self-fertilization. In *C. elegans*, the progeny can also be generated by outcrossing [77]. The third method to assessing the toxicity of toxicants and stresses across multiple generations is to make the exposure of the male nematodes to toxicants and stresses. After that, the offspring can be obtained by male mating. This form of exposure method can be used to evaluate the role of damage on male nematodes in causing transgenerational toxicity of toxicants and stresses. Also, this form of exposure method may reflect the induction of transgenerational toxicants and stresses in mammals to a certain degree.

1.13 Determination of Molecular Mechanisms for Observed Transgenerational Toxicity of Toxicants or Stresses

It has been well known that *C. elegans* is a powerful model for the study of molecular toxicology [13]. Meanwhile, *C. elegans* is also important for the study of target organs toxicology [14]. Nevertheless, for induction of transgenerational toxicity of toxicants and stresses, we assumed that the germline plays the pivotal role. Thus, the elucidation of molecular mechanisms of transgenerational toxicity of toxicants and stresses has been focused on the role of germline molecular signals. In the germline, some aspects of molecular signals have been identified to govern induction of transgenerational toxicity of pollutants. Firstly, transgenerational signaling communication is required for formation of transgenerational toxicity of pollutants. That is, in the germline, some secreted ligands at P0-G and their receptor(s) in the offspring acted together to regulate the induction of transgenerational pollutant toxicity. So far, at least germline insulin, Wnt, Ephrin, hedgehog, FGF, Notch, and neurotransmitter ligands at P0-G and their receptor in the offspring were identified to control transgenerational toxicity of nanoplastics or multi-walled carbon nanotubes in nematodes [32, 34, 78–82]. Secondly, epigenetic regulation related molecular signals regulated transgenerational toxicity of pollutants by existing their transgenerational memory and effect. During the control of transgenerational toxicity of nanoplastics or multi-walled carbon nanotubes, some microRNAs (such as *mir-38* and *mir-240*), histone methylation–related signals, and long non-coding RNAs acted in the germline to play the important role [26, 32, 33, 41]. Besides these molecular signals, some GPCR (such as CED-1) and NHRs (such as NHR-14) in the germline were also identified to regulate the transgenerational toxicity of nanoplastics by affecting downstream certain targets, such as secreted ligands [43, 44].

1.14 Determination of Metabolic Mechanisms Involved in Regulating Transgenerational Toxicity of Toxicants or Stresses

In *C. elegans*, toxicity of toxicants and stresses is not only under the control of certain molecular signals but also associated with alterations in certain biochemical metabolisms. With 6-PPD quinone (6-PPDQ) as example of emergent pollutants, exposure to 6-PPDQ could cause the disruption in dopamine, glucose, glycogen, and lipid metabolisms together with changes of expressions of related enzyme genes governing these biochemical metabolisms [6, 83–85]. More importantly, alterations of these biochemical metabolisms were associated with modulation of toxicity induction of 6-PPDQ [6, 83–85]. Moreover, alterations in glucose and glycogen metabolisms might directly mediate the 6-PPDQ toxicity induction [83, 84]. Additionally, accompanied with induction of pollutant toxicity across multiple generations, certain biochemical metabolisms (such as lipid metabolism) were also observed to be changed [86–87]. Therefore, metabolic signals can act together with molecular signals to modulate transgenerational toxicity of toxicants and stresses.

1.15 Perspectives

In this chapter, mainly based on the obtained data, we introduced and discussed research values of *C. elegans* for the study of transgenerational toxicology (Fig. 1.1). Due to short life cycle, we first discussed the convenience at this aspect for using *C. elegans* to assess transgenerational toxicity of toxicants and stresses. We further introduced the advantages of applying *C. elegans* to evaluate long-term effect of toxicants or stresses, to assess transgenerational toxicity of toxicants at ERCs or stresses at ERDs, to perform the exposure at different developmental stages, and to examine the physiological, molecular, and developmental changes during transgenerational process. After that, we discussed the usefulness to assess the transgenerational toxicity of toxicants or stresses under certain genetic, physiological, and pathological backgrounds, and to determine transgenerational toxicity of toxicants with different physical and chemical properties in *C. elegans*. Different forms of exposure for evaluating toxicity of toxicants or stresses across multiple generations in *C. elegans* were also introduced. Moreover, we discussed the determination of molecular and metabolic mechanisms for observed transgenerational toxicity of toxicants or stresses in *C. elegans*. Nevertheless, at present, the accumulated research data in this field is still somewhat limited. Definitely, as the related researches continue, some more research values of *C. elegans* for the study of transgenerational toxicology will be further raised in the future. For example, with the

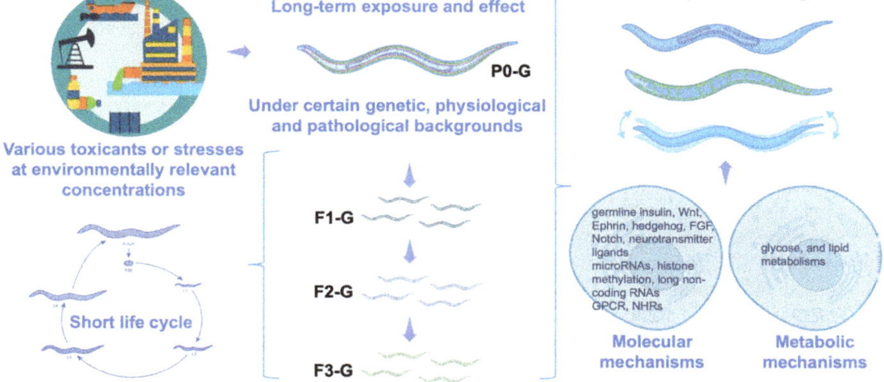

Fig. 1.1 Research values of *C. elegans* for the study of transgenerational toxicology

concern on human health, we suggest to perform large scale of transformation of wild-type and/or mutant human homologous genes into *C. elegans*. This may help us indirectly understand the transgenerational toxicity of toxicants and stresses on human health.

References

1. Wang D-Y (2018) Nanotoxicology in *Caenorhabditis elegans*. Springer Nature Singapore Pte Ltd, Singapore
2. Yu Y, Hua X, Chen H, Yang Y, Dang Y, Xiang M (2022) Tetrachlorobisphenol A mediates reproductive toxicity in *Caenorhabditis elegans* via DNA damage-induced apoptosis. Chemosphere 300:134588
3. Shi C, Wang C, Zeng L, Peng Y, Li Y, Hao H, Zheng Y, Chen C, Chen H, Zhang J, Xiang M, Huang Y, Li H (2023) Triphenyl phosphate induced reproductive toxicity through the JNK signaling pathway in *Caenorhabditis elegans*. J Hazard Mater 446:130643
4. Tejeda-Benitez L, Flegal R, Odigie K, Olivero-Verbel J (2016) Pollution by metals and toxicity assessment using *Caenorhabditis elegans* in sediments from the Magdalena River. Colombia Environ Pollut 212:238–250
5. Roh JY, Lee J, Choi J (2006) Assessment of stress-related gene expression in the heavy metal-exposed nematode *Caenorhabditis elegans*: a potential biomarker for metal-induced toxicity monitoring and environmental risk assessment. Environ Toxicol Chem 25:2946–2956
6. Li WH, Ju YR, Liao CM, Liao VH (2014) Assessment of selenium toxicity on the life cycle of *Caenorhabditis elegans*. Ecotoxicology 23:1245–1253
7. Hartman JH, Widmayer SJ, Bergemann CM, King DE, Morton KS, Romersi RF, Jameson LE, Leung MCK, Andersen EC, Taubert S, Meyer JN (2021) Xenobiotic metabolism and transport in *Caenorhabditis elegans*. J Toxicol Environ Health B Crit Rev 24:51–94
8. Nagar Y, Thakur RS, Parveen T, Patel DK, Ram KR, Satish A (2020) Toxicity assessment of parabens in *Caenorhabditis elegans*. Chemosphere 246:125730
9. Liu H-L, Tian L-J, Wang D-Y (2021) Notch receptor GLP-1 regulates toxicity of simulated microgravity stress by activating germline-intestine communication of insulin signaling in *C. elegans*. Biochem Biophys Res Commun 534:248–253
10. Sun L-M, Li D, Yuan Y-J, Wang D-Y (2021) Intestinal long non-coding RNAs in response to simulated microgravity stress in *Caenorhabditis elegans*. Sci Rep 11:1997
11. Zhao Y-Y, Dong S-S, Kong Y, Rui Q, Wang D-Y (2020) Molecular basis of intestinal canonical Wnt/β-catenin BAR-1 in response to simulated microgravity in *Caenorhabditis elegans*. Biochem Biophys Res Commun 522:198–204
12. Sun L-M, Li W-J, Li D, Wang D-Y (2020) microRNAs involved in the control of toxicity on locomotion behavior induced by simulated microgravity stress in *Caenorhabditis elegans*. Sci Rep 10:17510
13. Wang D-Y (2019) Molecular toxicology in *Caenorhabditis elegans*. Springer Nature Singapore Pte Ltd, Singapore
14. Wang D-Y (2019) Target organ toxicology in *Caenorhabditis elegans*. Springer Nature Singapore Pte Ltd, Singapore
15. Wang D-Y (2020) Exposure toxicology in *Caenorhabditis elegans*. Springer Nature Singapore Pte Ltd, Singapore
16. Wang D-Y (2022) Toxicology at environmentally relevant concentrations in *Caenorhabditis elegans*. Springer Nature Singapore Pte Ltd, Singapore
17. Wang Y, Xie W, Wang D-Y (2007) Transferable properties of multi-biological toxicity caused by cobalt exposure in *Caenorhabditis elegans*. Environ Toxicol Chem 26:2405–2412

18. Wang D-Y, Wang Y (2008) Nickel sulfate induces numerous defects in *Caenorhabditis elegans* that can also be transferred to progeny. Environ Pollut 151:585–592
19. Zhao Y-L, Lin Z-Q, Jia R-H, Li G-J, Xi Z-G, Wang D-Y (2014) Transgenerational effects of traffic-related fine particulate matter ($PM_{2.5}$) on nematode *Caenorhabditis elegans*. J Hazard Mater 274:106–114
20. Zhao L, Qu M, Wong G, Wang D-Y (2017) Transgenerational toxicity of nanopolystyrene particles in the range of μg/L in nematode *Caenorhabditis elegans*. Environ Sci Nano 4:2356–2366
21. Zhao Y-L, Chen J-Y, Wang R, Pu X-X, Wang D-Y (2023) A review of transgenerational and multigenerational toxicology in the *in vivo* model animal *Caenorhabditis elegans*. J Appl Toxicol 43:122–145
22. Qiu Y-X, Luo L-B, Yang Y-H, Kong Y, Li Y-H, Wang D-Y (2020) Potential toxicity of nanopolystyrene on lifespan and aging process of nematode *Caenorhabditis elegans*. Sci Total Environ 705:135918
23. Hua X, Wang D-Y (2023) Exposure to 6-PPD quinone at environmentally relevant concentrations inhibits both lifespan and healthspan in *C. elegans*. Environ Sci Technol 57:19295–19303
24. Liu H-L, Tian L-J, Wang S-T, Wang D-Y (2021) Size-dependent transgenerational toxicity induced by nanoplastics in nematode *Caenorhabditis elegans*. Sci Total Environ 790:148217
25. Sun L-M, Liao K, Wang D-Y (2021) Comparison of transgenerational reproductive toxicity induced by pristine and amino modified nanoplastics in *Caenorhabditis elegans*. Sci Total Environ 768:144362
26. Zhao Y-Y, Xu R-R, Hua X, Rui Q, Wang D-Y (2022) Multi-walled carbon nanotubes induce transgenerational toxicity associated with activation of germline long non-coding RNA *linc-7* in *C. elegans*. Chemosphere 301:134687
27. Zhang L, Wang S-T, Zhao Y-L, Nurdebek B, Bu Y-Q, Wang D-Y (2022) Long-term exposure to polystyrene nanoparticles causes transgenerational toxicity by affecting the function and expression of MEV-1 and DAF-2 signals in *Caenorhabditis elegans*. NanoImpact 26:100403
28. Nouara A, Wu Q-L, Li Y-X, Tang M, Wang H-F, Zhao Y-L, Wang D-Y (2013) Carboxylic acid functionalization prevents the translocation of multi-walled carbon nanotubes at predicted environmental relevant concentrations into targeted organs of nematode *Caenorhabditis elegans*. Nanoscale 5:6088–6096
29. Wu Q-L, Nouara A, Li Y-P, Zhang M, Wang W, Tang M, Ye B-P, Ding J-D, Wang D-Y (2013) Comparison of toxicities from three metal oxide nanoparticles at environmental relevant concentrations in nematode *Caenorhabditis elegans*. Chemosphere 90:1123–1131
30. Wu Q-L, Zhao Y-L, Li Y-P, Wang D-Y (2014) Susceptible genes regulate the adverse effects of TiO_2-NPs at predicted environmental relevant concentrations on nematode *Caenorhabditis elegans*. Nanomedicine 10:1263–1271
31. Xu R-R, Hua X, Rui Q, Wang D-Y (2022) Polystyrene nanoparticles caused dynamic alteration in mitochondrial unfolded protein response from parents to the offspring in *C. elegans*. Chemosphere 308:136154
32. Zhang L, Wang S-T, Zhao Y, Bi K, Wang D-Y (2022) Increase in germline methyltransferases governing methylation of histone H3K9 is associated with transgenerational nanoplastic toxicity in *Caenorhabditis elegans*. Environ Sci Nano 9:265–274
33. Hua X, Zhao Y, Yuan Y-J, Zhang L, Bian Q, Wang D-Y (2022) Nanoplastics cause transgenerational toxicity through inhibiting germline microRNA *mir-38* in *C. elegans*. J Hazard Mater 437:129302
34. Liu H-L, Zhao Y-L, Hua X, Wang D-Y (2022) Induction of transgenerational toxicity is associated with the activated germline insulin signals in nematodes exposed to nanoplastic at predicted environmental concentrations. Ecotoxicol Environ Saf 243:114022
35. Xing X-J, Guo Y-L, Wang D-Y (2009) Using the larvae nematode *Caenorhabditis elegans* to evaluate neurobehavioral toxicity to metallic salts. Ecotoxicol Environ Saf 72:1819–1823
36. Guo Y-L, Yang Y-C, Wang D-Y (2009) Induction of reproductive deficits in nematode *Caenorhabditis elegans* exposed to metals at different developmental stages. Reprod Toxicol 28:90–95

37. Li S, How CM, Liao VH (2018) Prolonged exposure of di(2-ethylhexyl) phthalate induces multigenerational toxic effects in *Caenorhabditis elegans*. Sci Total Environ 634:260–266
38. Chen H, Hua X, Li H, Wang C, Dang Y, Ding P, Yu Y (2021) Transgenerational neurotoxicity of polystyrene microplastics induced by oxidative stress in *Caenorhabditis elegans*. Chemosphere 272:129642
39. Kim SW, Moon J, An Y (2018) Matricidal hatching can induce multi-generational effects in nematode *Caenorhabditis elegans* after dietary exposure to nanoparticles. Environ Sci Pollut Res 25:36394–36402
40. Lu Q, Bu Y, Ma L, Liu R (2020) Transgenerational reproductive and developmental toxicity of tebuconazole in *Caenorhabditis elegans*. J Appl Toxicol 40:578–591
41. Hua X, Zhang L, Wang D-Y (2024) Nanoplastic at environmentally relevant concentrations activates germline *mir-240-rab-5* signaling cascade to affect secreted ligands associated with transgenerational toxicity induction in *C. elegans*. Environ Sci Nano 11:3615
42. Shao Y-T, Li Y-H, Wang D-Y (2024) Polylactic acid microplastics cause transgenerational reproductive toxicity associated with activation of insulin and hedgehog ligands in *C. elegans*. Sci Total Environ 942:173746
43. Wu J-W, Shao Y-T, Hua X, Wang Y-X, Wang D-Y (2024) Nanoplastic at environmentally relevant concentrations induces toxicity across multiple generations associated with inhibition in germline G protein-coupled receptor CED-1 in *Caenorhabditis elegans*. Chemosphere 364:143011
44. Liu Z-Y, Wang Y-X, Bian Q, Wang D-Y (2024) Transgenerational response of germline nuclear hormone receptor genes to nanoplastics at predicted environmental doses in *Caenorhabditis elegans*. Toxics 12:420
45. Song M-X, Ruan Q-L, Wang D-Y (2024) Comparison of transgenerational toxicity on locomotion behavior and development of D-type motor neurons between pristine and amino modified nanoplastics at environmentally relevant doses in *C. elegans*. Toxics 12:555
46. Shao H-M, Han Z-Y, Krasteva N, Wang D-Y (2019) Identification of signaling cascade in the insulin signaling pathway in response to nanopolystyrene particles. Nanotoxicology 13:174–188
47. Qu M, Liu Y-Q, Xu K-N, Wang D-Y (2019) Activation of p38 MAPK signaling-mediated endoplasmic reticulum unfolded protein response by nanopolystyrene particles. Adv Biosyst 3:1800325
48. Shao H-M, Kong Y, Wang D-Y (2020) Response of intestinal signaling communication between nucleus and peroxisome to nanopolystyrene at predicted environmental concentration. Environ Sci Nano 7:250–261
49. Zhao Y-L, Yang R-L, Rui Q, Wang D-Y (2016) Intestinal insulin signaling encodes two different molecular mechanisms for the shortened longevity induced by graphene oxide in *Caenorhabditis elegans*. Sci Rep 6:24024
50. Zhao Y-L, Zhi L-T, Wu Q-L, Yu Y-L, Sun Q-Q, Wang D-Y (2016) p38 MAPK-SKN-1/Nrf signaling cascade is required for intestinal barrier against graphene oxide toxicity in *Caenorhabditis elegans*. Nanotoxicology 10:1469–1479
51. Zhi L-T, Qu M, Ren M-X, Zhao L, Li Y-H, Wang D-Y (2017) Graphene oxide induces canonical Wnt/β-catenin signaling-dependent toxicity in *Caenorhabditis elegans*. Carbon 113:122–131
52. Ren M-X, Zhao L, Ding X-C, Krasteva N, Rui Q, Wang D-Y (2018) Developmental basis for intestinal barrier against the toxicity of graphene oxide. Part Fibre Toxicol 15:26
53. Zhi L-T, Fu W, Wang X, Wang D-Y (2016) ACS-22, a protein homologous to mammalian fatty acid transport protein 4, is essential for the control of toxicity and translocation of multi-walled carbon nanotubes in *Caenorhabditis elegans*. RSC Adv 6:4151–4159
54. Ding X-C, Rui Q, Wang D-Y (2018) Functional disruption in epidermal barrier enhances toxicity and accumulation of graphene oxide. Ecotoxicol Environ Saf 163:456–464
55. Zhao L, Kong J-T, Krasteva N, Wang D-Y (2018) Deficit in epidermal barrier induces toxicity and translocation of PEG modified graphene oxide in nematodes. Toxicol Res 7:1061–1070

56. Qu M, Xu K-N, Li Y-H, Wong G, Wang D-Y (2018) Using *acs-22* mutant *Caenorhabditis elegans* to detect the toxicity of nanopolystyrene particles. Sci Total Environ 643:119–126
57. Zhuang Z-H, Liu T-W, Liu Z-Y, Wang D-Y (2024) Polystyrene nanoparticles strengthen high glucose toxicity associated with alteration in insulin signaling pathway in *C. elegans*. Ecotoxicol Environ Saf 272:116056
58. Wang D-Y, Zhou X-H, Zheng X-D, Dai H (2024) Editorial: *Caenorhabditis elegans:* a model organism for pharmacological research of medicinal plants and their metabolites. Front Pharmacol 15:1472276
59. Yu Y-L, Zhi L-T, Wu Q-L, Jing L-N, Wang D-Y (2018) NPR-9 regulates innate immune response in *Caenorhabditis elegans* by antagonizing activity of AIB interneurons. Cell Mol Immunol 15:27–37
60. Zhi L-T, Yu Y-L, Li X-Y, Wang D-Y, Wang D-Y (2017) Molecular control of innate immune response to *Pseudomonas aeruginosa* infection by intestinal *let-7* in *Caenorhabditis elegans*. PLoS Pathog 13:e1006152
61. Zhang L, Wang Y-X, Wang D-Y (2023) Paeoniflorin increases the survival of *Pseudomonas aeruginosa* infected *Caenorhabditis elegans* at the immunosuppression stage by activating PMK-1, BAR-1, and EGL-1 signals. Arch Pharm Res 46:616–628
62. Wang Y-X, Zhang L, Yuan X-A, Wang D-Y (2023) Treatment with paeoniflorin increases lifespan of *Pseudomonas aeruginosa* infected *Caenorhabditis elegans* by inhibiting bacterial accumulation in intestinal lumen and biofilm formation. Front Pharmacol 14:1114219
63. Zhang L, Wang Y-X, Cao C, Zhu Y-K, Huang W, Yang Y, Qiu H-B, Liu S-Q, Wang D-Y (2022) Beneficial effect of Xuebijing against *Pseudomonas aeruginosa* infection in *Caenorhabditis elegans*. Front Pharmacol 13:949608
64. Li D, Deng Y-J, Wang S-T, Du H-H, Xiao G-S, Wang D-Y (2020) Assessment of nanopolystyrene toxicity under fungal infection condition in *Caenorhabditis elegans*. Ecotoxicol Environ Saf 197:110625
65. Zhao Y-L, Wang X, Wu Q-L, Li Y-P, Wang D-Y (2015) Translocation and neurotoxicity of CdTe quantum dots in RMEs motor neurons in nematode *Caenorhabditis elegans*. J Hazard Mater 283:480–489
66. Zhao Y-L, Wu Q-L, Li Y-P, Nouara A, Jia R-H, Wang D-Y (2014) *In vivo* translocation and toxicity of multi-walled carbon nanotubes are regulated by microRNAs. Nanoscale 6:4275–4284
67. Liu H-L, Kwak JI, Wang D-Y, An Y (2021) Multigenerational effects of polyethylene terephthalate microfibers in *Caenorhabditis elegans*. Environ Res 193:110569
68. Zhi L-T, Ren M-X, Qu M, Zhang H-Y, Wang D-Y (2016) Wnt ligands differentially regulate toxicity and translocation of graphene oxide through different mechanisms in *Caenorhabditis elegans*. Sci Rep 6:39261
69. Hua X, Wang D-Y (2024) Polyethylene nanoparticles at environmentally relevant concentrations enhances neurotoxicity and accumulation of 6-PPD quinone in *Caenorhabditis elegans*. Sci Total Environ 918:170760
70. Qu M, Nida A, Kong Y, Du H-H, Xiao G-S, Wang D-Y (2019) Nanopolystyrene at predicted environmental concentration enhances microcystin-LR toxicity by inducing intestinal damage in *Caenorhabditis elegans*. Ecotoxicol Environ Saf 183:109568
71. Dong S-S, Qu M, Rui Q, Wang D-Y (2018) Combinational effect of titanium dioxide nanoparticles and nanopolystyrene particles at environmentally relevant concentrations on nematodes *Caenorhabditis elegans*. Ecotoxicol Environ Saf 161:444–450
72. Cheng X, Dong S, Chen D, Rui Q, Guo J, Wang D, Jiang J (2020) Potential of esterase DmtH in transforming plastic additive dimethyl terephthalate to less toxic mono-methyl terephthalate. Ecotoxicol Environ Saf 187:109848
73. Wu Q-L, Li Y-X, Li Y-P, Zhao Y-L, Ge L, Wang H-F, Wang D-Y (2013) Crucial role of biological barrier at the primary targeted organs in controlling translocation and toxicity of multi-walled carbon nanotubes in nematode *Caenorhabditis elegans*. Nanoscale 5:11166–11178
74. Qu M, Qiu Y-X, Kong Y, Wang D-Y (2019) Amino modification enhances reproductive toxicity of nanopolystyrene on gonad development and reproductive capacity in nematode *Caenorhabditis elegans*. Environ Pollut 254:112978

75. Qu M, Wang D-Y (2020) Toxicity comparison between pristine and sulfonate modified nanopolystyrene particles in affecting locomotion behavior, sensory perception, and neuronal development in *Caenorhabditis elegans*. Sci Total Environ 703:134817
76. Moon J, Kwak JI, Kim SW, An Y (2017) Multigenerational effects of gold nanoparticles in *Caenorhabditis elegans*: continuous versus intermittent exposures. Environ Pollut 220:46–52
77. Morran LT, Cappy BJ, Anderson JL, Phillips PC (2009) Sexual partners for the stressed: facultative outcrossing in the self-fertilizing nematode *C. elegans*. Evolution 63:1473–1482
78. Hua X, Cao C, Zhang L, Wang D-Y (2023) Activation of FGF signal in germline mediates transgenerational toxicity of polystyrene nanoparticles at predicted environmental concentrations in *Caenorhabditis elegans*. J Hazard Mater 451:131174
79. He W-M, Gu A-H, Wang D-Y (2023) Sulfonate-modified polystyrene nanoparticle at precited environmental concentrations induces transgenerational toxicity associated with increase in germline notch signal of *Caenorhabditis elegans*. Toxics 11:511
80. Zhao Y-Y, Hua X, Rui Q, Wang D-Y (2023) Exposure to multi-walled carbon nanotubes causes suppression in octopamine signal associated with transgenerational toxicity induction in *C. elegans*. Chemosphere 318:137986
81. Xu R-R, Hua X, Rui Q, Wang D-Y (2022) Alteration in Wnt signaling mediates induction of transgenerational toxicity of polystyrene nanoplastics in *C. elegans*. NanoImpact 28:100425
82. Zhao Y, Hua X, Bian Q, Wang D-Y (2022) Nanoplastic exposure at predicted environmental concentrations induces activation of germline Ephrin signal associated with toxicity formation in the *Caenorhabditis elegans* offspring. Toxics 10:699
83. Wang Y-X, Wang D-Y (2024) Exposure to 6-PPD quinone enhances glycogen accumulation in *Caenorhabditis elegans*. Environ Pollut 359:124600
84. Liu Z-J, Li Y-H, Wang D-Y (2024) Exposure to 6-PPD quinone disrupts glucose metabolism associated with lifespan reduction by affecting insulin and AMPK signals in *Caenorhabditis elegans*. Chemosphere 363:142975
85. Wang Y-X, Hua X, Wang D-Y (2023) Exposure to 6-PPD quinone enhances lipid accumulation through activating metabolic sensors of SBP-1 and MDT-15 in *Caenorhabditis elegans*. Environ Pollut 333:121937
86. Zheng Y, Yu Z, Zhang J (2022) Multi-generational effects of enrofloxacin on lifespan and reproduction of *Caenorhabditis elegans* with SKN-1-mediated antioxidant responses and lipid metabolism disturbances. Sci Total Environ 804:150250
87. Wang L, Wu D, Yu Z, Huang S, Zhang J (2023) Hormone-mediated multi- and transgenerational reproductive toxicities of 1-ethyl-3-methylimidazolium hexafluorophosphate on *Caenorhabditis elegans*. Sci Total Environ 863:160958

Chapter 2
Transgenerational Toxicity at Various Aspects

Abstract Exposure to toxicants and stresses can cause multiple aspects of transgenerational toxicity in *C. elegans*. We first introduced transgenerational molecular alterations after exposure to toxicants or stresses, which is reflected by transgenerational genome instability, transgenerational gene expression plasticity, transgenerational alteration in epigenetic-related signals and germline ligand genes, and transgenerational alteration in protective molecular responses. After that, transgenerational developmental toxicity, transgenerational intestinal toxicity, transgenerational neurotoxicity, and transgenerational reproductive toxicity of toxicants were introduced. Moreover, we introduced transgenerational metabolic alterations and transgenerational physiological alterations after exposure to toxicants. Finally, transgenerational lifespan reduction after exposure to toxicants was further discussed.

Keywords Transgenerational toxicity · Endpoints · *Caenorhabditis elegans*

2.1 Introduction

Caenorhabditis elegans has become an important animal model for toxicological study, at least including both molecular toxicology and target organ toxicology of toxicants and stresses [1–3]. *C. elegans* can provide the toxicity assessment and the toxicology-related information at the whole animal level [4–7]. More importantly, because of its high sensitivity to environmental exposures [7–9], *C. elegans* has been proven to be helpful to detect toxicity of pollutants at environmentally relevant concentrations (ERCs) and to determine underlying molecular mechanisms [10–14].

In Chap. 1, we have discussed the research values of *C. elegans* for the study of transgenerational toxicology. Largely due to these research values, transgenerational toxicity of different pollutants has been assessed in nematodes [15–18]. Especially, transgenerational toxicity of pollutants at ERCs could be detected in this animal model [19–22]. The established endpoint systems directly determine which transgenerational toxicities can be evaluated in *C. elegans* [7]. In *C. elegans*, the

D. Wang, *Transgenerational Toxicology in Caenorhabditis elegans*,
https://doi.org/10.1007/978-981-92-1389-4_2

established endpoint systems contain basic endpoints, endpoints for assessing toxicity on primary and secondary targeted organs, endpoints for assessing toxicity on biochemical processes, and endpoints for assessing genotoxicity and genetic toxicity [7].

In this chapter, we first introduced transgenerational molecular alterations after pollutant exposure. After that, we introduced transgenerational developmental toxicity, transgenerational intestinal toxicity, transgenerational neurotoxicity, and transgenerational reproductive toxicity of pollutants. Moreover, transgenerational metabolic alterations, transgenerational physiological alterations, and transgenerational lifespan reduction after pollutant exposure were discussed. These obtained research progresses have laid an important foundation for further in-depth and extensive evaluation of transgenerational toxicity of toxicants and stresses and elucidation of underlying molecular mechanisms in *C. elegans*.

2.2 Transgenerational Molecular Alterations After Exposure to Toxicants or Stresses

We will introduce the transgenerational molecular alterations after exposure to toxicants or stresses from the following several aspects: (1) transgenerational genome instability, (2) transgenerational gene expression plasticity, (3) transgenerational alteration in germline ligand genes and epigenetic related signals, and (4) transgenerational alteration in protective molecular responses.

2.2.1 Transgenerational Genome Instability

The *C. elegans* germline is continuous. This suggests that the environmental pollutants can directly affect the germline and then influence the gametes that give rise to all subsequent generations, if the pollutants can be translocated into the gonads. After exposure to toxicants or stresses at P0-G, transgenerational molecular alterations are not only limited in changes of some certain molecular signals. The genome instability refers to a phenomenon that exposure to toxicants or stresses at parental generation (P0-G) can cause the increase in genetic mutations, chromosomal rearrangements, and other genomic alterations for multiple generations (F1, F2, F3, etc.). The transgenerational molecular alteration induced by toxicants or stresses can occur at the genome-wide level. The transgenerational genome instability is distinct from direct DNA damage of toxicants or stresses through a change in the DNA sequence. The observed transgenerational genome instability suggests that the "safe" levels of environmental pollutants need to be re-evaluated based on their potential to cause heritable damage on organisms. The obtained related data in *C. elegans* provides a foundational principle for understanding related human health that toxicants and stresses could potentially have consequences for the genomic integrity of our descendants.

The most common observed transgenerational genome instability induced by pollutants is the transgenerational genome mutation. In *C. elegans*, after multiple generations' (ten generations) exposure to different forms of silver nanoparticle (Ag-NPs), the total number of mutation variants was examined [23]. The increase in total number of mutation variants could be detected, and change of single nucleotide polymorphisms (SNPs) mainly contributed to this [23]. However, transgenerational alteration in number of mutation accumulation seemed to be not correlated with alteration in transgenerational toxicity induced by Ag-NPs [23]. Sulfidized Ag-NPs (sAg-NPs) caused higher transgenerational number of mutation accumulation than Ag-NP [23]. Nevertheless, transgenerational reproductive toxicity of sAg-NP was lower than that of Ag-NP [23], which implied that some of the formed mutations might be correlated to beneficial effects for pollutant exposure.

In *C. elegans*, exposure to flame retardants, such as TBBPA, TCEP, and TCPP, at P0-G caused reproductive toxicity, including enhanced germline apoptosis and inhibited oogenesis from F1-G to F6-G [24]. Accompanied with this, it was also found that double-increased mutation frequency could be detected in the offspring after flame retardant exposure, suggesting germ cell mutagenicity across multiple generations induced by flame retardant [24]. The transgenerational mutagenic actions caused by flame retardant mainly contained point mutations at A/T base pairs [24]. Nevertheless, normally, the exposure to toxicants and stresses at low doses or after short-term exposure would not cause transgenerational genome mutation at the whole genome level.

Besides the induction of transgenerational genome mutation, certain toxicants and stresses also resulted in other aspects of transgenerational genome instability. The *fog-2* mutants have both somatic hermaphrodites only generating female germline and males having normal male germline. In this sex-separated strain, feminized mutant *fog-2*, ionizing radiation led to transgenerational embryonic lethality [25]. Exposure to ionizing radiation in males caused transgenerational embryonic lethality till to F2-G [25]. Accompanied with this, the offspring of irradiated males exhibited phenotypes of DNA fragmentation, chromosomal rearrangement, and aneuploidy [25]. Depletion of histone H1.0/HIS-24 and heterochromatin protein/HPL-1 reversed transgenerational embryonic lethality induced by ionizing radiation, and removal of HIS-24 and HPL-1 enabled error-free homologous recombination repair of germline of F1-G of ionizing radiation-treated males [25]. This implied that ionizing radiation exposure caused transgenerational genome instability by activating HIS-24 and HPL-1. For the recovery of genome instability in the offspring of ionizing radiation-treated males, it was raised that maternally provided error-prone polymerase theta-mediated end joining (TMEJ) contributed to the repair of DNA double-strand breaks caused by ionizing radiation exposure [25]. This recovery mechanism may be conserved, since structural variants with signature microhomology of TMEJ present in both humans and natural *C. elegans* variants [25].

2.2.2 Transgenerational Gene Expression Plasticity

Transgenerational gene expression plasticity refers to the ability of nematodes to be inherited of response of genes ("turned on" or "turned off") induced by toxicants and stresses for multiple generations. In organisms including the *C. elegans*, exposure to toxicants and stresses normally induces widespread changes in gene expression. Nevertheless, pollutants at ERCs or stresses at environmentally relevant doses (ERDs) usually cause change in expression of a limited number of genes in nematodes [10]. Furthermore, some of the observed change in gene expressions induced by pollutants at ERCs or stresses at ERDs would be detected in the offspring of exposed nematodes.

The transgenerational gene expression plasticity induced by toxicants and stresses usually lasts for three to five generations or more before the eventually fading, implying the property of semi-stable but not permanent change. The severity and pattern for formation of transgenerational gene expression plasticity induced by toxicants and stresses are highly dependent on concentrations/doses and developmental stages at which the initial exposure occurred. Considering that some effects in the immediate offspring (F1-G) can be due to direct exposure of the embryo in utero or through paternal provisioning (e.g., pollutants in the yolk), transgenerational toxicity induced by pollutants is considered when effects are seen in the F3-G, where the F2-G germline was the first non-exposed one.

With the polylactic acid microplastics (PLA-MPs) as the example, transgenerational alteration in expression of genes controlling the reproductive process was induced by PLA-MPs exposure. As an important member of bioplastics, PLA is widely applied as the alternative to conventional fossil fuel-based plastics. Fragmentation of PLA-MPs can be accelerated by UV irradiation, and UV irradiation further enhanced PLA-MPs caused reproductive toxicity, such as inhibition in reproductive capacity and induction of germline apoptosis [26]. Moreover, transgenerational toxicity of PLA-MPs in inhibiting reproductive capacity, causing damage on gonad development, and inducing germline apoptosis was also enhanced by UV irradiation [27]. After exposure to 100 μg/L PLA-MPs, their transgenerational toxicity in inducing germline apoptosis could be detected till to F3-G [27]. CED-3, CED-4, CED-9, and EGL-1 are major regulators of *C. elegans* apoptosis [28–30]. Meanwhile, transgenerational increase in *ced-3*, *ced-4*, and *egl-1* expressions and decrease in *ced-9* expression were induced by photo-aged PLA-MPs (Fig. 2.1) [27], which provided the direct molecular basis for the observed transgenerational toxicity of photo-aged PLA-MPs. Considering that photo-aged PLA-MPs could only be accumulated in the body at P0-G and F1-G [27], transgenerational reproductive toxicity of photo-aged PLA-MPs was not mainly due to transgenerational accumulation of photo-aged PLA-MPs. Additionally, leachates of 1–100 μg/L photo-aged PLA-MPs did not cause reproductive toxicity [27], suggesting that the observed transgenerational reproductive toxicity of photo-aged PLA-MPs was also not related to their leachates. In *C. elegans*, DNA damage checkpoints (CLK-2, MTR-2, and HUS-1) act as important upstream regulators to drive EGL-1-mediated germline

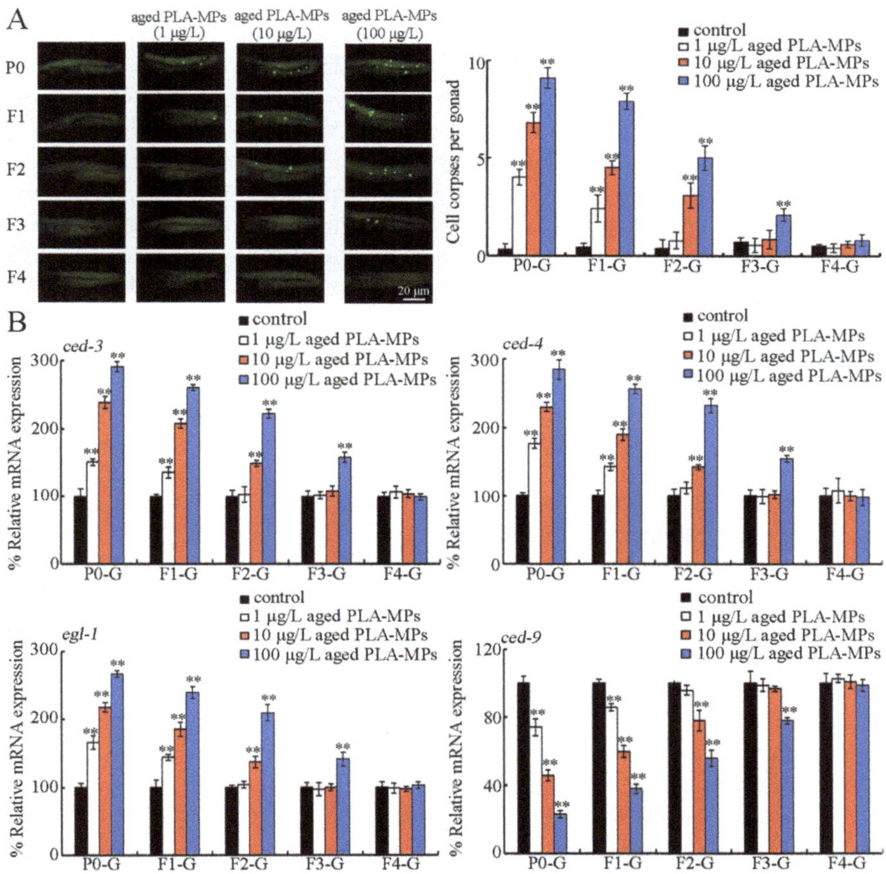

Fig. 2.1 Transgenerational effect of photo-aged PLA-MPs on induction of germline apoptosis [27]. (**a**) Transgenerational effect of photo-aged PLA-MPs on germline apoptosis induction based on AO staining. Thirty adult animals (adult day 1) were assessed for each treatment, and experiments were repeated three times. (**b**) Transgenerational effect of photo-aged PLA-MPs on expression of *ced-3*, *ced-4*, *egl-1*, and *ced-9*. Three independent experiments were performed. $^{**}P < 0.01$ vs control

apoptosis induction [31–33]. CLK-2, MTR-2, and HUS-1 could be activated by photo-aged PLA-MPs, and activated CLK-2, MTR-2, and HUS-1 could be detected till to F3-G after exposure to 100 μg/L photo-aged PLA-MPs [27]. RNAi of genes encoding these DNA damage checkpoints disrupted induction of germline apoptosis by photo-aged PLA-MPs [26]. Therefore, photo-aged PLA-MPs caused transgenerational reproductive toxicity by activating DNA damage checkpoints to result in EGL-1-mediated germline apoptosis.

2.2.3 Transgenerational Alteration in Epigenetic-Related Signals and Germline Ligand Genes

For the formation of transgenerational gene expression plasticity, at least the epigenetic machinery plays the crucial role, which is normally considered as the robust and long-lasting effects transmitted through the germline (eggs and sperm). At the P0-G after exposure to toxicants and stresses, the epigenetic changes established in germline cannot be fully erased, which would lead to altered gene expression in the F1-G, F2-G, and sometimes F3-G and F4-G. Epigenetic machineries identified in *C. elegans* mainly include (1) small non-coding RNAs (sncRNAs), (2) histone modifications, and (3) DNA methylation. That is, after exposure to toxicants or stresses, "memories" of small RNAs (like piRNAs and endogenous siRNAs) in the germline can be loaded into complexes and passed to subsequent generations, where they guide the silencing of specific genes. Similarly, at promoters of stress-responsive genes for toxicants and stresses, changes in histone marks (e.g., H3K4me3 for activation, H3K9me3 for repression) can be inherited to influence which genes are accessible for transcription in the offspring.

Further with the polylactic acid microplastics (PLA-MPs) as the example, among histone methylation transferase (HMT) genes expressed in the germline, expressions of *mes-2*, *blmp-1*, *set-27*, *set-13*, *set-24*, *set-28*, *set-31*, *set-32*, *set-2*, *set-16*, *set-23*, and *rbbp-5* in the germline could be decreased, and expressions of *met-2*, *set-17*, *set-4*, *set-3*, and *set-6* in the germline could be increased by 100 μg/L photo-aged PLA-MPs [27]. Moreover, after exposure to photo-aged PLA-MPs, only decrease in germline *mes-2*, *set-27*, *set-24*, *set-28*, *set-31*, and *set-2* expressions and increase in germline *met-2*, *set-3*, and *set-6* expressions were observed from P0-G to F3-G (Fig. 2.2) [27]. Meanwhile, photo-aged PLA-MPs induced germline apoptosis, as well as increase in *ced-4*, *ced-3*, and *egl-1* expressions and decrease in *ced-9* expression, could be strengthened by germline RNAi of *mes-2*, *set-27*, *set-24*, *set-28*, *set-31*, and *set-2* and suppressed by germline RNAi of *met-2*, *set-3*, and *set-6* [27]. Additionally, photo-aged PLA-MPs activated DNA damage checkpoint genes (*clk-1*, *mrt-2*, and *hus-1*) could also be increased by germline RNAi of *mes-2*, *set-27*, *set-24*, *set-28*, *set-31*, or *set-2* and decreased by germline RNAi of *met-2*, *set-3*, or *set-6* [27]. Therefore, change of HMTs-mediated histone methylation may drive the formation of transgenerational reproductive toxicity of photo-aged PLA-MPs.

Besides the epigenetic machinery, germline ligand genes also regulate transgenerational toxicity of toxicants and stresses. On the one hand, exposure to toxicants and stresses will cause heritable changes in the expression of germline ligand genes, which means that these changes can be passed down to multiple subsequent generations. On the other hand, as the role of ligands, the changed ligands in the germline can target to the corresponding receptors in the offspring of toxicants and stresses exposed nematodes to control the physiological processes and behaviors in the subsequent generations.

With the increasing production and consumption of plastics, the microplastics and especially the nanoplastics have been widely distributed in various environments. This suggests the severe exposure risk of microplastics and nanoplastic to

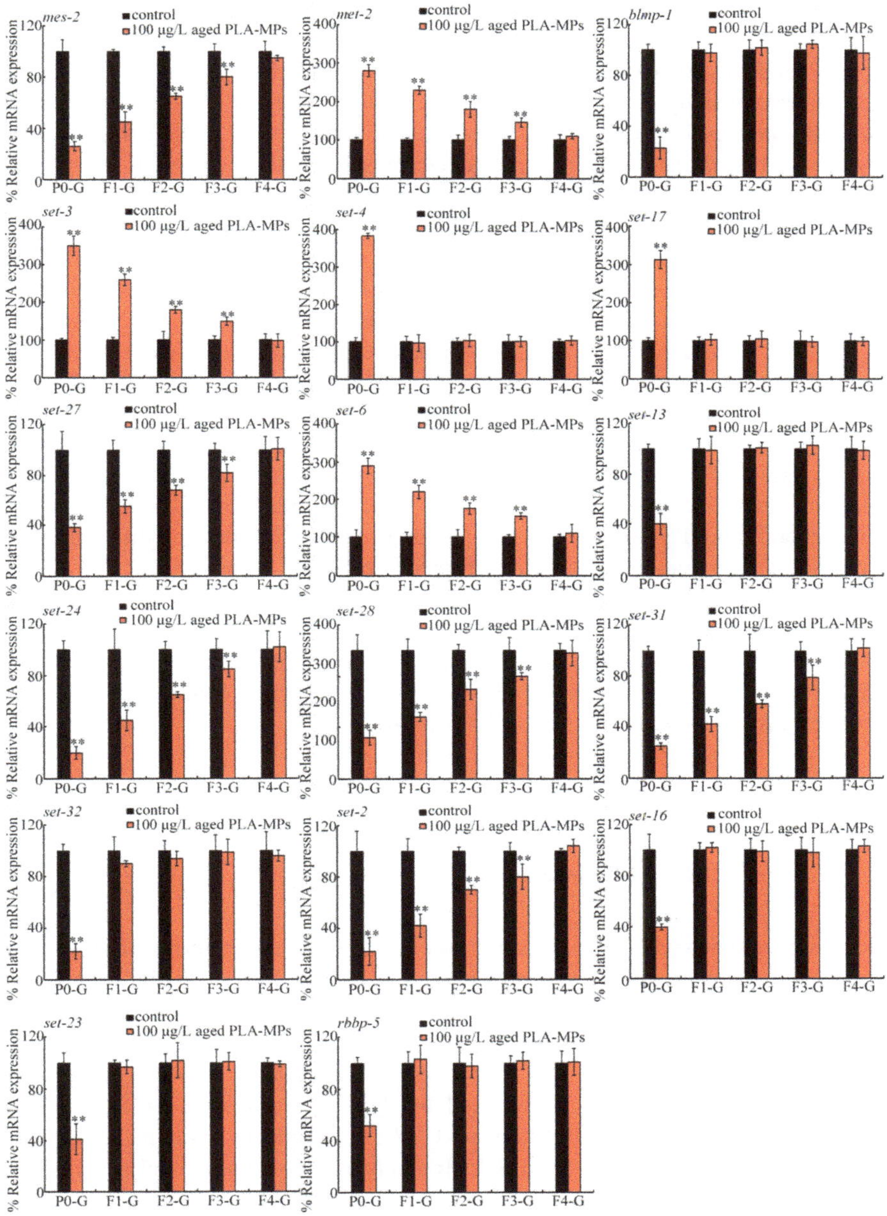

Fig. 2.2 Transgenerational expression of germline histone methylation transferase genes after exposure to photo-aged PLA-MPs [27]. A total of 30 intact gonads were used for the qRT-PCR assay. Exposure concentration of photo-aged PLA-MPs was 100 μg/L. Three independent experiments were performed. $^{**}P < 0.01$ vs control

environmental organisms and human health. In *C. elegans*, toxic effects of nanoplastics at different aspects have been described [34–38]. With polystyrene nanoparticle (PS-NP) as the example, exposure to 1–100 μg/L PS-NPs (20 nm) induced transgenerational upregulation of germline insulin ligands (INS-39, INS-3, and DAF-28) and insulin receptor DAF-2, and these three insulin ligands transgenerationally activated the DAF-2 to regulate the transgenerational PS-NP toxicity [39]. Exposure to PS-NP induced transgenerational increase in expression of germline Wnt ligand LIN-44 and decrease in expression of Wnt receptor MIG-1, and *mig-1* expression in the offspring could be increased by germline RNAi of *lin-44* [40]. The transgenerational PS-NP toxicity was inhibited by *lin-44* and enhanced by *mig-1* [40]. Exposure to PS-NP resulted in transgenerational activation of germline FGF ligand EGL-17 and FGF receptor EGL-15, and germline overexpression of EGL-17 increased EGL-15 expression in the offspring [41]. Transgenerational PS-NS toxicity in nematodes overexpressing germline EGL-17 was inhibited by RNAi of *egl-15* at F1-G [41]. Transgenerational expressions of hedgehog ligand genes (*grl-15*, *grl-16*, *qua-1*, and *wrt-1*) and hedgehog receptor genes (*ptr-23*, *scp-1*, *ptd-2*, and *ncr-1*) could be increased by PS-NP, and RNAi of them exhibited the resistance to transgenerational PS-NP toxicity [42]. Additionally, sulfonate-modified PS-NP induced transgenerational activation of germline Notch ligand LAG-2 and Notch receptor GLP-1, and germline LAG-2 could transgenerationally activate the GLP-1 to regulate the transgenerational nanoplastic toxicity [43]. Similarly, transgenerational activation of Ephrin ligand EFN-3 and Ephrin receptor VAB-1 was caused by PS-NP exposure, and Ephrin receptor VAB-1 in the offspring could be further activated by germline overexpression of EFN-3 [44].

2.2.4 Transgenerational Alteration in Protective Molecular Responses

Exposure to toxicants and stresses at P0-G can trigger certain forms of protective molecular responses, and these responses can be further detected in the subsequent multiple generations. The triggered protective responses by toxicants and stresses include antioxidation response, mitochondrial unfolded protein (mt UPR) response, endoplasmic reticulum UPR (ER UPR) response, innate immune response, etc. Antioxidation response can be reflected by dynamic expression of SOD-2 and SOD-3, mitochondrial Mn-SODs. The mt UPR can be reflected by dynamic expressions of HSP-6 and HSP-60. The ER UPR can be reflected by dynamic expression of HSP-4. The innate immune response can be reflected by dynamic expression of antimicrobial genes. Normally, low doses of toxicants and stresses would induce the activation of protective responses, and relatively high doses of toxicants and stresses would cause the suppression in protective responses. Further with different sizes of PS-NPs as the examples, exposure to 100 nm PS-NPs induced activation of SOD-3::GFP, HSP-6::GFP, and HSP-4::GFP expressions at both P0-G and their offspring (Fig. 2.3) [45]. After exposure to 1 μg/L PS-NPs (100 nm), the activation of SOD-3::GFP, HSP-6::GFP, and HSP-4::GFP expressions was also observed at

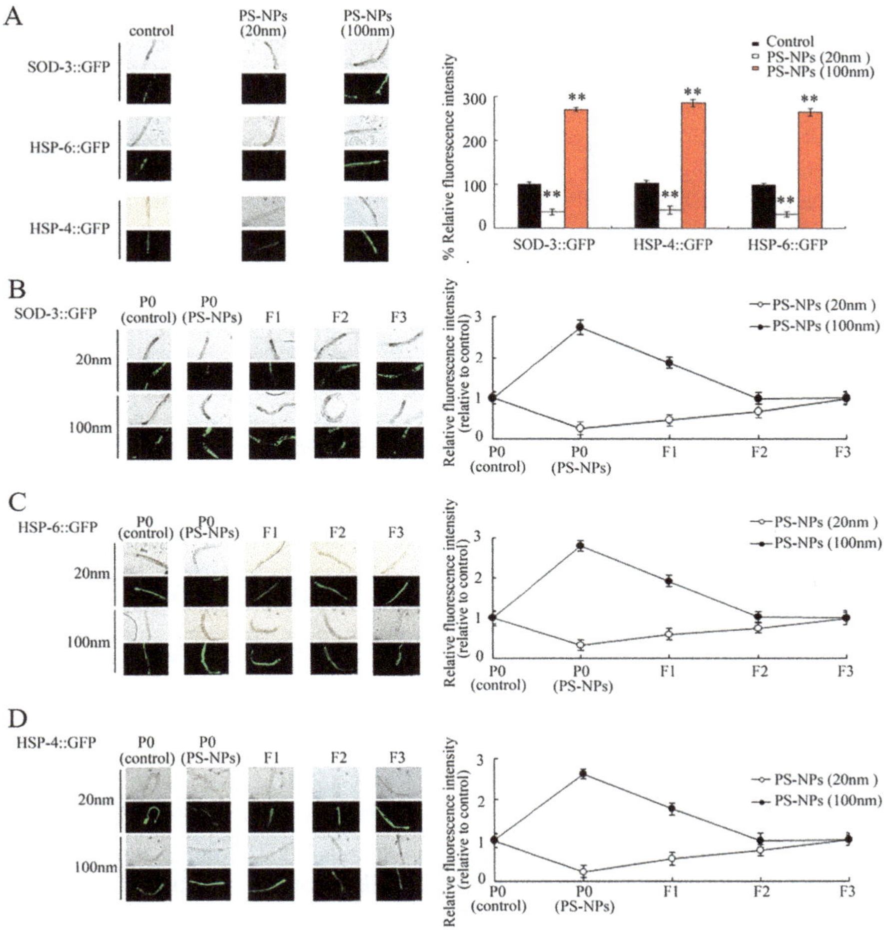

Fig. 2.3 Comparison of toxicity between 20 nm PS-NPs and 100 nm PS-NPs in affecting expressions of SOD-3::GFP, HSP-6::GFP, and HSP-4::GFP [45]. (**a**) Comparison of toxicity between 20 nm PS-NPs and 100 nm PS-NPs in affecting expressions of SOD-3::GFP, HSP-6::GFP, and HSP-4::GFP at the P0 generation. Bars represent means ± SD. $^{**}P < 0.01$ vs. control. (**b**) Effect of exposure to 20 nm PS-NPs or 100 nm PS-NPs on transgenerational expression of SOD-3::GFP. The data was normalized to the control (control = 1). Bars represent means ± SD. (**c**) Effect of exposure to 20 nm PS-NPs or 100 nm PS-NPs on transgenerational expression of HSP-6::GFP. The data was normalized to the control (control = 1). Bars represent means ± SD. (**d**) Effect of exposure to 20 nm PS-NPs or 100 nm PS-NPs on transgenerational expression of HSP-4::GFP. The data was normalized to the control (control = 1). Bars represent means ± SD. Exposure concentration of 20 nm or 10 nm PS-NPs was 1 μg/L

F1-G (Fig. 2.3) [45], suggesting the activation of antioxidation response, mt UPR, and ER UPR to protect the nematodes from the damage of 100 nm PS-NPs in the offspring. Different from these, exposure to 20 nm PS-NPs caused inhibition in SOD-3::GFP, HSP-6::GFP, and HSP-4::GFP expressions at P0-G, and the inhibition in antioxidation response, mt UPR, and ER UPR was also observed in the offspring

(Fig. 2.3) [45]. After exposure to 1 μg/L PS-NPs (100 nm), the inhibition in SOD-3::GFP, HSP-6::GFP, and HSP-4::GFP expressions was also detected at F1-G (Fig. 2.3) [45], suggesting the suppression in antioxidation response, mt UPR, and ER UPR to mediate the damage of 20 nm PS-NPs on the offspring. Besides these, exposure to 6-PPD quinone (6-PPDQ) could decrease expression of *lys-7* and *spp-1*, two antimicrobial genes [46]. The decrease in these two antimicrobial gene expressions caused by 10 μg/L 6-PPDQ could last till to F4-G [46], suggesting the immunosuppression induction to mediate the damage of 6-PPDQ on nematodes.

2.3 Transgenerational Developmental Toxicity of Toxicants

Transgenerational developmental toxicity of toxicants and stresses is usually evaluated by the endpoint of body length to reflect the *C. elegans* growth [7]. Exposure to pollutants, such as heavy metals, could cause the transgenerational inhibition in body length [47–50]. For example, toxicity in inhibiting body length by exposure to Fe, Ni, Pb, and Co could be transferred from heavy metal exposed nematodes to their offspring [47–50]. Transgenerational developmental toxicity of toxicants and stresses can also be assessed by change of neuronal and reproductive development during the transgenerational process after exposure to toxicants and stresses, which is described in 2.5 and 2.6.

2.4 Transgenerational Intestinal Toxicity of Toxicants

The intestine is the primary target organs of pollutants [3]. This is the biggest organs in *C. elegans*, and it is composed of only 20 cells, forming a simple tube. Despite simplicity of intestine, its functions are complex and conserved. The intestine is the primary site of nutrient absorption and storage, major detoxification center, core of innate immunity, sensor for environmental pollutants and stress, and important source of signaling molecules.

Usually, oxidative damage is used to reflect the intestinal toxicity of toxicants and stresses using endpoints of reactive oxygen species (ROS) production and intestinal autofluorescence. Meanwhile, some transgenic markers, such as SKN-1::GFP, SOD-3::GFP, and GST-4::GFP, can also be used indirectly to reflect the induction of oxidative damage after exposure to toxicants and stresses. With 6-PPDQ as the example, transgenerational induction of intestinal ROS production could be caused by 6-PPDQ (Fig. 2.4) [46]. After exposure to 10 μg/L 6-PPDQ, induction of intestinal ROS production could be observed till to the F5-G (Fig. 2.4) [46], suggesting the formation of severe transgenerational oxidative damage.

Intestinal permeability reflects the functional state of *C. elegans* intestinal barrier, and this can be observed after staining with the dyes, such as erioglaucine disodium [3]. Transgenerational enhancement in intestinal permeability could be

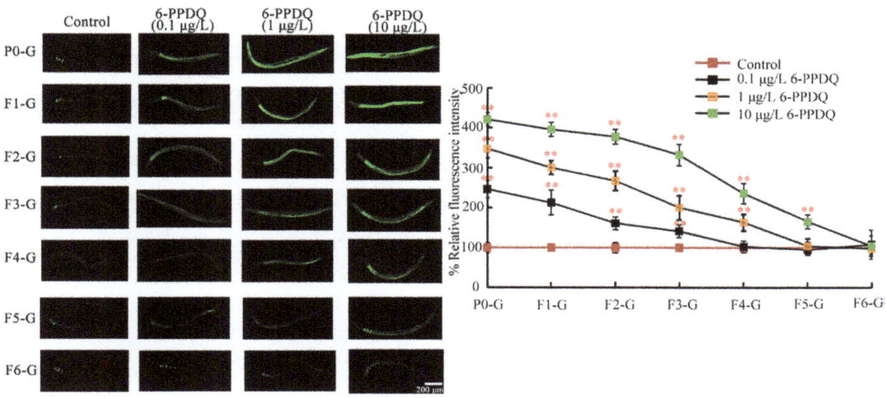

Fig. 2.4 Transgenerational induction of intestinal ROS production after 6-PPDQ exposure [46]. $^{**}P < 0.01$ vs control

induced by exposure to 6-PPDQ. After exposure to 10 μg/L 6-PPDQ, enhancement in intestinal permeability could be observed till to the F4-G, and severe 6-PPDQ accumulation in body cavity was formed at F1-G and F2-G (Fig. 2.5) [46]. Accompanied with this, transgenerational inhibition in expression of *acs-22*, *hmp-2*, and *pkc-3* governing functional state of intestinal barrier was caused by 6-PPDQ [46].

2.5 Transgenerational Neurotoxicity of Toxicants

The nervous system is one of the important secondary target systems for pollutants in nematodes [3]. *C. elegans* has a simple yet well-mapped nervous system (302 neurons in a hermaphrodite). The *C. elegans* connectivity and molecular regulatory pathways are highly conserved. In *C. elegans*, exposure to toxicants and stresses can alter some locomotion behaviors, perception behaviors, and learning and memory behaviors [51–56]. Additionally, development of nervous system and some aspects of certain neurons could be affected by toxicants and stresses [57–59]. All these aspects of neurotoxicity have the possibility to be transferred from toxicants and stresses exposed nematodes to their offspring.

With the 6-PPDQ as the example, transgenerational inhibition in locomotion behavior reflected by change of endpoints of head thrashing and body bending was caused by 6-PPDQ (Fig. 2.6) [60]. After exposure to 10 μg/L 6-PPDQ, transgenerational inhibition in locomotion behavior lasted till to the F5-G and was recovered at F6-G (Fig. 2.6) [60]. The transgenerational neurotoxicity of 6-PPDQ on nervous system development was not so severe as on the locomotion behavior [60]. Exposure to 6-PPDQ only caused transgenerational increase in neuronal loss and ventral cord gap formation in GABAergic nervous system, and this transgenerational damage on development of GABAergic nervous system by 10 μg/L 6-PPDQ could be observed till to F2-G (Fig. 2.6) [60].

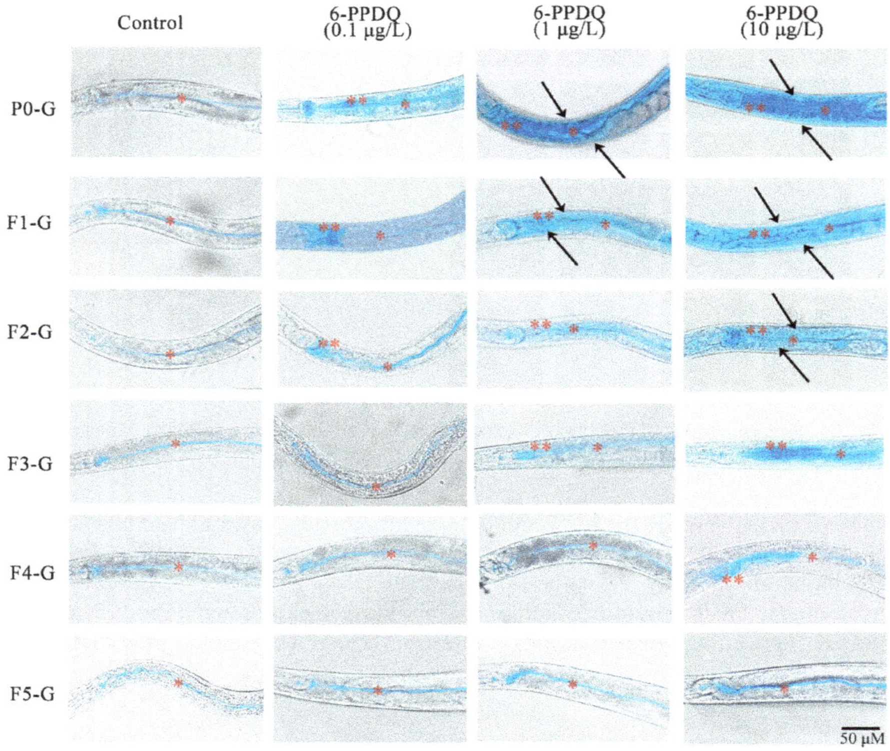

Fig. 2.5 Transgenerational intestinal permeability after 6-PPDQ exposure [46]. The intestinal cells (**) and intestinal lumen (*) were labeled by asterisks, and the body cavity was labeled by arrowheads

2.6 Transgenerational Reproductive Toxicity of Toxicants

In *C. elegans*, the reproductive system is another important secondary target system for pollutants [3]. Transgenerational impact of pollutants on *C. elegans* germline and reproductive outcomes can serve as a critical warning for the potential ecological and health consequences of environmental pollutants. The *C. elegans* germline is a syncytial structure that produces the sperm and the oocytes. Exposure to pollutants can cause toxicity on reproductive capacity and gonad development in hermaphrodites and behavior and development of males [35, 61–64]. Normally, long-term or high dose of pollutant exposure would result in transgenerational reproductive toxicity.

With PLA-MPs as the example, exposure to PLA-MPs induced transgenerational reduction in reproductive capacity reflected by decrease in both brood size and number of fertilized eggs in uterus [9]. After exposure to 100 μg/L PLA-MPs, transgenerational reduction in reproductive capacity lasted till to the F2-G [9]. Besides this, exposure to PLA-MPs also caused transgenerational damage on gonad development, and damage on gonad development induced by 100 μg/L PLA-MPs could

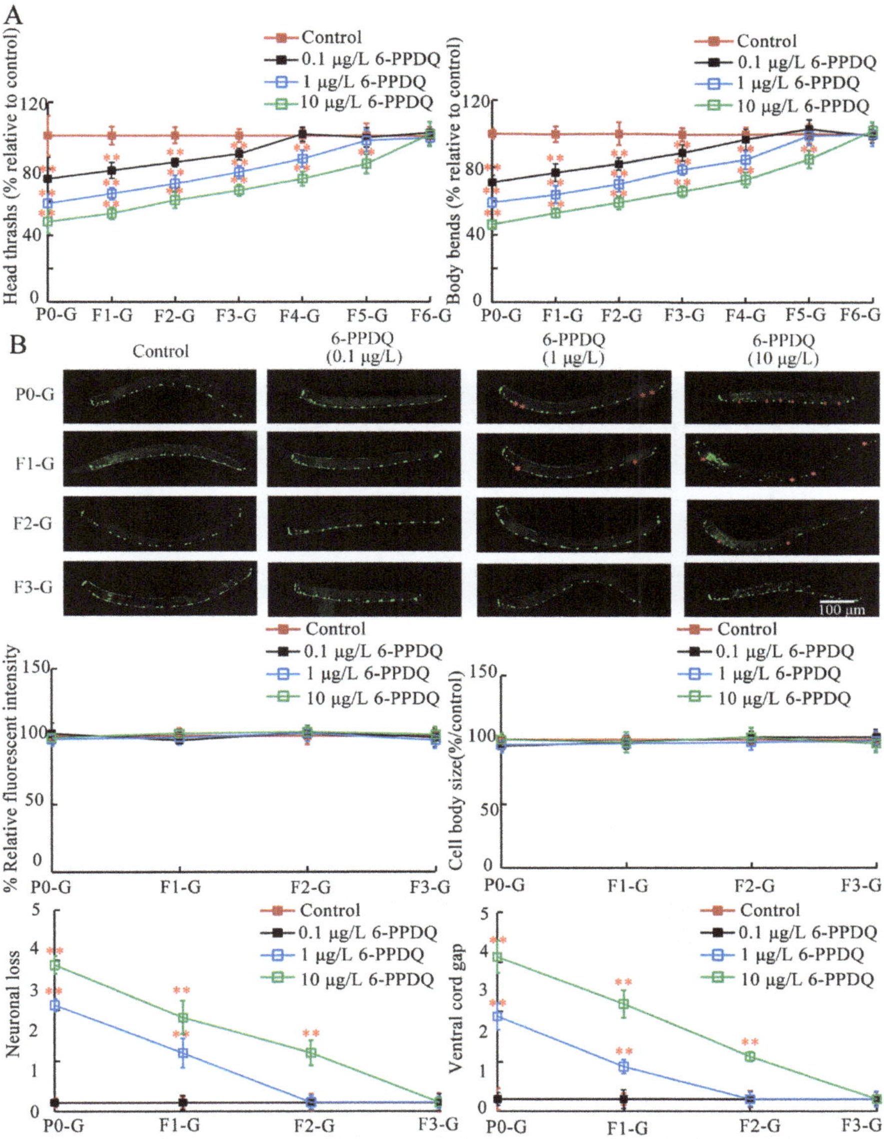

Fig. 2.6 Transgenerational neurotoxicity induced by 6-PPDQ exposure [60]. (**a**) Transgenerational toxicity of 6-PPDQ on locomotion behavior. (**b**) Transgenerational effect of 6-PPDQ on development of D-type motor neurons on GABAergic system. In images, asterisks indicate the position with neuronal loss. $^{**}P < 0.01$ vs control

be observed till to the F2-G (Fig. 2.7) [9]. Additionally, exposure to 100 μg/L PLA-MPs could further enhance transgenerational germline apoptosis and change in expression of genes (*ced-3*, *ced-4*, *egl-1*, and *ced-9*) governing apoptosis from P0-G to F2-G [9].

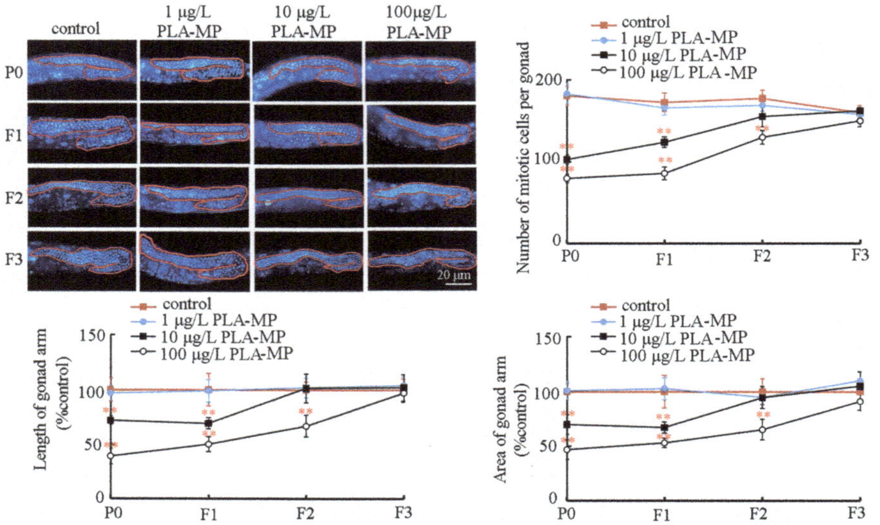

Fig. 2.7 Transgenerational effect of exposure to PLA-MP on gonad development [9] $^{**}P < 0.01$ vs control

2.7 Transgenerational Metabolic Alterations After Exposure to Toxicants

Accompanied with the toxicity on target organs, one of the normally observed toxicity is disruption in metabolisms by toxicants and stresses. In *C. elegans*, some of the metabolisms were observed to be disrupted by pollutants, such as metabolisms of amino acids, vitamin D3, lipid, glycogen, glucose, and citric acid cycle [4, 65–71]. The disruption in metabolisms by pollutants could be further observed in their offspring, suggesting the existence of transgenerational disruption in biological metabolisms [72–76].

With the 6-PPDQ as the example of pollutants, exposure to 6-PPDQ induced enhanced lipid accumulation, and the enhanced lipid accumulation could be observed till to F4-G after exposure to 10 μg/L 6-PPDQ at P0-G (Fig. 2.8) [60]. Accompanied with this induced transgenerational lipid accumulation by 6-PPDQ, a transgenerational increase in expressions of *fasn-1* and *pod-2* encoding enzymes governing fatty acid synthesis and expressions of *fat-5*, *fat-6*, and *fat-7* encoding delta-9 desaturases governing monounsaturated fatty acyl-CoAs synthesis was observed [60]. Meanwhile, exposure to 6-PPDQ also resulted in transgenerational decrease in expressions of *acs-2* and *ech-2* encoding enzymes governing mitochondrial fatty acid β-oxidation and expressions of *acs-1* and *ech-3* encoding enzymes controlling peroxisomal fatty acid β-oxidation [60].

Moreover, exposure to 6-PPDQ also induced transgenerational elevation of glucose synthesis. After exposure to 10 μg/L 6-PPDQ, elevation of glucose content could be found till to F3-G [77]. Nevertheless, expressions of *F47B8.10* and *fbp-1*

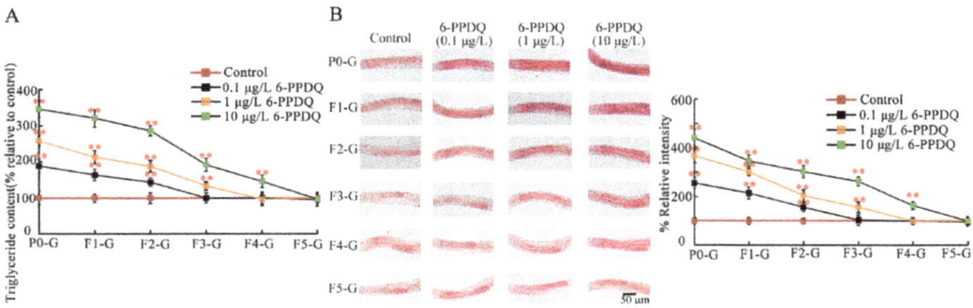

Fig. 2.8 Transgenerational lipid accumulation after exposure to 6-PPDQ at P0-G [60]. (**a**) Transgenerational change of triglyceride content after exposure to 6-PPDQ at P0-G. (**b**) Transgenerational change of lipid accumulation after exposure to 6-PPDQ at P0-G based on oil red O staining. $^{**}P < 0.01$ vs control

encoding enzymes governing gluconeogenesis did not show transgenerational increase in their expressions [77]. In contrast, transgenerational decrease in expressions of *hxk-1*, *hxk-3*, *pyk-1*, and *pyk-2* encoding enzymes governing glycolysis was induced by 6-PPDQ (Fig. 2.9) [77], which suggested that transgenerational inhibition in glycolysis contributed to formation of transgenerational elevation of glucose synthesis after 6-PPDQ exposure.

2.8 Transgenerational Physiological Alterations After Exposure to Toxicants

Accompanied with the toxicity on target organs, another usually detected toxicity is the alterations in physiological behavior by toxicants and stresses. Exposure to toxicants and stresses can alter some of the physiological behaviors, such as pumping rate and defecation behavior [78–83]. Some of these physiology-related endpoints were used to assess transgenerational effect of pollutants. For example, mean defecation cycle length and development of AVL and DVB neurons controlling defecation behavior were used to evaluate transgenerational toxicity of traffic-related fine particulate matter on physiological state [17].

2.9 Transgenerational Lifespan Reduction After Exposure to Toxicants

In *C. elegans*, lifespan is a valuable endpoint to reflect the long-term effect of toxicants and stresses [7]. Besides this, healthspan is used to reflect physiological state during the aging process [7]. Additionally, immunosenescence reflects the association between immunosuppression and lifespan reduction. Exposure to pollutants could cause lifespan reduction, inhibition in healthspan, and immunosenescence

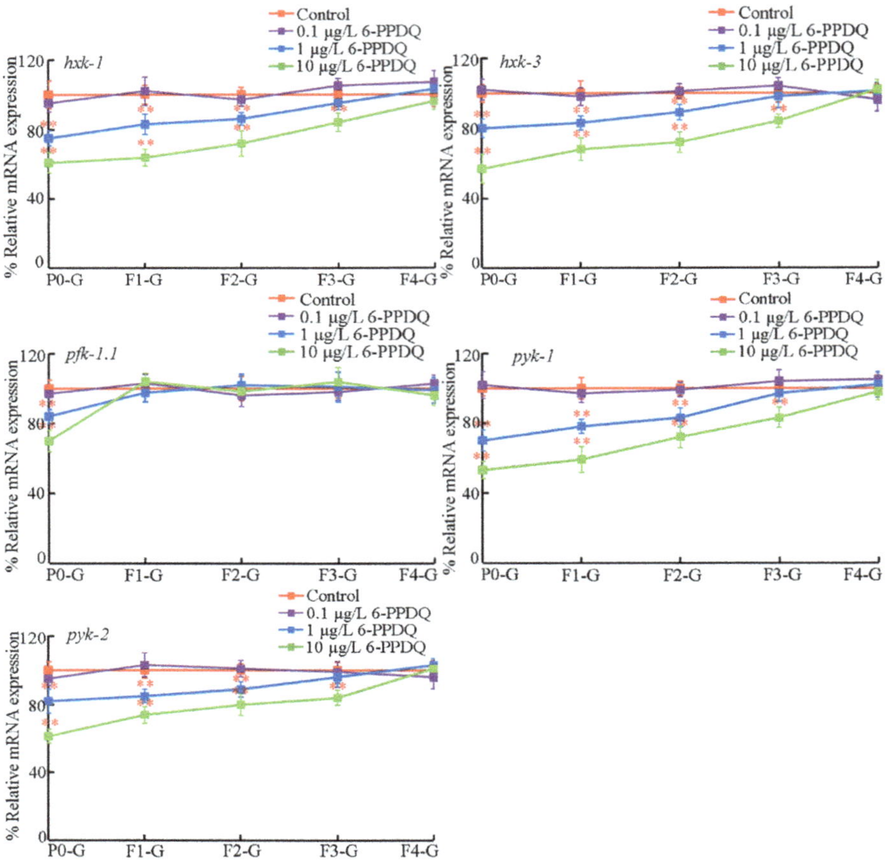

Fig. 2.9 Transgenerational changes of enzyme genes governing glycolysis after 6-PPDQ exposure at P0-G [77]. $^{**}P < 0.01$ vs control

[78, 84–86]. The endpoint of lifespan was also used to assess transgenerational toxicity of traffic-related fine particulate matter and TiO_2-NPs [17, 87].

2.10 Perspectives

In this chapter, we have introduced and discussed several aspects of transgenerational toxicity induced by toxicants and stresses, including transgenerational molecular alterations, transgenerational developmental toxicity and transgenerational damage on target organs, transgenerational metabolic and physiological alterations, and transgenerational toxicity on longevity. Nevertheless, very limited studies have been performed to determine transgenerational toxicity of toxicants and stresses on biochemical processes and in inducing genotoxicity and genetic toxicity. In addition, with transgenerational lifespan reduction by toxicants as the example, still very limited studies have been performed to investigate the effect of pollutants in inhibiting healthspan and in causing immunosenescence.

References

1. Wang D-Y (2018) Nanotoxicology in *Caenorhabditis elegans*. Springer Nature Singapore Pte Ltd., Singapore
2. Wang D-Y (2019) Molecular toxicology in *Caenorhabditis elegans*. Springer Nature Singapore Pte Ltd., Singapore
3. Wang D-Y (2019) Target organ toxicology in *Caenorhabditis elegans*. Springer Nature Singapore Pte Ltd., Singapore
4. Wang Y-X, Hua X, Wang D-Y (2023) Exposure to 6-PPD quinone enhances lipid accumulation through activating metabolic sensors of SBP-1 and MDT-15 in *Caenorhabditis elegans*. Environ Pollut 333:121937
5. Hua X, Wang D-Y (2023) Disruption of dopamine metabolism by exposure to 6-PPD quinone in *Caenorhabditis elegans*. Environ Pollut 337:122649
6. Liu Z-Y, Hua X, Zhao Y, Bian Q, Wang D-Y (2024) Polyethylene nanoplastics cause reproductive toxicity associated with activation of both estrogenic hormone receptor NHR-14 and DNA damage checkpoints in *C. elegans*. Sci Total Environ 906:167471
7. Wang D-Y (2020) Exposure toxicology in *Caenorhabditis elegans*. Springer Nature Singapore Pte Ltd, Singapore
8. Tang M, Ding G, Li L, Xiao G, Wang D (2023) Exposure to polystyrene nanoparticles at predicted environmental concentrations enhances toxic effects of *Acinetobacter johnsonii* AC15 infection on *Caenorhabditis elegans*. Ecotoxicol Environ Saf 262:115131
9. Shao Y-T, Li Y-H, Wang D-Y (2024) Polylactic acid microplastics cause transgenerational reproductive toxicity associated with activation of insulin and hedgehog ligands in *C. elegans*. Sci Total Environ 942:173746
10. Wang D-Y (2022) Toxicology at environmentally relevant concentrations in *Caenorhabditis elegans*. Springer Nature Singapore Pte Ltd, Singapore
11. Hua X, Liang G-Y, Chao J, Wang D-Y (2024) Exposure to 6-PPD quinone causes damage on mitochondrial complex I/II associated with lifespan reduction in *Caenorhabditis elegans*. J Hazard Mater 472:134598
12. Wang Y-X, Yuan X-A, Zhou R, Bu Y-Q, Wang D-Y (2023) Combinational exposure to hydroxyatrazine increases neurotoxicity of polystyrene nanoparticles on *Caenorhabditis elegans*. Sci Total Environ 880:163283
13. Wang Y-X, Liang G-Y, Chao J, Wang D-Y (2024) Comparison of intestinal toxicity in enhancing intestinal permeability and in causing ROS production of six PPD quinones in *Caenorhabditis elegans*. Sci Total Environ 927:172306
14. Hua X, Wang D-Y (2024) Polyethylene nanoparticles at environmentally relevant concentrations enhances neurotoxicity and accumulation of 6-PPD quinone in *Caenorhabditis elegans*. Sci Total Environ 918:170760
15. Ni S, Zhang H, Sun L, Zhao Y, Pei C, Nie Y, Liu X, Wu L, Xu A (2022) Transgenerational reproductive toxicity of 2,4,6-trinitrotoluene (TNT) and its metabolite 4-ADNT in *Caenorhabditis elegans*. Environ Toxicol Pharmacol 92:103865
16. Zhao Y, Chen J, Wang R, Pu X, Wang D (2023) A review of transgenerational and multigenerational toxicology in the in vivo model animal *Caenorhabditis elegans*. J Appl Toxicol 43:122–145

17. Zhao Y, Lin Z, Jia R, Li G, Xi Z, Wang D (2014) Transgenerational effects of traffic-related fine particulate matter ($PM_{2.5}$) on nematode *Caenorhabditis elegans*. J Hazard Mater 274:106–114
18. Wang L, Wu D, Yu Z, Huang S, Zhang J (2023) Hormone-mediated multi- and transgenerational reproductive toxicities of 1-ethyl-3-methylimidazolium hexafluorophosphate on *Caenorhabditis elegans*. Sci Total Environ 863:160958
19. Chen H, Chen X, Gu Y, Jiang Y, Guo H, Chen J, Yu J, Wang C, Chen C, Li H (2024) Transgenerational reproductive toxicity induced by carboxyl and amino charged microplastics at environmental concentrations in *Caenorhabditis elegans*: involvement of histone methylation. Sci Total Environ 949:175132
20. Zhang L, Wang S-T, Zhao Y-L, Nurdebek B, Bu Y-Q, Wang D-Y (2022) Long-term exposure to polystyrene nanoparticles causes transgenerational toxicity by affecting the function and expression of MEV-1 and DAF-2 signals in *Caenorhabditis elegans*. NanoImpact 26:100403
21. Xu R-R, Hua X, Rui Q, Wang D-Y (2022) Polystyrene nanoparticles caused dynamic alteration in mitochondrial unfolded protein response from parents to the offspring in *C. elegans*. Chemosphere 308:136154
22. Zhao Y-Y, Hua X, Rui Q, Wang D-Y (2023) Exposure to multi-walled carbon nanotubes causes suppression in octopamine signal associated with transgenerational toxicity induction in *C. elegans*. Chemosphere 318:137986
23. Wamuho A, Unrine JM, Kieran TJ, Glenn TC, Schultz CL, Farman M, Svendsen C, Spurgeon DJ, Tsyusko OV (2019) Genomic mutations after multigenerational exposure of *Caenorhabditis elegans* to pristine and sulfidized silver nanoparticles. Environ Pollut 254:113078
24. Li J, Dai L, Feng Y, Cao Z, Ding Y, Xu H, Xu A, Du H (2024) Multigenerational effects and mutagenicity of three flame retardants on germ cells in *Caenorhabditis elegans*. Ecotoxicol Environ Saf 269:115815
25. Wang S, Meyer DH, Schumacher B (2023) Inheritance of paternal DNA damage by histone-mediated repair restriction. Nature 613:365–374
26. Shao Y-T, Hua X, Li Y-H, Wang D-Y (2024) Comparison of reproductive toxicity between pristine and aged polylactic acid microplastics in *Caenorhabditis elegans*. J Hazard Mater 466:133545
27. Wu J-W, Shao Y-T, Hua X, Li Y-H, Wang D-Y (2025) Photo-aged polylactic acid microplastics causes severe transgenerational decline in reproductive capacity in *C. elegans*: insight into activation of DNA damage checkpoints affected by multiple germline histone methyltransferases. Environ Pollut 382:126697
28. Chen F, Hersh BM, Conradt B, Zhou Z, Riemer D, Gruenbaum Y, Horvitz HR (2000) Translocation of *C. elegans* CED-4 to nuclear membranes during programmed cell death. Science 287:1485–1489
29. Conradt B, Horvitz HR (1998) The *C. elegans* protein EGL-1 is required for programmed cell death and interacts with the Bcl-2-like protein CED-9. Cell 93:519–529
30. Reddien PW, Cameron S, Horvitz HR (2001) Phagocytosis promotes programmed cell death in *C. elegans*. Nature 412:198–202
31. Ahmed S, Alpi A, Hengartner MO, Gartner A (2001) *C. elegans* RAD-5/CLK-2 defines a new DNA damage checkpoint protein. Curr Biol 11:1934–1944
32. Hofmann ER, Milstein S, Boulton SJ, Ye M, Hofmann JJ, Stergiou L, Gartner A, Vidal M, Hengartner MO (2002) *Caenorhabditis elegans* HUS-1 is a DNA damage checkpoint protein required for genome stability and EGL-1-mediated apoptosis. Curr Biol 12:1908–1918
33. Gartner A, Milstein S, Ahmed S, Hodgkin J, Hengartner MO (2000) A conserved checkpoint pathway mediates DNA damage--induced apoptosis and cell cycle arrest in *C. elegans*. Mol Cell 5:435–443
34. Zhao Y-Y, Li D, Rui Q, Wang D-Y (2020) Toxicity induction of nanopolystyrene under microgravity stress condition in *Caenorhabditis elegans*. Sci Total Environ 703:135623
35. Qu M, Qiu Y-X, Kong Y, Wang D-Y (2019) Amino modification enhances reproductive toxicity of nanopolystyrene on gonad development and reproductive capacity in nematode *Caenorhabditis elegans*. Environ Pollut 254:112978

36. Qu M, Nida A, Kong Y, Du H-H, Xiao G-S, Wang D-Y (2019) Nanopolystyrene at predicted environmental concentration enhances microcystin-LR toxicity by inducing intestinal damage in *Caenorhabditis elegans*. Ecotoxicol Environ Saf 183:109568
37. Qu M, Wang D-Y (2020) Toxicity comparison between pristine and sulfonate modified nanopolystyrene particles in affecting locomotion behavior, sensory perception, and neuronal development in *Caenorhabditis elegans*. Sci Total Environ 703:134817
38. Li D, Deng Y-J, Wang S-T, Du H-H, Xiao G-S, Wang D-Y (2020) Assessment of nanopolystyrene toxicity under fungal infection condition in *Caenorhabditis elegans*. Ecotoxicol Environ Saf 197:110625
39. Liu H-L, Zhao Y-L, Hua X, Wang D-Y (2022) Induction of transgenerational toxicity is associated with the activated germline insulin signals in nematodes exposed to nanoplastic at predicted environmental concentrations. Ecotoxicol Environ Saf 243:114022
40. Xu R-R, Hua X, Rui Q, Wang D-Y (2022) Alteration in Wnt signaling mediates induction of transgenerational toxicity of polystyrene nanoplastics in *C. elegans*. NanoImpact 28:100425
41. Hua X, Cao C, Zhang L, Wang D-Y (2023) Activation of FGF signal in germline mediates transgenerational toxicity of polystyrene nanoparticles at predicted environmental concentrations in *Caenorhabditis elegans*. J Hazard Mater 451:131174
42. Wu J-W, Shao Y-T, Hua X, Wang D-Y (2024) Activated hedgehog and insulin ligands by decreased transcriptional factor DAF-16 mediate transgenerational nanoplastic toxicity in *Caenorhabditis elegans*. J Hazard Mater 480:135909
43. He W-M, Gu A-H, Wang D-Y (2023) Sulfonate-modified polystyrene nanoparticle at precited environmental concentrations induces transgenerational toxicity associated with increase in germline notch signal of *Caenorhabditis elegans*. Toxics 11:511
44. Zhao Y, Hua X, Bian Q, Wang D-Y (2022) Nanoplastic exposure at predicted environmental concentrations induces activation of germline Ephrin signal associated with toxicity formation in the *Caenorhabditis elegans* offspring. Toxics 10:699
45. Liu H-L, Tian L-J, Wang S-T, Wang D-Y (2021) Size-dependent transgenerational toxicity induced by nanoplastics in nematode *Caenorhabditis elegans*. Sci Total Environ 790:148217
46. Wang Y-X, Wang D-Y (2024) Transgenerational intestinal toxicity of 6-PPD quinone in causing ROS production, enhancement in intestinal permeability and suppression in innate immunity in *C. elegans*. Environ Pollut 363:125208
47. Hu Y-O, Wang Y, Ye B-P, Wang D-Y (2008) Phenotypic and behavioral defects induced by iron exposure can be transferred to progeny in *Caenorhabditis elegans*. Biomed Environ Sci 21:467–473
48. Wang D-Y, Wang Y (2008) Nickel sulfate induces numerous defects in *Caenorhabditis elegans* that can also be transferred to progeny. Environ Pollut 151:585–592
49. Wang D-Y, Yang P (2007) The multi-biological defects caused by lead exposure exhibit transferable properties from exposed parents to their progeny in *Caenorhabditis elegans*. J Environ Sci 19:1367–1372
50. Wang Y, Xie W, Wang D-Y (2007) Transferable properties of multi-biological toxicity caused by cobalt exposure in *Caenorhabditis elegans*. Environ Toxicol Chem 26:2405–2412
51. Wang D-Y, Xing X-J (2008) Assessment of locomotion behavioral defects induced by acute toxicity from heavy metal exposure in nematode *Caenorhabditis elegans*. J Environ Sci 20:1132–1137
52. Li Y-X, Yu S-H, Wu Q-L, Tang M, Wang D-Y (2013) Transmissions of serotonin, dopamine and glutamate are required for the formation of neurotoxicity from Al_2O_3-NPs in nematode *Caenorhabditis elegans*. Nanotoxicology 7:1004–1013
53. Xing X-J, Du M, Zhang Y-F, Wang D-Y (2009) Adverse effects of metal exposure on chemotaxis towards water-soluble attractants regulated mainly by ASE sensory neuron in nematode *Caenorhabditis elegans*. J Environ Sci 21:1684–1694
54. Zhang Y-F, Ye B-P, Wang D-Y (2010) Effects of metal exposure on associative learning behavior in nematode *Caenorhabditis elegans*. Arch Environ Contamin Toxicol 59:129–136

55. Li Y-P, Li Y-X, Wu Q-L, Ye H-Y, Sun L-M, Ye B-P, Wang D-Y (2013) High concentration of vitamin E decreases thermosensation and thermotaxis learning and the underlying mechanisms in nematode *Caenorhabditis elegans*. PLoS One 8:e71180
56. Ye H-Y, Ye B-P, Wang D-Y (2008) Trace administration of vitamin E can retrieve and prevent UV-irradiation- and metal exposure-induced memory deficits in nematode *Caenorhabditis elegans*. Neurobiol Learn Mem 90:10–18
57. Du M, Wang D-Y (2009) The neurotoxic effects of heavy metal exposure on GABAergic system in nematode *Caenorhabditis elegans*. Environ Toxicol Pharmacol 27:314–320
58. Xing X-J, Du M, Xu X-M, Rui Q, Wang D-Y (2009) Exposure to metals induces morphological and functional alteration of AFD neurons in nematode *Caenorhabditis elegans*. Environ Toxicol Pharmacol 28:104–110
59. Zhao Y-L, Wang X, Wu Q-L, Li Y-P, Wang D-Y (2015) Translocation and neurotoxicity of CdTe quantum dots in RMEs motor neurons in nematode *Caenorhabditis elegans*. J Hazard Mater 283:480–489
60. Wang Y-X, Wu J-W, Wang D-Y (2025) 6-PPD quinone causes lipid accumulation across multiple generations differentially affected by metabolic sensors and components of COMPASS complex in *Caenorhabditis elegans*. Environ Pollut 366:125539
61. Liu Z-Y, Bian Q, Wang D-Y (2025) 6-PPD quinone induces response of nuclear hormone receptors in the germline associated with formation of reproductive toxicity in *Caenorhabditis elegans*. J Hazard Mater 495:138815
62. Hua X, Feng X, Liang G-Y, Chao J, Wang D-Y (2023) Long-term exposure to 6-PPD quinone reduces reproductive capacity by enhancing germline apoptosis associated with activation of both DNA damage and cell corpse engulfment in *Caenorhabditis elegans*. J Hazard Mater 454:131495
63. Liu P-D, He K-W, Li Y-X, Wu Q-L, Yang P, Wang D-Y (2012) Exposure to mercury causes formation of male-specific structural deficits by inducing oxidative damage in nematodes. Ecotoxicol Environ Saf 79:90–100
64. Ruan Q, Ju J, Li Y, Li X, Liu R, Liang G, Zhang J, Pu Y, Wang D, Yin L (2012) Chlorpyrifos exposure reduces reproductive capacity owing to a damaging effect on gametogenesis in the nematode *Caenorhabditis elegans*. J Appl Toxicol 32:527–535
65. Wan X, Liang G-Y, Wang D-Y (2025) 6-PPD quinone at environmentally relevant concentrations disrupts citric acid cycle in *Caenorhabditis elegans*: role of reduction in acetyl CoA and pyruvate contents. Environ Chem Ecotoxicol 7:1119–1129
66. Wang W, Li Y-H, Wang D-Y (2025) Long-term exposure to 6-PPD quinone inhibits glutamate synthesis and glutamate receptor function associated with its toxicity induction in *Caenorhabditis elegans*. Toxics 13:434
67. Wang W, Li Y-H, Wang D-Y (2025) Exposure to 6-PPD quinone disrupts adsorption and catabolism of leucine to cause mitochondrial dysfunction in *Caenorhabditis elegans*. Toxics 13:544
68. Wu J-W, Li L-E, Hu D-Y, Liu R, Bian Q, Wang D-Y (2025) Environmentally relevant concentrations of 6-PPDQ disrupts vitamin D3 adsorption and receptor function in *Caenorhabditis elegans*. Environ Sci Processes Impacts 27:2798–2808
69. Wang Y-X, Hu G-C, Wang D-Y (2025) Increased S-adenosyl methionine strengthens the suppression in mitochondrial unfolded protein response induced by 6-PPD quinone at environmentally relevant concentrations in *Caenorhabditis elegans*. Environ Pollut 386:127231
70. Wang Y-X, Wang D-Y (2024) Exposure to 6-PPD quinone enhances glycogen accumulation in *Caenorhabditis elegans*. Environ Pollut 359:124600
71. Liu Z-J, Li Y-H, Wang D-Y (2024) Exposure to 6-PPD quinone disrupts glucose metabolism associated with lifespan reduction by affecting insulin and AMPK signals in *Caenorhabditis elegans*. Chemosphere 363:142975
72. He W, Chen X, Zhou S, Fu Z, Xu J, Liu Q, Cubero FJ, Li Z, Zhou Y, Wang D, Jiang G, Gu A (2025) Bisphenol S induced transgenerational lipid accumulation by triggering intestine-to-germline histone H3K4me3 modification in *Caenorhabditis elegans*. Environ Sci Technol 59:14957–14968

73. Chow Y, Sato F (2019) Transgenerational lipid-reducing activity of benzylisoquinoline alkaloids in *Caenorhabditis elegans*. Genes Cells 24:70–81
74. Li Z, Yu Z, Yin D (2021) Multi- and trans-generational disturbances of perfluorobutane sulfonate and perfluorohexane sulfonate on lipid metabolism in *Caenorhabditis elegans*. Chemosphere 280:130666
75. Li Z, Yu Z, Cui C, Ai F, Yin D (2020) Multi-generational obesogenic effects of sulfomethoxazole on *Caenorhabditis elegans* through epigenetic regulation. J Hazard Mater 382:121061
76. Li Z, Yu Z, Gao P, Yin D (2020) Multigenerational effects of perfluorooctanoic acid on lipid metabolism of *Caenorhabditis elegans* and its potential mechanism. Sci Total Environ 703:134762
77. Liu Z-J, Li Y-H, Wang D-Y (2024) Transgenerational response of glucose metabolism in *Caenorhabditis elegans* exposed to 6-PPD quinone. Chemosphere 367:143653
78. Hua X, Wang D-Y (2023) Exposure to 6-PPD quinone at environmentally relevant concentrations inhibits both lifespan and healthspan in *C. elegans*. Environ Sci Technol 57:19295–19303
79. Zhao Y-L, Wu Q-L, Tang M, Wang D-Y (2014) The *in vivo* underlying mechanism for recovery response formation in nano-titanium dioxide exposed *Caenorhabditis elegans* after transfer to the normal condition. Nanomedicine: Nanotechnol. Biol Med 10:89–98
80. Wei J, Zhang Y, Shi W, Lu L, Zhou Q, Pu Y, Yin L (2025) Copper exposure induces neurotoxicity through ferroptosis in *C. elegans*. Chem Biol Interact 407:111369
81. Wu Q, Qu Y, Li X, Wang D (2012) Chromium exhibits adverse effects at environmental relevant concentrations in chronic toxicity assay system of nematode *Caenorhabditis elegans*. Chemosphere 87:1281–1287
82. How CM, Yen PL, Wei CC, Li SW, Liao VH (2019) Early life exposure to di(2-ethylhexyl) phthalate causes age-related declines associated with insulin/IGF-1-like signaling pathway and SKN-1 in *Caenorhabditis elegans*. Environ Pollut 251:871–878
83. Wu Q-L, Li Y-X, Tang M, Wang D-Y (2012) Evaluation of environmental safety concentrations of DMSA coated Fe_2O_3-NPs using different assay systems in nematode *Caenorhabditis elegans*. PLoS One 7:e43729
84. Wu J-W, Shen S-H, Wang D-Y (2024) 6-PPD quinone at environmentally relevant concentrations induces immunosenescence by causing immunosuppression during the aging process. Chemosphere 368:143719
85. Zhao Y-L, Yang R-L, Rui Q, Wang D-Y (2016) Intestinal insulin signaling encodes two different molecular mechanisms for the shortened longevity induced by graphene oxide in *Caenorhabditis elegans*. Sci Rep 6:24024
86. Zhao Y-L, Zhi L-T, Wu Q-L, Yu Y-L, Sun Q-Q, Wang D-Y (2016) p38 MAPK-SKN-1/Nrf signaling cascade is required for intestinal barrier against graphene oxide toxicity in *Caenorhabditis elegans*. Nanotoxicology 10:1469–1479
87. Hu C, Hou J, Zhu Y, Lin D (2020) Multigenerational exposure to TiO2 nanoparticles in soil stimulates stress resistance and longevity of survived *C. elegans* via activating insulin/IGF-like signaling. Environ Pollut 263:114376

Chapter 3
Factors Affecting Transgenerational Toxicity

Abstract Induction of transgenerational toxicity after exposure to stresses and pollutants can be affected by multiple factors. In this chapter, we first discussed the contribution of exposure dose, exposure duration, developmental stage, and genetic background to transgenerational toxicity of stresses and pollutants. Moreover, influence of physical properties and chemical properties on transgenerational toxicity induction of pollutants was discussed.

Keywords Transgenerational toxicity · Factors · *Caenorhabditis elegans*

3.1 Introduction

Caenorhabditis elegans was initially developed as a classic model animal for the study of life science. After that, due to its property of high sensitivity to stresses and pollutants, *C. elegans* was applied in the field of environmental science and toxicology. In *C. elegans*, knowledge systems of molecular toxicology, target toxicology, exposure toxicology, and toxicology at environmentally relevant concentrations were established [1–5]. These backgrounds provide important foundation for further systematically performing the study on transgenerational toxicology in this animal model.

In Chap. 2, we have introduced and discussed different endpoints useful for assessing transgenerational toxicity of stresses and pollutants. These raised endpoints can be employed to detect transgenerational toxicity of stresses and pollutants at different aspects. Additionally, different endpoints show different sensitivity to toxicity of stresses and pollutants. Besides this, induction of transgenerational toxicity of stresses and pollutants can be affected by some other factors. Thus, in this chapter, we introduced and discussed the contribution of exposure dose, exposure duration, developmental stage, genetic background, physical property, and chemical property to formation of transgenerational toxicity of stresses and pollutants.

D. Wang, *Transgenerational Toxicology in Caenorhabditis elegans*,
https://doi.org/10.1007/978-981-92-1389-4_3

3.2 Exposure Dose

The exposure dose is usually the first factor to be considered to affect the toxicity induction of stresses and pollutants. During the control of transgenerational toxicity of stresses and pollutants, normally, exposure to the higher doses of stresses and pollutants will induce more severe transgenerational toxicity [6–8]. At the P0-G, exposure to polystyrene nanoparticle (PS-NP), an important member of nanoplastics, resulted in several aspects of toxicity, such as neurotoxicity, reproductive toxicity, and intestinal damage [9–17]. With PS-NP as the example of pollutants, exposure to 100–10,000 μg/L at P0-G caused intestinal oxidative stress, neurotoxicity, and reproductive toxicity at the F1-G (Fig. 3.1) [8]. More importantly, PS-NP-induced toxicity at the F1-G was dose dependent, and 10,000 μg/L PS-NP caused intestinal oxidative stress, neurotoxicity, and reproductive toxicity, which were more severe than those induced by 100 μg/L PS-NP at the F1-G (Fig. 3.1) [8].

Besides this, exposure to different doses of stresses and pollutants may induce different transgenerational responses. The 6-PPD quinone (6-PPDQ) is a tire-derived emergent contaminant [18], and its exposure caused multiple aspects of damage on *C. elegans*, such as neurotoxicity, lifespan reduction, metabolic disruption, and intestinal damage, as well as transgenerational toxicity [19–28]. Accompanied with the toxicity induction, 0.1–1 μg/L 6-PPDQ induced activation of mitochondrial unfolded protein response (mt UPR) and mitophagy, whereas 10 μg/L 6-PPDQ caused suppression in mt UPR and mitophagy [29, 30]. These observations implied that exposure to low doses of stresses and pollutants may induce certain forms of transgenerational protective responses and exposure to high doses of stresses and pollutants may cause severe transgenerational toxicity.

3.3 Exposure Duration

In *C. elegans*, longer-duration exposure to stresses and pollutants will cause more severe toxicity than relatively short exposure duration [4, 31, 32]. For this reason, to detect the possible transgenerational toxicity of stresses and pollutants, the long-term exposure duration is usually performed. For example, exposure to PS-NPs from L1-larvar for 6.5 days could cause transgenerational intestinal damage, transgenerational neurotoxicity, and transgenerational reproductive toxicity [33–39]. Also, exposure to 6-PPDQ from L1-larvar for 6.5 days could result in transgenerational intestinal damage and transgenerational lipid accumulation [40, 41].

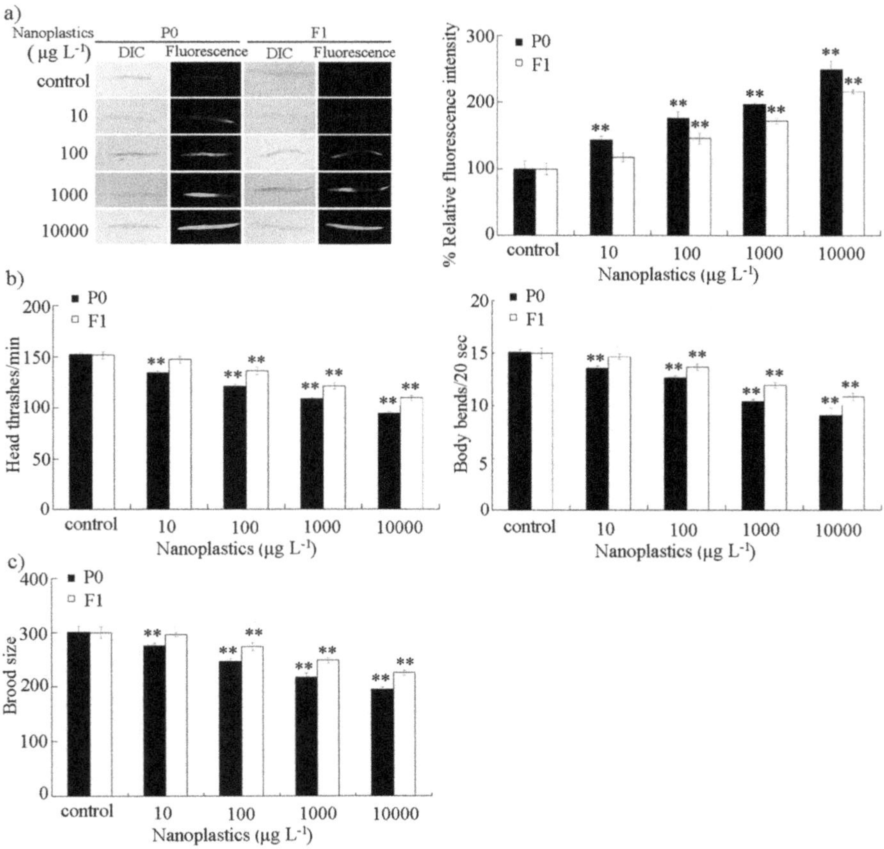

Fig. 3.1 Transgenerational toxicity of nanopolystyrene particles labeled with Rho B after prolonged exposure [8]. (**a**) Transgenerational toxicity of nanopolystyrene particles labeled with Rho B in inducing intestinal ROS production after prolonged exposure. (**b**) Transgenerational toxicity of nanopolystyrene particles labeled with Rho B in decreasing locomotion behavior after prolonged exposure. (**c**) Transgenerational toxicity of nanopolystyrene particles labeled with Rho B in reducing brood size after prolonged exposure. Prolonged exposure was performed from L1-larvae to adult day 1. Bars represent means ± SD. $^{**}P < 0.05$ vs. control

3.4 Developmental Stage

In *C. elegans*, larvae are more sensitive to toxicity of stresses and pollutants than young adults or adults [4, 42–44]. That is, in larval nematodes, it will be more possible to detect the potential toxicity of stresses and pollutants at low doses. Therefore, the detection of transgenerational pollutants toxicity was normally performed from larval stages. For example, exposure to PS-NP from L1-larvae till to adult day 3 caused transgenerational suppression in mt UPR [45]. Exposure to multi-walled carbon nanotubes (MWCNTs) from L1-larvae till to adult day 1 resulted in

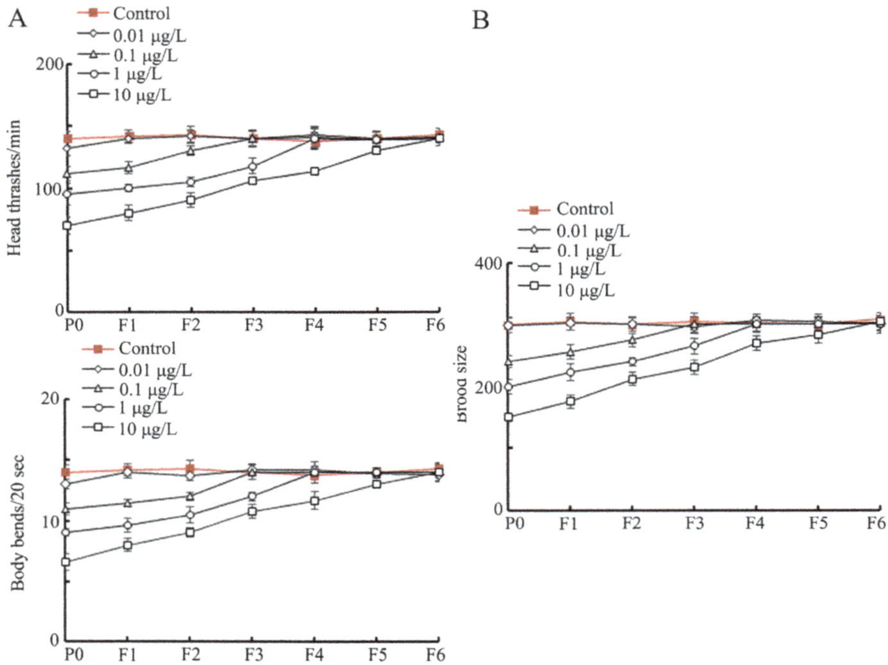

Fig. 3.2 Transgenerational toxicity induced by MWCNT exposure in wild-type nematodes [46]. (**a**) MWCNT exposure induced transgenerational toxicity in decreasing locomotion behavior. (**b**) MWCNT exposure induced transgenerational toxicity in reducing brood size. Control, nematodes without MWCNT exposure. Bars represent means ± SD

transgenerational neurotoxicity and transgenerational reproductive toxicity (Fig. 3.2) [46].

3.5 Genetic Background

In *C. elegans*, different genetic backgrounds can induce susceptibility or resistance to toxicity of stresses and pollutants [4]. Especially, mutation of some genes can induce the susceptible property of nematodes to stresses and pollutants [47, 48], which is helpful for detecting potential toxicity of stresses and pollutants at low doses. During the control of transgenerational toxicity of stresses and pollutants, susceptibility or resistance to transgenerational toxicity will also be induced by RNAi of certain genes. For example, RNAi of *mev-1*, *cbp-1*, *taf-1, sir-2.1*, *hda-3*, *ced-1*, *rab-5*, and *nhr-12* led to susceptibility to transgenerational neurotoxicity and/or transgenerational reproductive toxicity of PS-NP [49–53]. In contrast, RNAi of *daf-2*, *daf-12*, *nhr-14*, *nhr-47*, hedgehog ligand genes (*grl-15*, *grl-16*, *qua-1*, and *wrt-1*), and hedgehog receptor genes (*ptr-23*, *scp-1*, *ptd-2*, and *ncr-1*) conferred resistance to transgenerational neurotoxicity and transgenerational reproductive toxicity of PS-NP [49, 53, 54]. Additionally, transgenerational reproductive toxicity of photo-aged polylactic acid microplastics could be increased by RNAi of *mes-2*, *set-27*, *set-24*, *set-28*, *set-31*, and *set-2* and decreased by RNAi of *met-2*, *set-3*, and *set-6* RNAi [55].

3.6 Physical Property

3.6.1 Size

Among the physical properties of particle pollutants, size may be the most important one affecting their toxicity induction. Normally, the smaller particle pollutants will induce more severe toxicity at different aspects than larger size of particle pollutants [56–58]. With the PS-NPs as the example of particle pollutants, after exposure at the concentration of 100 μg/L, 100 nm PS-NPs caused transgenerational neurotoxicity and transgenerational reproductive toxicity till to the F3-G, whereas 20 nm PS-NPs could result in transgenerational neurotoxicity and transgenerational reproductive toxicity till to the F6-G (Fig. 3.3) [59].

3.6.2 Leachate or Ion Release

For particle pollutants, such as nanoparticles, the leachates will affect their toxicity induction to a certain degree [1, 4]. For metal nanoparticles, metal ion released will influence their toxicity formation to a certain degree [1, 4]. For example, the leachate of PS-NPs at a concentration of 10 mg/L could result in intestinal oxidative stress, locomotion inhibition, and reproductive reduction [8]. Therefore, during assessing transgenerational toxicity of particle pollutants, the possible contribution of leachate to transgenerational toxicity induction needs to be examined. Or at least this aspect should be excluded. In addition, the possible contribution of leachates from aged particle pollutants to their transgenerational toxicity induction should be paid attention to carefully, since relatively a large amount of leachates would be formed for these particle pollutants.

3.6.3 Aging

For particle pollutants, aging process, such as UV induced photo-aging, can strengthen their transgenerational toxicity induction. With PS-NP as the example of particle pollutants, after exposure at the concentration of 100 μg/L, pristine PS-NPs

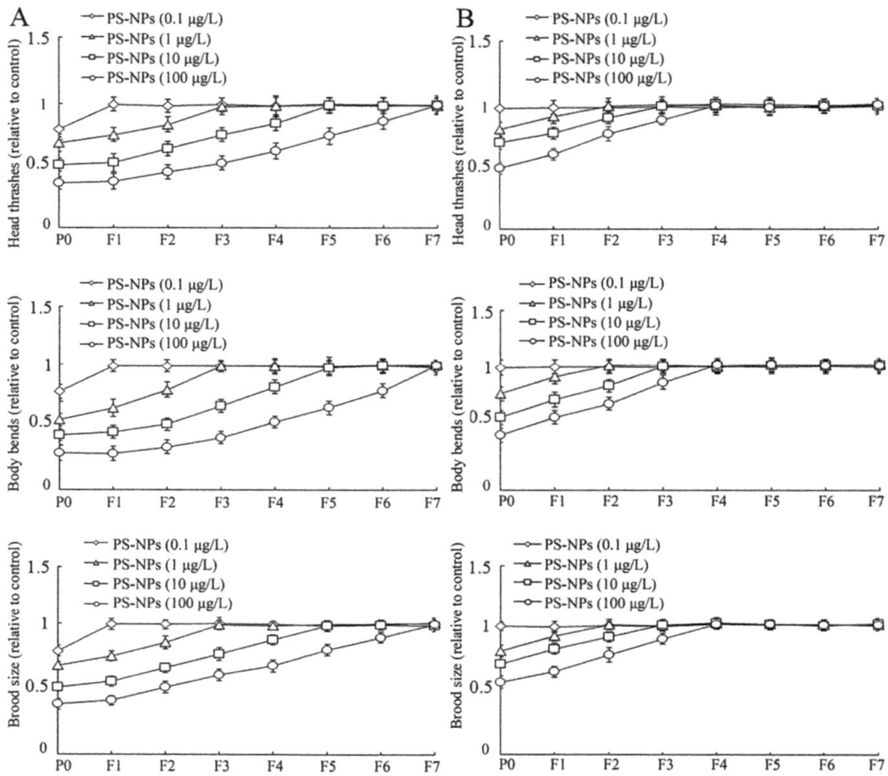

Fig. 3.3 Comparison of transgenerational toxicity between 20 nm PS-NPs and 100 nm PS-NPs [59]. (**a**) Transgenerational toxicity of 20 nm PS-NPs in decreasing locomotion behavior and in reducing brood size. The data was normalized to the control (control = 1). (**b**) Transgenerational toxicity of 100 nm PS-NPs in decreasing locomotion behavior and in reducing brood size. The data was normalized to the control (control = 1). The nematodes were exposed to PS-NPs from L1-larvae for 6.5 days at the P0 generation, and the nematodes were grown on normal NGM plates without PS-NPs exposure from the F1 to the F7 generations. Bars represent means ± SD

caused transgenerational reproductive toxicity till to the F1-G; however, photoaged PS-NPs could induce transgenerational reproductive toxicity till to the F2-G [60].

3.7 Chemical Property

Some chemical modifications, such as amino modification and sulfonate modification, can enhance toxicity induction of pollutants, such as the particle pollutants [10, 15]. In contrast, some modification, such as PEG modification, carboxylic acid modification, and ZnS coating, can reduce the toxicity induction of pollutants, such as the particle pollutants [61–65]. In *C. elegans*, transgenerational reproductive

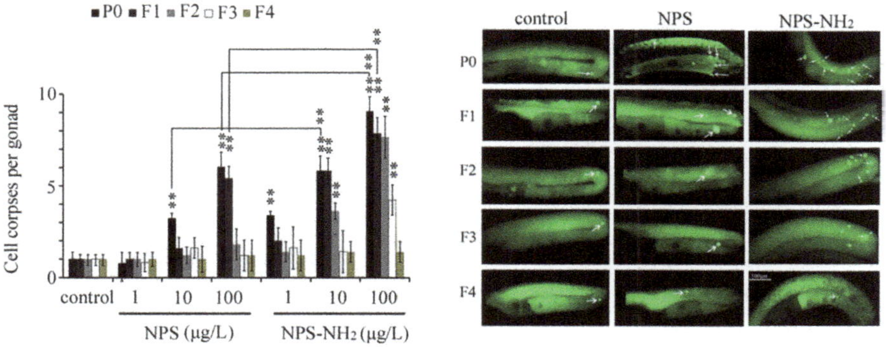

Fig. 3.4 Comparison of the effects between NPS and NPS-NH_2 in inducing germline apoptosis transgenerationally [67]. (**a**) Comparison of the effects between NPS and NPS-NH_2 in inducing transgenerational toxicity to cause germline apoptosis. N = 50. **$P < 0.01$ vs control (if not specially indicated). (**b**) Images showing the induction of germline apoptosis. Arrow indicates the germline apoptosis signals. Exposure concentration of NPS and NPS-NH_2 was 100 μg/L. Control, K-medium

toxicity induced by pristine and amino-modified PS-NPs was compared [66]. Transgenerational PS-NPs toxicity can be strengthened by different chemical modifications [66]. For example, after exposure at the same concentration, amino-modified PS-NPs caused more severe transgenerational reproductive toxicity (such as transgenerational inhibition in reproductive capacity) than pristine PS-NPs [67]. After exposure at the concentration of 100 μg/L, transgenerational germline apoptosis induction induced by pristine PS-NPs was observed till to the F1-G; however, transgenerational germline apoptosis induction induced by amino-modified PS-NPs could last till to the F3-G (Fig. 3.4) [67].

Besides this, amino modification also enhances the PS-NPs-induced transgenerational neurotoxicity. After exposure at the concentration of 10 μg/L at the P0-G, pristine PS-NPs-induced inhibition in locomotion could be observed till to F3-G, and pristine PS-NPs caused damage on development of D-type motor neurons could be observed till to F1-G (Fig. 3.5) [68]. In contrast, amino-modified PS-NPs-induced inhibition in locomotion could be detected till to F4-G, and damage caused by amino-modified PS-NPs on development of D-type motor neurons could be detected till to F2-G (Fig. 3.5) [68]. The difference in transgenerational neurotoxicity caused by pristine and amino-modified PS-NPs was associated with the difference in transgenerational activation of *mec-4*, *crt-1*, *itr-1*, and *tra-3* governing neurodegeneration; transgenerational inhibition in *mpk-1*, *jnk-1*, *dbl-1*, and *daf-7* expressions; and transgenerational response of ligand genes (*ins-3*, *ins-39*, *daf-28*, *lin-44*, *egl-17*, *efn-3*, and *lag-2*) (Fig. 3.6) [68].

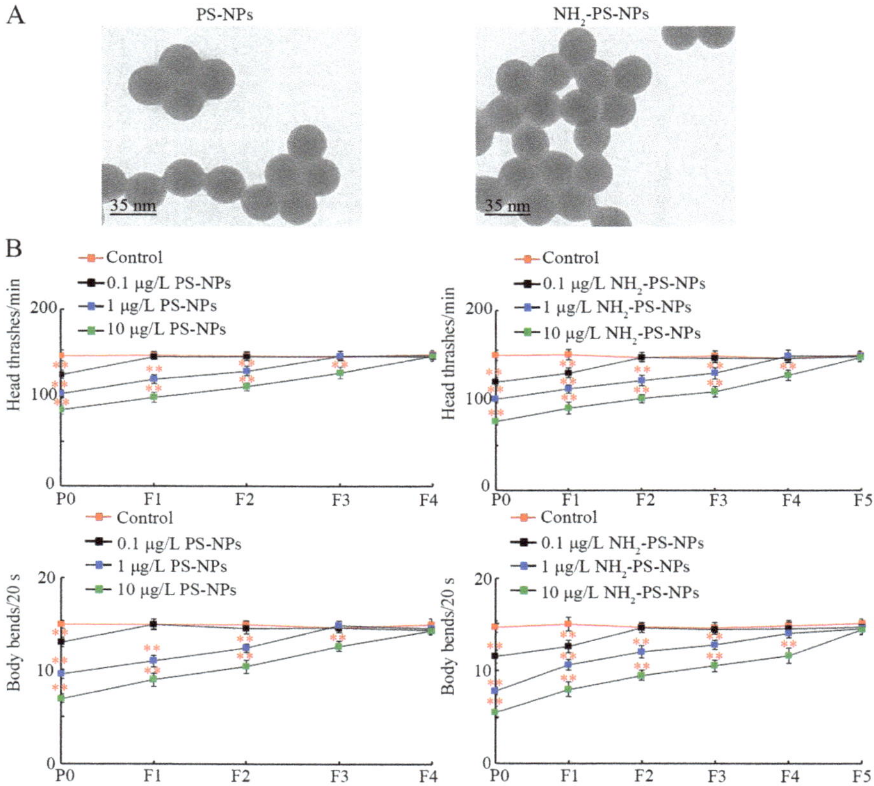

Fig. 3.5 Comparison of transgenerational effect between pristine and amino-modified PS-NPs on locomotion behavior [68]. (**a**) TEM images of pristine and amino-modified PS-NPs before sonication. (**b**) Comparison of transgenerational effect between pristine and amino-modified PS-NPs on head thrash and body bend. $^{**}P < 0.01$ vs. control

3.8 Perspectives

In this chapter, we discussed the factors possible affecting the induction of transgenerational toxicity after exposure to stresses and pollutants. Actually, these factors are somewhat similar to those exerting their effects on toxicity induction of stresses and pollutants at the P0-G. Considering that the factors discussed are mainly focused on those supported by published references, some other factors confirmed at the P0-G need to be also examined for their possible role in affecting transgenerational toxicity induction of stresses and pollutants. In addition, the available published references discussed in this chapter are mainly on pollutants. The factors affecting transgenerational toxicity of different stresses need to be further carefully determined, since toxicity induction of certain stresses is somewhat or very different from that of pollutants.

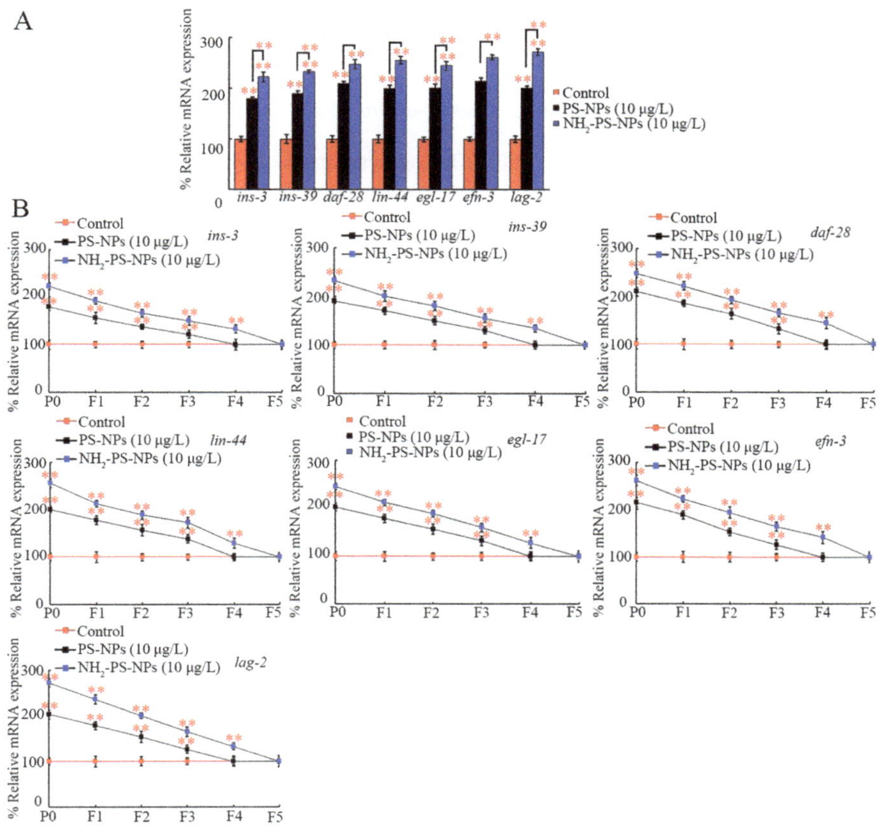

Fig. 3.6 Comparison of transgenerational effect between pristine and amino-modified PS-NPs on expressions of *ins-3*, *ins-39*, *daf-28*, *lin-44*, *egl-17*, *efn-3*, and *lag-2* [68]. (**a**) Comparison of effect between pristine and amino-modified PS-NPs on expressions of *ins-3*, *ins-39*, *daf-28*, *lin-44*, *egl-17*, *efn-3*, and *lag-2* at P0-G. (**b**) Comparison of transgenerational effect between pristine and amino-modified PS-NPs on expressions of *ins-3*, *ins-39*, *daf-28*, *lin-44*, *egl-17*, *efn-3*, and *lag-2*. $^{**}P < 0.01$ vs. control (if not specially indicated)

References

1. Wang D-Y (2018) Nanotoxicology in *Caenorhabditis elegans*. Springer Nature Singapore Pte Ltd, Singapore
2. Wang D-Y (2019) Molecular toxicology in *Caenorhabditis elegans*. Springer Nature Singapore Pte Ltd., Singapore
3. Wang D-Y (2019) Target organ toxicology in *Caenorhabditis elegans*. Springer Nature Singapore Pte Ltd, Singapore
4. Wang D-Y (2020) Exposure toxicology in *Caenorhabditis elegans*. Springer Nature Singapore Pte Ltd, Singapore
5. Wang D-Y (2022) Toxicology at environmentally relevant concentrations in *Caenorhabditis elegans*. Springer Nature Singapore Pte Ltd., Singapore

6. Wang M, Nie Y, Liu Y, Dai H, Wang J, Si B, Yang Z, Cheng L, Liu Y, Chen S, Xu A (2019) Transgenerational effects of diesel particulate matter on *Caenorhabditis elegans* through maternal and multigenerational exposure. Ecotoxicol Environ Saf 170:635–643
7. Contreras EQ, Cho M, Zhu H, Puppala HL, Escalera G, Zhong W, Colvin VL (2013) Toxicity of quantum dots and cadmium salt to *Caenorhabditis elegans* after multigenerational exposure. Environ Sci Technol 47:1148–1154
8. Zhao L, Qu M, Wong G, Wang D-Y (2017) Transgenerational toxicity of nanopolystyrene particles in the range of μg/L in nematode *Caenorhabditis elegans*. Environ Sci Nano 4:2356–2366
9. Qu M, Kong Y, Yuan Y-J, Wang D-Y (2019) Neuronal damage induced by nanopolystyrene particles in nematode *Caenorhabditis elegans*. Environ Sci Nano 6:2591–2601
10. Qu M, Qiu Y-X, Kong Y, Wang D-Y (2019) Amino modification enhances reproductive toxicity of nanopolystyrene on gonad development and reproductive capacity in nematode *Caenorhabditis elegans*. Environ Pollut 254:112978
11. Qu M, Nida A, Kong Y, Du H-H, Xiao G-S, Wang D-Y (2019) Nanopolystyrene at predicted environmental concentration enhances microcystin-LR toxicity by inducing intestinal damage in *Caenorhabditis elegans*. Ecotoxicol Environ Saf 183:109568
12. Qu M, Xu K-N, Li Y-H, Wong G, Wang D-Y (2018) Using *acs-22* mutant *Caenorhabditis elegans* to detect the toxicity of nanopolystyrene particles. Sci Total Environ 643:119–126
13. Qiu Y-X, Liu Y-Q, Li Y-H, Li G-J, Wang D-Y (2020) Effect of chronic exposure to nanopolystyrene on nematode *Caenorhabditis elegans*. Chemosphere 256:127172
14. Qiu Y-X, Luo L-B, Yang Y-H, Kong Y, Li Y-H, Wang D-Y (2020) Potential toxicity of nanopolystyrene on lifespan and aging process of nematode *Caenorhabditis elegans*. Sci Total Environ 705:135918
15. Qu M, Wang D-Y (2020) Toxicity comparison between pristine and sulfonate modified nanopolystyrene particles in affecting locomotion behavior, sensory perception, and neuronal development in *Caenorhabditis elegans*. Sci Total Environ 703:134817
16. Li D, Deng Y-J, Wang S-T, Du H-H, Xiao G-S, Wang D-Y (2020) Assessment of nanopolystyrene toxicity under fungal infection condition in *Caenorhabditis elegans*. Ecotoxicol Environ Saf 197:110625
17. Hua X, Wang D-Y (2022) Cellular uptake, transport, and organelle response after exposure to microplastics and nanoplastics: current knowledge and perspectives for environmental and health risks. Rev Environ Contam Toxicol 260:12
18. Hua X, Wang D-Y (2023) Tire-rubber related pollutant 6-PPD quinone: a review of its transformation, environmental distribution, bioavailability, and toxicity. J Hazard Mater 459:132265
19. Hua X, Feng X, Liang G-Y, Chao J, Wang D-Y (2023) Long-term exposure to tire-derived 6-PPD quinone causes intestinal toxicity by affecting functional state of intestinal barrier in *Caenorhabditis elegans*. Sci Total Environ 861:160591
20. Hua X, Feng X, Liang G-Y, Chao J, Wang D-Y (2023) Exposure to 6-PPD quinone at environmentally relevant concentrations causes abnormal locomotion behaviors and neurodegeneration in *Caenorhabditis elegans*. Environ Sci Technol 57:4940–4950
21. Hua X, Feng X, Liang G-Y, Chao J, Wang D-Y (2023) Long-term exposure to 6-PPD quinone reduces reproductive capacity by enhancing germline apoptosis associated with activation of both DNA damage and cell corpse engulfment in *Caenorhabditis elegans*. J Hazard Mater 454:131495
22. Hua X, Wang D-Y (2023) Exposure to 6-PPD quinone at environmentally relevant concentrations inhibits both lifespan and healthspan in *C. elegans*. Environ Sci Technol 57:19295–19303
23. Liu Z-Y, Bian Q, Wang D-Y (2024) Exposure to 6-PPD quinone causes ferroptosis activation associated with induction of reproductive toxicity in *Caenorhabditis elegans*. J Hazard Mater 471:134356
24. Wu J-W, Shen S-H, Wang D-Y (2024) 6-PPD quinone at environmentally relevant concentrations induces immunosenescence by causing immunosuppression during the aging process. Chemosphere 368:143719

25. Liu Z-J, Li Y-H, Wang D-Y (2024) Transgenerational response of glucose metabolism in *Caenorhabditis elegans* exposed to 6-PPD quinone. Chemosphere 367:143653
26. Wang Y-X, Hu G-C, Wang D-Y (2025) Increased S-adenosyl methionine strengthens the suppression in mitochondrial unfolded protein response induced by 6-PPD quinone at environmentally relevant concentrations in *Caenorhabditis elegans*. Environ Pollut 386:127231
27. Wang Y-X, Wang D-Y (2024) Exposure to 6-PPD quinone enhances glycogen accumulation in *Caenorhabditis elegans*. Environ Pollut 359:124600
28. Hua X, Wang D-Y (2025) 6-PPD quinone causes alteration in ubiquinone-mediated complex III associated with toxicity on mitochondrial function and longevity in *Caenorhabditis elegans*. J Environ Chem Eng 13:116571
29. Hua X, Wang D-Y (2024) 6-PPD quinone at environmentally relevant concentrations induced damage on longevity in *C. elegans*: mechanistic insight from inhibition in mitochondrial UPR response. Sci Total Environ 954:176275
30. Hua X, Wang D-Y (2025) Environmentally relevant concentration of 6-PPD quinone inhibits two types of mitophagy to cause mitochondrial dysfunction and lifespan reduction in *Caenorhabditis elegans*. Environ Sci: Processes Impacts 27:1928–1940
31. Wu Q-L, Li Y-P, Tang M, Wang D-Y (2012) Evaluation of environmental safety concentrations of DMSA coated Fe_2O_3-NPs using different assay systems in nematode *Caenorhabditis elegans*. PLoS One 7:e43729
32. Wu S, Lu J-H, Rui Q, Yu S-H, Cai T, Wang D-Y (2011) Aluminum nanoparticle exposure in L1 larvae results in more severe lethality toxicity than in L4 larvae or young adults by strengthening the formation of stress response and intestinal lipofuscin accumulation in nematodes. Environ Toxicol Pharmacol 31:179–188
33. Wang Y-X, Wang D-Y (2025) Effect of disruption in intestinal barrier function during transgenerational process on nanoplastic toxicity induction in *Caenorhabditis elegans*. Environ Sci Nano 12:2741–2749
34. Hua X, Cao C, Zhang L, Wang D-Y (2023) Activation of FGF signal in germline mediates transgenerational toxicity of polystyrene nanoparticles at predicted environmental concentrations in *Caenorhabditis elegans*. J Hazard Mater 451:131174
35. Zhang L, Wang S-T, Zhao Y, Bi K, Wang D-Y (2022) Increase in germline methyltransferases governing methylation of histone H3K9 is associated with transgenerational nanoplastic toxicity in *Caenorhabditis elegans*. Environ Sci Nano 9:265–274
36. Hua X, Zhao Y, Yuan Y-J, Zhang L, Bian Q, Wang D-Y (2022) Nanoplastics cause transgenerational toxicity through inhibiting germline microRNA *mir-38* in *C. elegans*. J Hazard Mater 437:129302
37. Liu H-L, Zhao Y-L, Hua X, Wang D-Y (2022) Induction of transgenerational toxicity is associated with the activated germline insulin signals in nematodes exposed to nanoplastic at predicted environmental concentrations. Ecotoxicol Environ Saf 243:114022
38. Xu R-R, Hua X, Rui Q, Wang D-Y (2022) Alteration in Wnt signaling mediates induction of transgenerational toxicity of polystyrene nanoplastics in *C. elegans*. NanoImpact 28:100425
39. Zhao Y, Hua X, Bian Q, Wang D-Y (2022) Nanoplastic exposure at predicted environmental concentrations induces activation of germline Ephrin signal associated with toxicity formation in the *Caenorhabditis elegans* offspring. Toxics 10:699
40. Wang Y-X, Wu J-W, Wang D-Y (2025) 6-PPD quinone causes lipid accumulation across multiple generations differentially affected by metabolic sensors and components of COMPASS complex in *Caenorhabditis elegans*. Environ Pollut 366:125539
41. Wang Y-X, Wang D-Y (2024) Transgenerational intestinal toxicity of 6-PPD quinone in causing ROS production, enhancement in intestinal permeability and suppression in innate immunity in *C. elegans*. Environ Pollut 363:125208
42. Xing X-J, Guo Y-L, Wang D-Y (2009) Using the larvae nematode *Caenorhabditis elegans* to evaluate neurobehavioral toxicity to metallic salts. Ecotoxicol Environ Saf 72:1819–1823

43. Guo Y-L, Yang Y-C, Wang D-Y (2009) Induction of reproductive deficits in nematode *Caenorhabditis elegans* exposed to metals at different developmental stages. Reprod Toxicol 28:90–95
44. Xing X-J, Rui Q, Du N, Wang D-Y (2009) Exposure to lead and mercury in young larvae induces more severe deficits in neuronal survival and synaptic function than in adult nematodes. Arch Environ Contam Toxicol 56:732–741
45. Xu R-R, Hua X, Rui Q, Wang D-Y (2022) Polystyrene nanoparticles caused dynamic alteration in mitochondrial unfolded protein response from parents to the offspring in *C. elegans*. Chemosphere 308:136154
46. Zhao Y-Y, Xu R-R, Hua X, Rui Q, Wang D-Y (2022) Multi-walled carbon nanotubes induce transgenerational toxicity associated with activation of germline long non-coding RNA *linc-7* in *C. elegans*. Chemosphere 301:134687
47. Rui Q, Zhao Y-L, Wu Q-L, Tang M, Wang D-Y (2013) Biosafety assessment of titanium dioxide nanoparticles in acutely exposed nematode *Caenorhabditis elegans* with mutations of genes required for oxidative stress or stress response. Chemosphere 93:2289–2296
48. Wu Q-L, Zhao Y-L, Li Y-P, Wang D-Y (2014) Susceptible genes regulate the adverse effects of TiO_2-NPs at predicted environmental relevant concentrations on nematode *Caenorhabditis elegans*. Nanomedicine 10:1263–1271
49. Zhang L, Wang S-T, Zhao Y-L, Nurdebek B, Bu Y-Q, Wag D-Y (2022) Long-term exposure to polystyrene nanoparticles causes transgenerational toxicity by affecting the function and expression of MEV-1 and DAF-2 signals in *Caenorhabditis elegans*. NanoImpact 26:100403
50. Hua X, Wang D-Y (2024) Transgenerational response of germline histone acetyltransferases and deacetylases to nanoplastics at predicted environmental doses in *Caenorhabditis elegans*. Sci Total Environ 952:175903
51. Wu J-W, Shao Y-T, Hua X, Wang Y-X, Wang D-Y (2024) Nanoplastic at environmentally relevant concentrations induces toxicity across multiple generations associated with inhibition in germline G protein-coupled receptor CED-1 in *Caenorhabditis elegans*. Chemosphere 364:143011
52. Hua X, Zhang L, Wang D-Y (2024) Nanoplastic at environmentally relevant concentrations activates germline *mir-240-rab-5* signaling cascade to affect secreted ligands associated with transgenerational toxicity induction in *C. Elegans*. Environ Sci Nano 11:3615
53. Liu Z-Y, Wang Y-X, Bian Q, Wang D-Y (2024) Transgenerational response of germline nuclear hormone receptor genes to nanoplastics at predicted environmental doses in *Caenorhabditis elegans*. Toxics 12:420
54. Wu J-W, Shao Y-T, Hua X, Wang D-Y (2024) Activated hedgehog and insulin ligands by decreased transcriptional factor DAF-16 mediate transgenerational nanoplastic toxicity in *Caenorhabditis elegans*. J Hazard Mater 480:135909
55. Wu J-W, Shao Y-T, Hua X, Li Y-H, Wang D-Y (2025) Photo-aged polylactic acid microplastics causes severe transgenerational decline in reproductive capacity in *C. elegans*: insight into activation of DNA damage checkpoints affected by multiple germline histone methyltransferases. Environ Pollut 382:126697
56. Li Y-X, Wang W, Wu Q-L, Li Y-P, Tang M, Ye B-P, Wang D-Y (2012) Molecular control of TiO_2-NPs toxicity formation at predicted environmental relevant concentrations by Mn-SODs proteins. PLoS One 7:e44688
57. Wu Q-L, Wang W, Li Y-X, Li Y-P, Ye B-P, Tang M, Wang D-Y (2012) Small sizes of TiO_2-NPs exhibit adverse effects at predicted environmental relevant concentrations on nematodes in a modified chronic toxicity assay system. J Hazard Mater 243:161–168
58. Li Y-X, Yu S-H, Wu Q-L, Tang M, Pu Y-P, Wang D-Y (2012) Chronic Al_2O_3-nanoparticle exposure causes neurotoxic effects on locomotion behaviors by inducing severe ROS production and disruption of ROS defense mechanisms in nematode *Caenorhabditis elegans*. J Hazard Mater 219-220:221–230
59. Liu H-L, Tian L-J, Wang S-T, Wang D-Y (2021) Size-dependent transgenerational toxicity induced by nanoplastics in nematode *Caenorhabditis elegans*. Sci Total Environ 790:148217

60. Chen H, Gu Y, Jiang Y, Yu J, Chen C, Shi C, Li H (2023) Photoaged polystyrene nanoplastics result in transgenerational reproductive toxicity associated with the methylation of histone H3K4 and H3K9 in *Caenorhabditis elegans*. Environ Sci Technol 57:19341–19351
61. Nouara A, Wu Q-L, Li Y-X, Tang M, Wang H-F, Zhao Y-L, Wang D-Y (2013) Carboxylic acid functionalization prevents the translocation of multi-walled carbon nanotubes at predicted environmental relevant concentrations into targeted organs of nematode *Caenorhabditis elegans*. Nanoscale 5:6088–6096
62. Wu Q-L, Li Y-X, Li Y-P, Zhao Y-L, Ge L, Wang H-F, Wang D-Y (2013) Crucial role of biological barrier at the primary targeted organs in controlling translocation and toxicity of multi-walled carbon nanotubes in nematode *Caenorhabditis elegans*. Nanoscale 5:11166–11178
63. Wu Q-L, Zhao Y-L, Fang J-P, Wang D-Y (2014) Immune response is required for the control of *in vivo* translocation and chronic toxicity of graphene oxide. Nanoscale 6:5894–5906
64. Zhao Y-L, Wang X, Wu Q-L, Li Y-P, Tang M, Wang D-Y (2015) Quantum dots exposure alters both development and function of D-type GABAergic motor neurons in nematode *Caenorhabditis elegans*. Toxicol Res 4:399–408
65. Zhao Y-L, Wang X, Wu Q-L, Li Y-P, Wang D-Y (2015) Translocation and neurotoxicity of CdTe quantum dots in RMEs motor neurons in nematode *Caenorhabditis elegans*. J Hazard Mater 283:480–489
66. Qu M, Miao L, Chen H, Zhang X, Wang Y (2023) SKN-1/Nrf2-dependent regulation of mitochondrial homeostasis modulates transgenerational toxicity induced by nanoplastics with different surface charges in *Caenorhabditis elegans*. J Hazard Mater 457:131840
67. Sun L-M, Liao K, Wang D-Y (2021) Comparison of transgenerational reproductive toxicity induced by pristine and amino modified nanoplastics in *Caenorhabditis elegans*. Sci Total Environ 768:144362
68. Song M-X, Ruan Q-L, Wang D-Y (2024) Comparison of transgenerational toxicity on locomotion behavior and development of D-type motor neurons between pristine and amino modified nanoplastics at environmentally relevant doses in *C. elegans*. Toxics 12:555

Chapter 4
Roles of Transgenerational Translocation and Biological Barrier in Controlling Transgenerational Toxicity of Pollutants

Abstract After exposure to pollutants, their transgenerational translocation could be formed in *Caenorhabditis elegans*. Transgenerational translocation of particle pollutants may be somewhat different from transgenerational translocation of organic pollutants. ACS-22, CLK-1, ISP-1, and DAF-2 were identified as important regulators for the control of transgenerational translocation of pollutants. Transgenerational damage on intestinal barrier function could be induced by exposure to particle and organic pollutants. ACS-22, ERM-1, HMP-2, PKC-3, and immunity response proteins (such as LYS-7 and SPP-1) functioned to regulate the transgenerational damage on intestinal barrier function by pollutants.

Keywords Transgenerational toxicity · Transgenerational translocation · Biological barrier · *Caenorhabditis elegans*

4.1 Introduction

Caenorhabditis elegans has high sensitivity to pollutants, which makes it suitable to examine toxicity and accumulation of pollutants after exposure at environmentally relevant concentrations [1–3]. The established knowledge system of molecular toxicology and target organ toxicology provides the important foundation for the further study of transgenerational toxicology after pollutant exposure at the parental generation (P0-G) [4, 5].

We have introduced in Chap. 2 the endpoints useful for assessing transgenerational toxicity of stresses and pollutants. We further discussed the factors affecting induction of transgenerational toxicity of stresses and pollutants in Chap. 3. Actually, transgenerational translocation is also an important factor affecting the transgenerational pollutants toxicity. Thus, in this chapter, we first introduced transgenerational translocation of pollutants and discussed the molecular control of transgenerational translocation of pollutants. Moreover, we introduced the transgenerational damage on intestinal barrier function after pollutant exposure and discussed the molecular control of transgenerational damage on intestinal barrier function by pollutants.

D. Wang, *Transgenerational Toxicology in Caenorhabditis elegans*,
https://doi.org/10.1007/978-981-92-1389-4_4

4.2 Transgenerational Translocation of Pollutants

4.2.1 Transgenerational Translocation of Particle Pollutants

After exposure to particle pollutants, they could be accumulated in the body of P0-G nematodes and their offspring [6, 7]. In *C. elegans*, nanoplastic exposure resulted in several aspects of damage, including organ toxicity, disrupted metabolism, and inhibited longevity [8–20]. In addition, nanoplastic exposure caused transgenerational neurotoxicity and transgenerational reproductive toxicity [21–23]. Polystyrene nanoparticle (PS-NP), a member of nanoplastics, was first selected as the example of particle pollutants. Accompanied with formation of transgenerational PS-NPs toxicity, fluorescence dye Rho B labeled PS-NPs could be severely accumulated in intestinal lumen, pharynx, intestinal cells, and gonad at the P0-G (Fig. 4.1) [24]. In

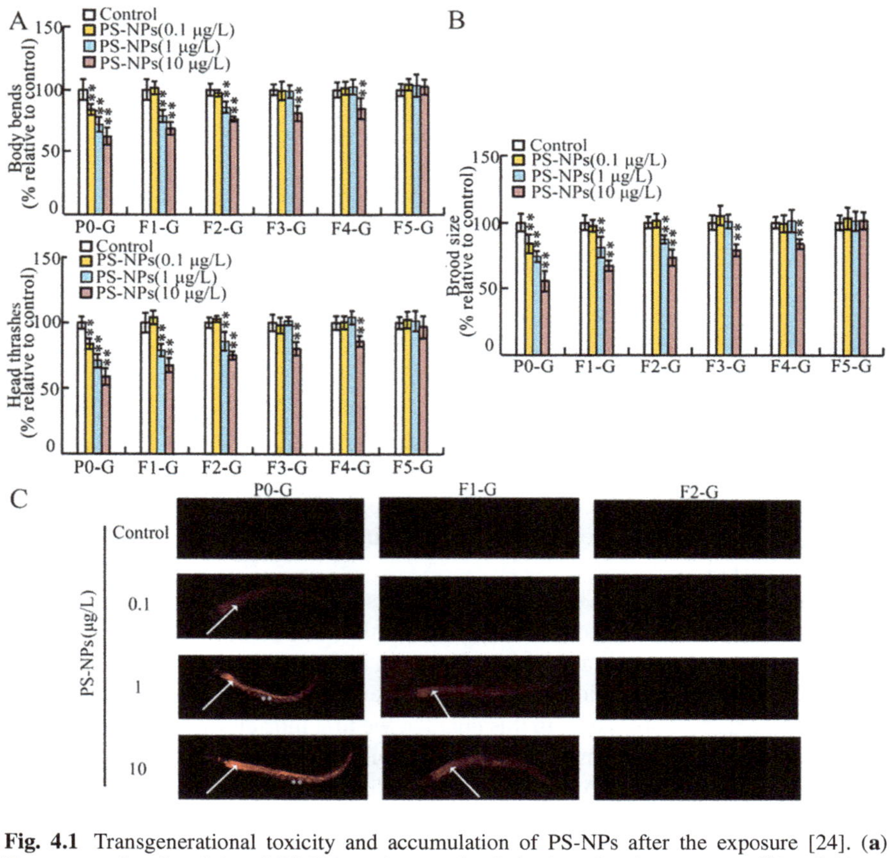

Fig. 4.1 Transgenerational toxicity and accumulation of PS-NPs after the exposure [24]. (**a**) Transgenerational toxicity of PS-NPs on locomotion behavior after the exposure. $^{**}P < 0.01$ vs. control. (**b**) Transgenerational toxicity of PS-NPs on the brood size after the exposure. $^{**}P < 0.01$ vs. control. (**c**) Accumulation of PS-NPs from P0-G to F2-G. Arrowheads indicate PS-NP accumulation in the intestine, and asterisks indicate PS-NP accumulation in the gonad of nematodes

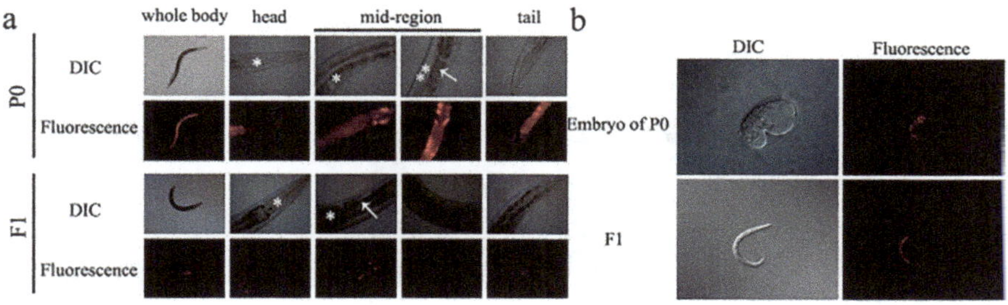

Fig. 4.2 Distribution and translocation of CdTe QDs in exposed wild-type nematodes and their progeny [28]. (**a**) CdTe QDs distribution in exposed wild-type nematodes and their progeny. Arrowheads indicate the intestine. Pharynx (*) in the head, embryos (*) in the mid-region, and gonad (**) in the mid-region were also indicated. (**b**) CdTe QDs distribution in embryos of exposed wild-type nematodes and the animals developed from these embryos. Exposure concentration of CdTe QDs was 20 mg L^{-1}. P0, Parents; F1, the filial generation

contrast, only moderate accumulation of Rho B–labeled PS-NPs in intestinal lumen and pharynx could be observed at the F1-G, and no accumulation of Rho B–labeled PS-NPs was observed from F2-G (Fig. 4.1) [24]. CdTe quantum dots (CdTe QDs) also caused multiple toxicity on *C. elegans*, such as neurotoxicity and lipid accumulation [25–27]. CdTe QD was selected as another example of particle pollutants. Accompanied with induction of transgenerational neurotoxicity and transgenerational intestinal toxicity, CdTe QDs could be accumulated at both P0-G and F1-G (Fig. 4.2) [28]. Only moderate CdTe QDs accumulation was observed in the pharynx and intestine at F1-G, and no CdTe QDs accumulation could be detected at F2-G (Fig. 4.2) [28]. Polylactic acid (PLA) is a kind of biodegradable plastic. PLA microplastics (PLA-MPs) exposure caused reproductive toxicity, which could be enhanced by UV photo-aging [29, 30]. Accompanied with induction of transgenerational reproductive toxicity, both pristine and photo-aged PLA-MPs could be severely accumulated at both anterior and posterior in *C. elegans* body at the P0-G [31, 32]. In contrast, only a small number of particles for pristine and photo-aged PLA-MPs were observed at the F1-G, and no particles for pristine and photo-aged PLA-MPs were observed from the F2-G [31, 32].

4.2.2 Transgenerational Translocation of Organic Pollutants

After the exposure, transgenerational translocation of organic pollutants could also be observed. The 6-PPD quinone (6-PPDQ) is tire-rubber-derived organic pollutants, and its exposure caused multiple toxicities on *C. elegans*, such as organ damage, lifespan reduction, mitochondrial dysfunction, and metabolic dysfunction [33–51]. Exposure to 6-PPDQ further induced transgenerational disruption in metabolism and transgenerational intestinal damage [52–54]. The 6-PPDQ was selected as the example of organic pollutants. Accompanied with these

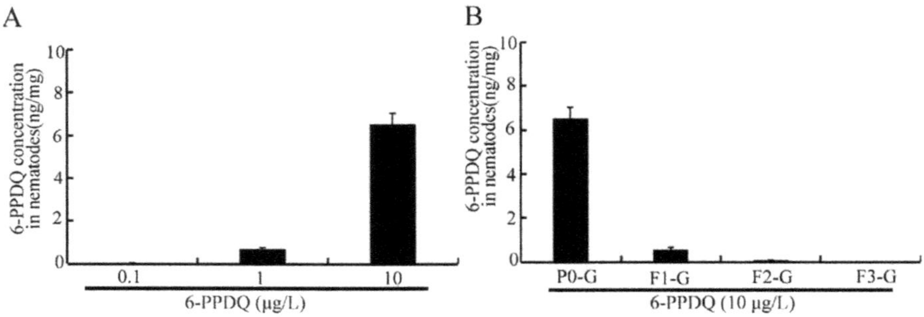

Fig. 4.3 Transgenerational accumulation of 6-PPDQ in nematodes [54]. (**a**) 6-PPDQ accumulation at P0-G after exposure to 0.1–10 μg/L 6-PPDQ. (**b**) Transgenerational 6-PPDQ accumulation after exposure to 10 μg/L 6-PPDQ at P0-G

transgenerational toxicities, 6-PPDQ accumulation was not only caused by 1–10 μg/L 6-PPDQ exposure but also be observed at F1-G and F2-G after exposure to 10 μg/L 6-PPDQ at the P0-G (Fig. 4.3) [54]. Nevertheless, 6-PPDQ accumulation could not be detected from F3-G after exposure to 10 μg/L 6-PPDQ at the P0-G (Fig. 4.3) [54].

4.3 Molecular Control of Transgenerational Translocation of Pollutants

4.3.1 Function of ACS-22 in Regulating Transgenerational Translocation of Pollutants

In *C. elegans*, ACS-22 is required for maintenance of functional state of intestinal barrier [12, 34, 55]. Mutation or RNAi of *acs-22* increased intestinal permeability and thus affects pollutants translocation [12, 34, 56]. With the PS-NPs as the example of pollutants, RNAi of *acs-22* at the P0-G caused the susceptibility to transgenerational neurotoxicity and transgenerational reproductive toxicity induced by 10 μg/L PS-NPs (Fig. 4.4) [24]. Compared to detection of transgenerational PS-NPs toxicity till to the F4-G in wild-type nematodes, transgenerational PS-NPs toxicity could be detected till to the F6-G in *acs-22(RNAi)* nematodes (Fig. 4.4) [24]. Meanwhile, compared to translocation of Rho B–labeled PS-NPs till to F1-G in wild-type nematodes, translocation of Rho B–labeled PS-NPs could be observed till to F2-G in *acs-22(RNAi)* nematodes (Fig. 4.4) [24].

If the *acs-22* RNAi was performed at the F1-G, the susceptibility to transgenerational neurotoxicity and transgenerational reproductive toxicity of 10 μg/L PS-NPs was also induced (Fig. 4.5) [24]. Compared to detection of transgenerational PS-NPs toxicity till to the F4-G in wild-type nematodes, transgenerational PS-NPs toxicity could be further found till to the F6-G in *acs-22(RNAi)* nematodes (Fig. 4.5) [24].

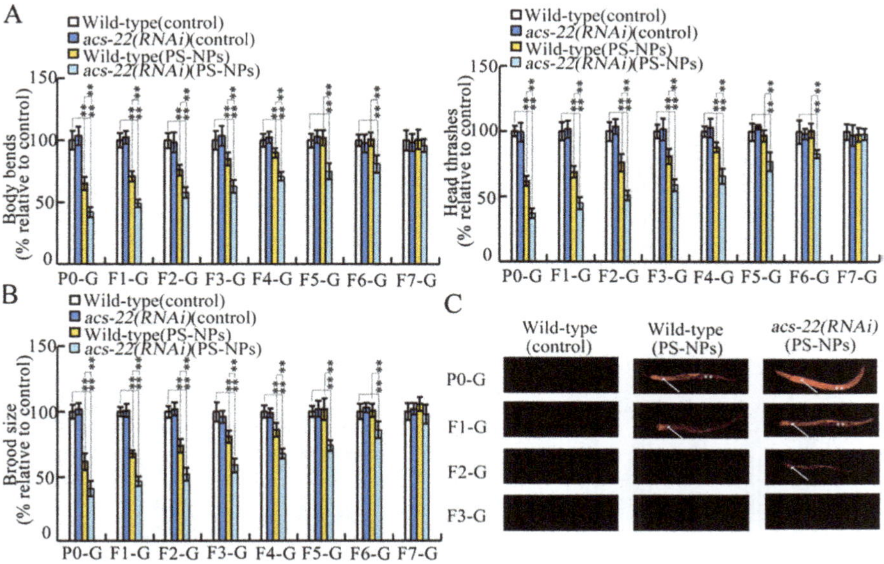

Fig. 4.4 Effect of *acs-22* RNAi at P0-G on transgenerational PS-NP toxicity and accumulation [24]. (**a**) Effect of *acs-22* RNAi at P0-G on transgenerational PS-NP toxicity on locomotion behavior. $^{**}P < 0.01$. (**b**) Effect of *acs-22* RNAi at P0-G on transgenerational PS-NP toxicity on brood size. $^{**}P < 0.01$. (**c**) Effect of *acs-22* RNAi at P0-G on transgenerational PS-NP accumulation. Arrowheads indicate PS-NP accumulation in the intestine, and asterisks indicate PS-NP accumulation in the gonad of nematodes

Additionally, compared to translocation and accumulation of Rho B–labeled PS-NPs till to F1-G in wild-type nematodes, translocation and accumulation of Rho B–labeled PS-NPs was further observed till to F2-G in *acs-22(RNAi)* nematodes (Fig. 4.5) [24].

4.3.2 Function of CLK-1, ISP-1, and DAF-2 in Regulating Transgenerational Translocation of Pollutants

In *C. elegans*, CLK-1 and ISP-1 are components of mitochondrial complex III, and DAF-2 is insulin receptor. Mutation or RNAi of *clk-1*, *isp-1*, and *daf-2* could confer the resistance to toxicity of pollutants, such as CdTe QDs [4, 5, 28, 48, 57–59]. Compared with the observed accumulation of CdTe QDs at both P0-G and F1-G in wild-type nematodes, no accumulation of CdTe QDs could be observed at both P0-G and F1-G in *clk-1(RNAi)*, *isp-1(RNAi)*, and *daf-2(RNAi)* nematodes (Fig. 4.6) [28]. That is, the transgenerational translocation and accumulation of CdTe QDs could be inhibited by RNAi of *clk-1*, *isp-1*, and *daf-2*.

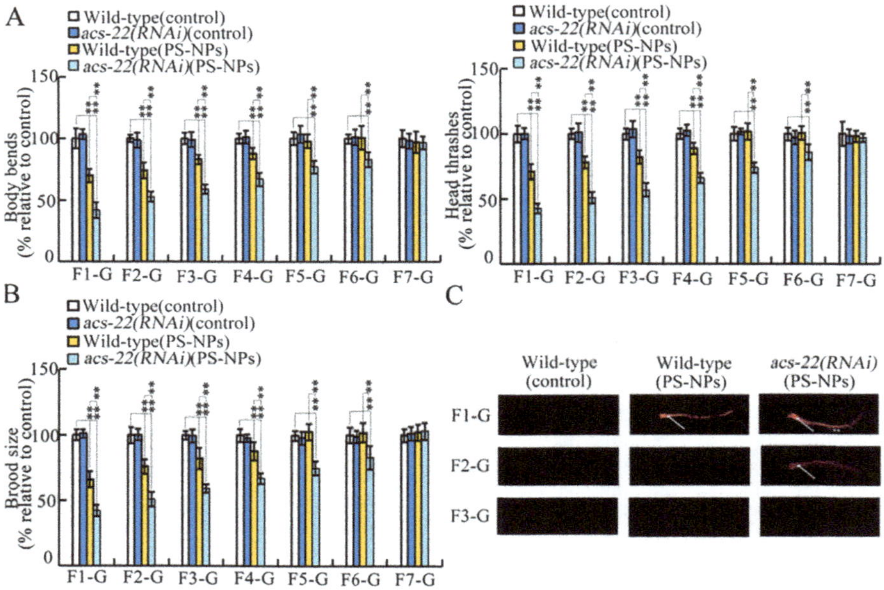

Fig. 4.5 Effect of *acs-22* RNAi at F1-G on transgenerational PS-NP toxicity and accumulation [24]. (**a**) Effect of *acs-22* RNAi at F1-G on transgenerational PS-NP toxicity on locomotion behavior. $^{**}P < 0.01$. (**b**) Effect of *acs-22* RNAi at F1-G on transgenerational PS-NP toxicity on brood size. $^{**}P < 0.01$. (**c**) Effect of *acs-22* RNAi at F1-G on transgenerational PS-NP accumulation. Arrowheads indicate PS-NP accumulation in the intestine, and asterisks indicate PS-NP accumulation in the gonad of nematodes

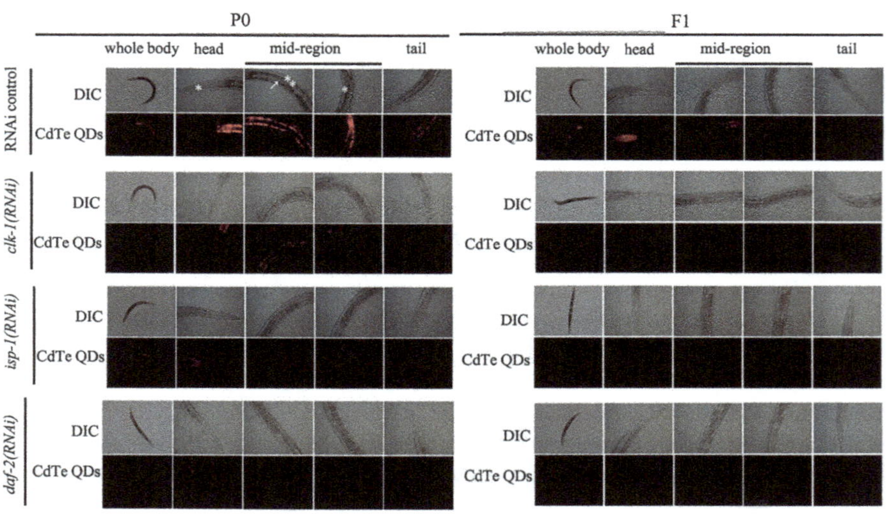

Fig. 4.6 Effects of intestine-specific RNAi of *clk-1*, *isp-1*, or *daf-2* gene on the distribution and translocation of CdTe QDs in nematodes [28]. Arrowheads indicate the intestine. Pharynx (*) in the head, embryos (*) in the mid-region, and gonad (**) in the mid-region were also indicated. RNAi control, VP303. Exposure concentration of CdTe QDs was 20 mg L^{-1}. P0, Parents; F1, the filial generation. QDs, quantum dots

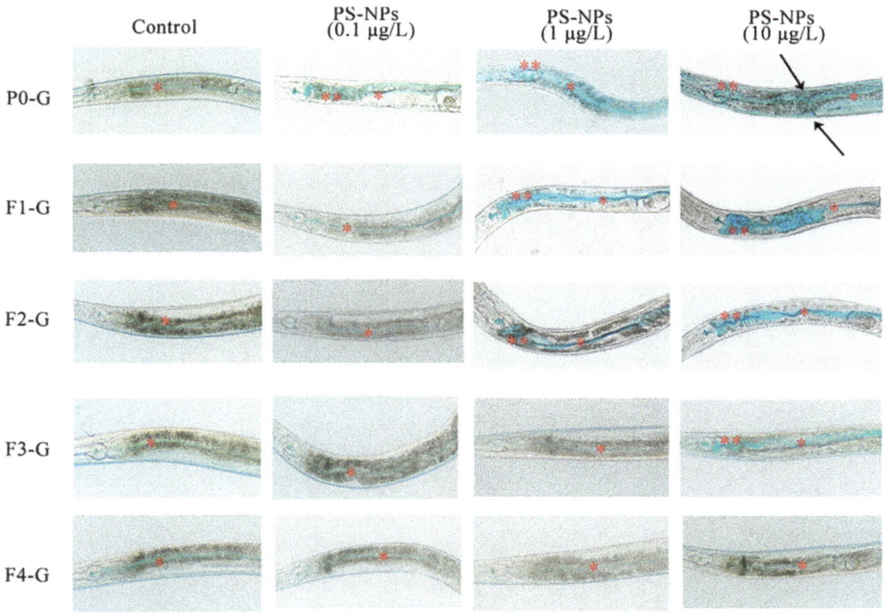

Fig. 4.7 Effect of PS-NP exposure on intestinal permeability at P0-G and the offspring after exposure based on blue dye staining [24]. Single asterisk and double asterisks indicate intestinal lumen and intestinal cells, respectively. Arrowheads indicate the body cavity

4.4 Transgenerational Damage on Intestinal Barrier Function by Pollutants

4.4.1 Transgenerational Damage on Intestinal Barrier Function by Particle Pollutants

With PS-NP as the example of particle pollutants, transgenerational damage on intestinal barrier function was detected after particle pollutant exposure at P0-G. Blue dye staining was used to indicate the change of intestinal permeability. Although 0.1 μg/L did not cause transgenerational damage on intestinal barrier function, 1–10 μg/L PS-NPs resulted in severe transgenerational damage on intestinal barrier function (Fig. 4.7) [24]. After exposure to 10 μg/L PS-NPs at the P0-G, transgenerational damage on intestinal barrier function could be detected till to the F3-G (Fig. 4.7) [24].

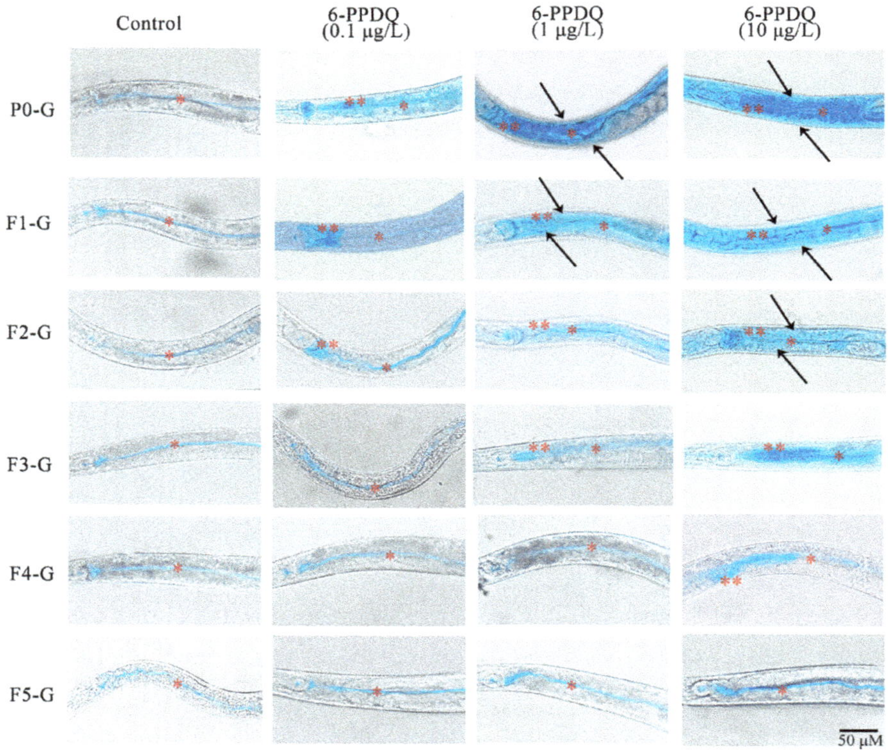

Fig. 4.8 Transgenerational intestinal permeability after 6-PPDQ exposure [54]. The intestinal cells (**) and intestinal lumen (*) were labeled by asterisks, and the body cavity was labeled by arrowheads

4.4.2 Transgenerational Damage on Intestinal Barrier Function by Organic Pollutants

With 6-PPDQ as the example of organic pollutants, transgenerational damage on intestinal barrier function could also be observed after organic pollutant exposure at P0-G. Exposure to 0.1–10 μg/L 6-PPDQ all caused severe transgenerational damage on intestinal barrier function (Fig. 4.8). Transgenerational damage on intestinal barrier function caused by 0.1, 1, and 10 μg/L 6-PPDQ could be observed till to F2-G, F3-G, and F4-G, respectively (Fig. 4.8) [54]. Moreover, translocation and accumulation of blue dye signals in the body cavity could be even observed at F1-G after exposure to 1 μg/L 6-PPDQ and at F1-G and F2-G after exposure to 10 μg/L 6-PPDQ (Fig. 4.8) [54].

4.5 Molecular Control of Transgenerational Damage on Intestinal Barrier Function by Pollutants

4.5.1 Function of ACS-22 in Regulating Transgenerational Damage on Intestinal Barrier Function by Pollutants

RNAi of *acs-22* caused increased intestinal permeability under the normal condition (Fig. 4.9) [24], suggesting the requirement for ACS-22 for maintenance of normal intestinal barrier function. Exposure to PS-NPs caused inhibition in ACS-22, and transgenerational inhibition in ACS-22 could be found till to F3-G after exposure to 10 μg/L PS-NPs at the P0-G (Fig. 4.9) [24]. These observations suggested the involvement of transgenerational inhibition in ACS-22 in mediating transgenerational damage on intestinal barrier function induced by pollutants.

4.5.2 Function of ERM-1, HMP-2, and PKC-3 in Regulating Transgenerational Damage on Intestinal Barrier Function by Pollutants

Besides the ACS-22, ERM-1, HMP-2, and PKC-3 were also involved in controlling transgenerational damage on intestinal barrier function by pollutants. Among genes regulating intestinal barrier functions [12, 34, 55, 60], 6-PPDQ exposure caused

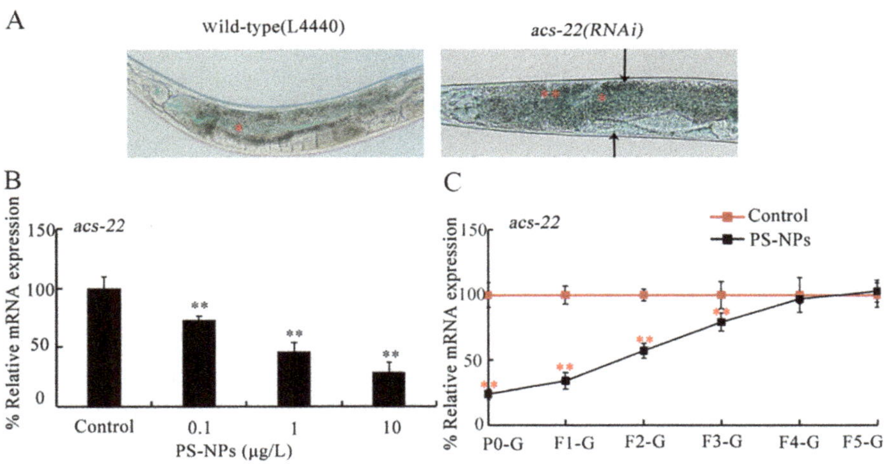

Fig. 4.9 Transgenerational expression of *acs-22* after PS-NP exposure [24]. (**a**) Effect of *acs-22* RNAi on intestinal permeability under the condition without PS-NP exposure based on blue dye staining. Single asterisk and double asterisks indicate intestinal lumen and intestinal cells, respectively. Arrowheads indicate the body cavity. (**B**) Effect of PS-NP exposure on *acs-22* expression at P0-G. $^{**}P < 0.01$ vs. control. (**c**) Effect of PS-NP (10 μg L^{-1}) exposure on transgenerational *acs-22* expression. $^{**}P < 0.01$ vs. control

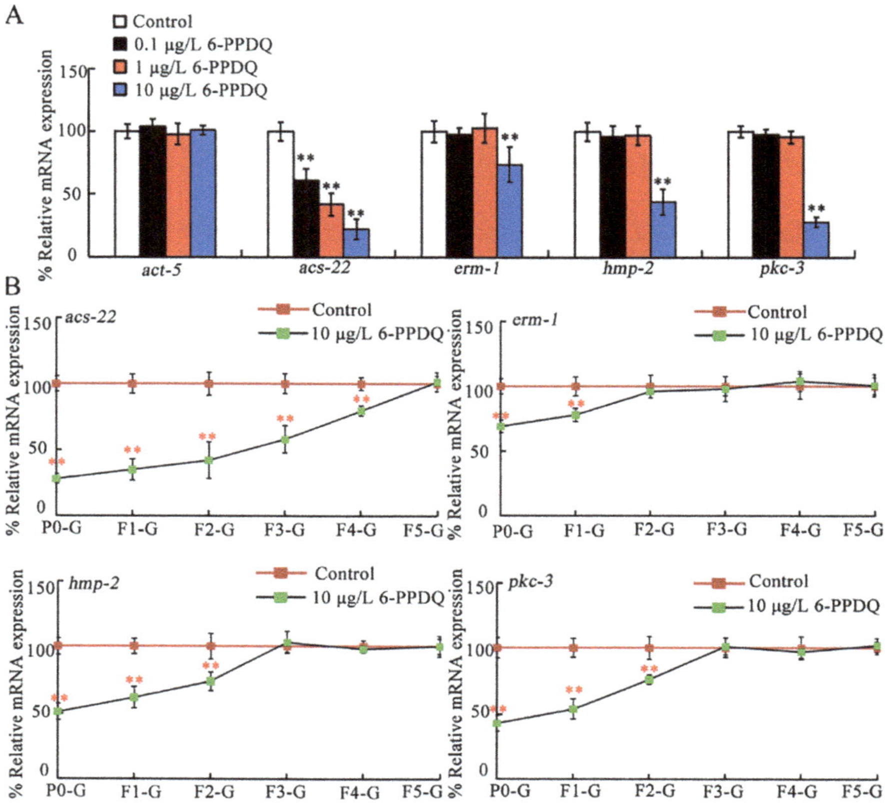

Fig. 4.10 Transgenerational expressions of genes governing functional state of intestinal barrier after 6-PPDQ exposure [54]. (**a**) Effect of 6-PPDQ exposure on expressions of *act-5*, *acs-22*, *erm-1*, *hmp-2*, or *pkc-3* at P0-G. (**b**) Transgenerational expressions of *acs-22*, *erm-1*, *hmp-2*, or *pkc-3* after exposure to 10 μg/L 6-PPDQ at P0-G. $^{**}P < 0.01$ vs control

inhibition in ACS-22, ERM-1, HMP-2, and PKC-3 (Fig. 4.10) [54]. Moreover, after exposure to 10 μg/L 6-PPDQ, transgenerational inhibition in ACS-22 was detected till to F4-G, transgenerational inhibition in HMP-2 and PKC-3 could be observed till to the F2-G, and inhibition in ERM-1 was further found at the F1-G (Fig. 4.10) [54]. Therefore, ACS-22, ERM-1, HMP-2, and PKC-3 regulate the transgenerational damage on intestinal barrier function by pollutants to different degrees.

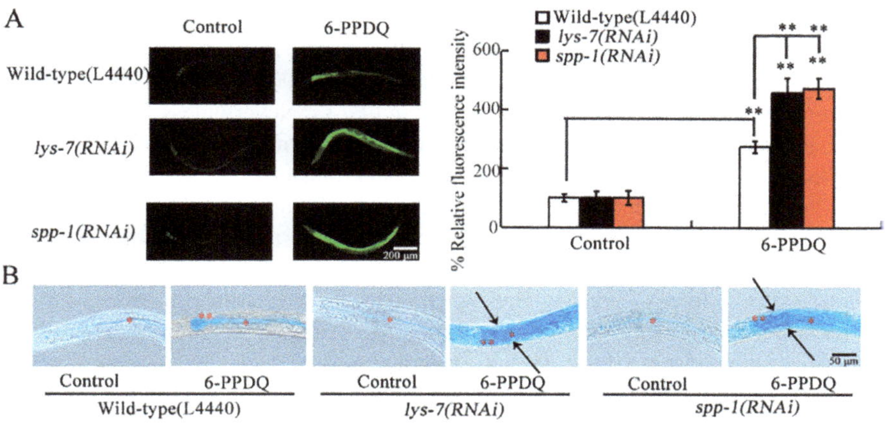

Fig. 4.11 Effect of RNAi of *lys-7* and *spp-1* on ROS production (**a**) and intestinal permeability (**b**) in 6-PPDQ-exposed nematodes [54]. The intestinal cells (**) and intestinal lumen (*) were labeled by asterisks, and the body cavity was labeled by arrowheads. Exposure concentration of 6-PPDQ was 0.1 μg/L. $^{**}P < 0.01$

4.5.3 Function of Immunity Response Proteins in Regulating Transgenerational Damage on Intestinal Barrier Function by Pollutants

In *C. elegans*, innate immunity is mainly assessed by change of antimicrobial genes expression [61–63]. With 6-PPDQ as the example of pollutants, inhibited expression of antimicrobial genes (*lys-7* and *spp-1*) was caused by 1–10 μg/L 6-PPDQ at the P0-G [54]. Moreover, after exposure to 10 μg/L 6-PPDQ, transgenerational inhibition in LYS-7 and SPP-1 could be detected till to the F4-G [54]. That is, exposure to relatively high concentrations of 6-PPDQ induced the transgenerational immunosuppression.

For the functions of LYS-7 and SPP-1, it was noted that RNAi of *lys-7* and *spp-1* caused susceptibility to 6-PPDQ-induced intestinal oxidative damage (Fig. 4.11) [54]. Additionally, the damage caused by 6-PPDQ on intestinal barrier function was further strengthened by RNAi of *lys-7* and *spp-1* (Fig. 4.11) [54]. Therefore, transgenerational immunosuppression mediates transgenerational damage on intestinal barrier function to a certain degree.

4.6 Perspectives

In *C. elegans*, both particle pollutants and organic pollutants could cause the transgenerational translocation. For particle pollutants, after P0-G exposure, they can be normally accumulated at the P0-G and F1-G, and no further accumulation would be detected from the F2-G. This may be largely due to the particulate property of this kind of pollutants, which makes them difficult to form transgenerational translocation for more generations. For organic pollutants, after P0-G exposure, they can be accumulated in the offspring, such as F1-G and F2-G. Nevertheless, with 6-PPDQ as the example of organic pollutants, only very small amount of 6-PPDQ could be detected at the F2-G. Therefore, transgenerational translocation can provide limited contribution to the induction of transgenerational pollutants toxicity. Usually, from F2-G, the observed toxicity in the offspring may be not due to the pollutants accumulation.

During the control of transgenerational translocation of pollutants, two groups of proteins were identified as the corresponding regulators. One group is the proteins with function governing intestinal barrier function, such as ACS-22. Another group is the proteins governing stress response, such as CLK-1, ISP-1, and DAF-2.

In organisms, disruption in biological barriers directly accounts for the observed accumulation of pollutants in the body. Both particle pollutants and organic pollutants could result in transgenerational damage on intestinal barrier function. Moreover, after exposure to particle and organic pollutants, the generation times for the observed transgenerational damage on intestinal barrier function were much longer than those for the observed transgenerational translocation. This suggested that, on the one hand, transgenerational damage on intestinal barrier function contributed to formation of transgenerational translocation of pollutants to a certain degree. On the other hand, the formed transgenerational damage on intestinal barrier function by pollutant exposure might more tend to be associated with induction of transgenerational immunosuppression, which further led to the induction of transgenerational toxicity on other aspects.

For the biological barriers, besides the transgenerational damage on intestinal barrier function, the possible transgenerational damage of different forms of pollutants on other biological barriers needs to be further examined. Among the other biological barriers, the transgenerational effect of pollutants on development and function of reproductive barrier, such as the gonad barrier, and the role of reproductive barrier in modulating transgenerational translocation of pollutants are at least suggested to be further carefully determined.

References

1. Wang D-Y (2018) Nanotoxicology in *Caenorhabditis elegans*. Springer Nature Singapore Pte Ltd, Singapore
2. Wang D-Y (2020) Exposure toxicology in *Caenorhabditis elegans*. Springer Nature Singapore Pte Ltd, Singapore
3. Wang D-Y (2022) Toxicology at environmentally relevant concentrations in *Caenorhabditis elegans*. Springer Nature Singapore Pte Ltd., Singapore
4. Wang D-Y (2019) Molecular toxicology in *Caenorhabditis elegans*. Springer Nature Singapore Pte Ltd, Singapore
5. Wang D-Y (2019) Target organ toxicology in *Caenorhabditis elegans*. Springer Nature Singapore Pte Ltd., Singapore

6. Hu C, Hou J, Zhu Y, Lin D-H (2020) Multigenerational exposure to TiO_2 nanoparticles in soil stimulates stress resistance and longevity of survived *C. elegans* via activating insulin/IGF-like signaling. Environ Pollut 263:114376
7. Kim SW, Moon J, An Y (2018) Matricidal hatching can induce multi-generational effects in nematode *Caenorhabditis elegans* after dietary exposure to nanoparticles. Environ Sci Pollut Res 25:36394–36402
8. Hua X, Wang D-Y (2022) Cellular uptake, transport, and organelle response after exposure to microplastics and nanoplastics: current knowledge and perspectives for environmental and health risks. Rev Environ Contam Toxicol 260:12
9. Qu M, Kong Y, Yuan Y-J, Wang D-Y (2019) Neuronal damage induced by nanopolystyrene particles in nematode *Caenorhabditis elegans*. Environ Sci Nano 6:2591–2601
10. Qu M, Qiu Y-X, Kong Y, Wang D-Y (2019) Amino modification enhances reproductive toxicity of nanopolystyrene on gonad development and reproductive capacity in nematode *Caenorhabditis elegans*. Environ Pollut 254:112978
11. Qu M, Nida A, Kong Y, Du H-H, Xiao G-S, Wang D-Y (2019) Nanopolystyrene at predicted environmental concentration enhances microcystin-LR toxicity by inducing intestinal damage in *Caenorhabditis elegans*. Ecotoxicol Environ Saf 183:109568
12. Qu M, Xu K-N, Li Y-H, Wong G, Wang D-Y (2018) Using *acs-22* mutant *Caenorhabditis elegans* to detect the toxicity of nanopolystyrene particles. Sci Total Environ 643:119–126
13. Qiu Y-X, Liu Y-Q, Li Y-H, Li G-J, Wang D-Y (2020) Effect of chronic exposure to nanopolystyrene on nematode *Caenorhabditis elegans*. Chemosphere 256:127172
14. Qiu Y-X, Luo L-B, Yang Y-H, Kong Y, Li Y-H, Wang D-Y (2020) Potential toxicity of nanopolystyrene on lifespan and aging process of nematode *Caenorhabditis elegans*. Sci Total Environ 705:135918
15. Li D, Deng Y-J, Wang S-T, Du H-H, Xiao G-S, Wang D-Y (2020) Assessment of nanopolystyrene toxicity under fungal infection condition in *Caenorhabditis elegans*. Ecotoxicol Environ Saf 197:110625
16. Liu H-L, Shao H-M, Guo Z-J, Wang D-Y (2020) Nanopolystyrene exposure activates a fat metabolism related signaling-mediated protective response in *Caenorhabditis elegans*. NanoImpact 17:100204
17. Dong S-S, Qu M, Rui Q, Wang D-Y (2018) Combinational effect of titanium dioxide nanoparticles and nanopolystyrene particles at environmentally relevant concentrations on nematodes *Caenorhabditis elegans*. Ecotoxicol Environ Saf 161:444–450
18. Zhao Y-Y, Li D, Rui Q, Wang D-Y (2020) Toxicity induction of nanopolystyrene under microgravity stress condition in *Caenorhabditis elegans*. Sci Total Environ 703:135623
19. Liu H-L, Wang D-Y (2021) Intestinal mitochondrial unfolded protein response induced by nanoplastic particles in *Caenorhabditis elegans*. Chemosphere 267:128917
20. Tang M, Ding G, Lu X, Huang Q, Du H, Xiao G, Wang D (2022) Exposure to nanoplastic particles enhances *Acinetobacter* survival, biofilm formation, and serum resistance. Nano 12:4222
21. Sun L-M, Liao K, Wang D-Y (2021) Comparison of transgenerational reproductive toxicity induced by pristine and amino modified nanoplastics in *Caenorhabditis elegans*. Sci Total Environ 768:144362
22. Liu H-L, Tian L-J, Wang S-T, Wang D-Y (2021) Size-dependent transgenerational toxicity induced by nanoplastics in nematode *Caenorhabditis elegans*. Sci Total Environ 790:148217
23. Zhao L, Qu M, Wong G, Wang D-Y (2017) Transgenerational toxicity of nanopolystyrene particles in the range of μg/L in nematode *Caenorhabditis elegans*. Environ Sci Nano 4:2356–2366
24. Wang Y-X, Wang D-Y (2025) Effect of disruption in intestinal barrier function during transgenerational process on nanoplastic toxicity induction in *Caenorhabditis elegans*. Environ Sci Nano 12:2741–2749
25. Zhao Y-L, Wang X, Wu Q-L, Li Y-P, Wang D-Y (2015) Translocation and neurotoxicity of CdTe quantum dots in RMEs motor neurons in nematode *Caenorhabditis elegans*. J Hazard Mater 283:480–489
26. Zhao Y-L, Wang X, Wu Q-L, Li Y-P, Tang M, Wang D-Y (2015) Quantum dots exposure alters both development and function of D-type GABAergic motor neurons in nematode *Caenorhabditis elegans*. Toxicol Res 4:399–408

27. Wu Q-L, Zhi L-T, Qu Y-Y, Wang D-Y (2016) Quantum dots increased fat storage in intestine of *Caenorhabditis elegans* by influencing molecular basis for fatty acid metabolism. Nanomedicine 12:1175–1184
28. Liu Z-F, Zhou X-F, Wu Q-L, Zhao Y-L, Wang D-Y (2015) Crucial role of intestinal barrier in the formation of transgenerational toxicity in quantum dot exposed nematodes *Caenorhabditis elegans*. RSC Adv 5:94257–94266
29. Shao Y-T, Wang Y-X, Hua X, Li Y-H, Wang D-Y (2023) Polylactic acid microparticles in the range of μg/L reduce reproductive capacity by affecting the gonad development and the germline apoptosis in *Caenorhabditis elegans*. Chemosphere 336:139193
30. Shao Y-T, Hua X, Li Y-H, Wang D-Y (2024) Comparison of reproductive toxicity between pristine and aged polylactic acid microplastics in *Caenorhabditis elegans*. J Hazard Mater 466:133545
31. Shao Y-T, Li Y-H, Wang D-Y (2024) Polylactic acid microplastics cause transgenerational reproductive toxicity associated with activation of insulin and hedgehog ligands in *C. elegans*. Sci Total Environ 942:173746
32. Wu J-W, Shao Y-T, Hua X, Li Y-H, Wang D-Y (2025) Photo-aged polylactic acid microplastics causes severe transgenerational decline in reproductive capacity in *C. elegans*: insight into activation of DNA damage checkpoints affected by multiple germline histone methyltransferases. Environ Pollut 382:126697
33. Hua X, Wang D-Y (2023) Tire-rubber related pollutant 6-PPD quinone: a review of its transformation, environmental distribution, bioavailability, and toxicity. J Hazard Mater 459:132265
34. Hua X, Feng X, Liang G-Y, Chao J, Wang D-Y (2023) Long-term exposure to tire-derived 6-PPD quinone causes intestinal toxicity by affecting functional state of intestinal barrier in *Caenorhabditis elegans*. Sci Total Environ 861:160591
35. Hua X, Feng X, Liang G-Y, Chao J, Wang D-Y (2023) Exposure to 6-PPD quinone at environmentally relevant concentrations causes abnormal locomotion behaviors and neurodegeneration in *Caenorhabditis elegans*. Environ Sci Technol 57:4940–4950
36. Hua X, Feng X, Liang G-Y, Chao J, Wang D-Y (2023) Long-term exposure to 6-PPD quinone reduces reproductive capacity by enhancing germline apoptosis associated with activation of both DNA damage and cell corpse engulfment in *Caenorhabditis elegans*. J Hazard Mater 454:131495
37. Hua X, Wang D-Y (2023) Exposure to 6-PPD quinone at environmentally relevant concentrations inhibits both lifespan and healthspan in *C. elegans*. Environ Sci Technol 57:19295–19303
38. Liu Z-Y, Bian Q, Wang D-Y (2024) Exposure to 6-PPD quinone causes ferroptosis activation associated with induction of reproductive toxicity in *Caenorhabditis elegans*. J Hazard Mater 471:134356
39. Wu J-W, Shen S-H, Wang D-Y (2024) 6-PPD quinone at environmentally relevant concentrations induces immunosenescence by causing immunosuppression during the aging process. Chemosphere 368:143719
40. Wang Y-X, Hu G-C, Wang D-Y (2025) Increased S-adenosyl methionine strengthens the suppression in mitochondrial unfolded protein response induced by 6-PPD quinone at environmentally relevant concentrations in *Caenorhabditis elegans*. Environ Pollut 386:127231
41. Hu D-Y, Wang Y-X, Hu G-C, Liu R, Wang D-Y (2025) 6-PPD quinone inhibited retinoic acid synthesis mediates toxicity through feedback loop between ALH-3/DHS-19-SEX-1 axis and intestinal signals in *Caenorhabditis elegans*. J Environ Expo Assess 4:40
42. Wang W, Li Y-H, Wang D-Y (2025) Long-term exposure to 6-PPD quinone inhibits glutamate synthesis and glutamate receptor function associated with its toxicity induction in *Caenorhabditis elegans*. Toxics 13:434
43. Wang Y-X, Liang G-Y, Chao J, Wang D-Y (2024) Comparison of intestinal toxicity in enhancing intestinal permeability and in causing ROS production of six PPD quinones in *Caenorhabditis elegans*. Sci Total Environ 927:172306
44. Liu Z-J, Li Y-H, Wang D-Y (2024) Exposure to 6-PPD quinone disrupts glucose metabolism associated with lifespan reduction by affecting insulin and AMPK signals in *Caenorhabditis elegans*. Chemosphere 363:142975

45. Wang Y-X, Wang D-Y (2024) Exposure to 6-PPD quinone enhances glycogen accumulation in *Caenorhabditis elegans*. Environ Pollut 359:124600
46. Wan X, Liang G-Y, Wang D-Y (2025) 6-PPD quinone at environmentally relevant concentrations disrupts citric acid cycle in *Caenorhabditis elegans*: role of reduction in acetyl CoA and pyruvate contents. Environ Chem Ecotoxicol 7:1119–1129
47. Hua X, Liang G-Y, Chao J, Wang D-Y (2024) Exposure to 6-PPD quinone causes damage on mitochondrial complex I/II associated with lifespan reduction in *Caenorhabditis elegans*. J Hazard Mater 472:134598
48. Hua X, Wang D-Y (2025) 6-PPD quinone causes alteration in ubiquinone-mediated complex III associated with toxicity on mitochondrial function and longevity in *Caenorhabditis elegans*. J Environ Chem Eng 13:116571
49. Hua X, Wang D-Y (2025) 6-PPD quinone at environmentally relevant concentrations activates feedback response of electron transport chain to mediate damage on mitochondrial function and longevity in *Caenorhabditis elegans*. Environ. Chem. Ecotoxicol. 7:2356–2365
50. Hua X, Wang D-Y (2024) 6-PPD quinone at environmentally relevant concentrations induced damage on longevity in *C. elegans*: mechanistic insight from inhibition in mitochondrial UPR response. Sci Total Environ 954:176275
51. Hua X, Wang D-Y (2025) Environmentally relevant concentration of 6-PPD quinone inhibits two types of mitophagy to cause mitochondrial dysfunction and lifespan reduction in *Caenorhabditis elegans*. Environ Sci: Process Impacts 27:1928–1940
52. Liu Z-J, Li Y-H, Wang D-Y (2024) Transgenerational response of glucose metabolism in *Caenorhabditis elegans* exposed to 6-PPD quinone. Chemosphere 367:143653
53. Wang Y-X, Wu J-W, Wang D-Y (2025) 6-PPD quinone causes lipid accumulation across multiple generations differentially affected by metabolic sensors and components of COMPASS complex in *Caenorhabditis elegans*. Environ Pollut 366:125539
54. Wang Y-X, Wang D-Y (2024) Transgenerational intestinal toxicity of 6-PPD quinone in causing ROS production, enhancement in intestinal permeability and suppression in innate immunity in *C. elegans*. Environ Pollut 363:125208
55. Liu H-L, Guo D-Q, Kong Y, Rui Q, Wang D-Y (2019) Damage on functional state of intestinal barrier by microgravity stress in nematode *Caenorhabditis elegans*. Ecotoxicol Environ Saf 183:109554
56. Zhi L-T, Fu W, Wang X, Wang D-Y (2016) ACS-22, a protein homologous to mammalian fatty acid transport protein 4, is essential for the control of toxicity and translocation of multi-walled carbon nanotubes in *Caenorhabditis elegans*. RSC Adv 6:4151–4159
57. Liu H-L, Zhao Y-L, Hua X, Wang D-Y (2022) Induction of transgenerational toxicity is associated with the activated germline insulin signals in nematodes exposed to nanoplastic at predicted environmental concentrations. Ecotoxicol Environ Saf 243:114022
58. Shao H-M, Han Z-Y, Krasteva N, Wang D-Y (2019) Identification of signaling cascade in the insulin signaling pathway in response to nanopolystyrene particles. Nanotoxicology 13:174–188
59. Zhao Y-L, Yang R-L, Rui Q, Wang D-Y (2016) Intestinal insulin signaling encodes two different molecular mechanisms for the shortened longevity induced by graphene oxide in *Caenorhabditis elegans*. Sci Rep 6:24024
60. Ren M-X, Zhao L, Ding X-C, Krasteva N, Rui Q, Wang D-Y (2018) Developmental basis for intestinal barrier against the toxicity of graphene oxide. Part Fibre Toxicol 15:26
61. Millet AC, Ewbank JJ (2004) Immunity in *Caenorhabditis elegans*. Curr Opin Immunol 16:4–9
62. Kurz CL, Tan MW (2004) Regulation of aging and innate immunity in *C. elegans*. Aging Cell 3:185–193
63. Kim DH, Ausubel FM (2005) Evolutionary perspectives on innate immunity from the study of *Caenorhabditis elegans*. Curr Opin Immunol 17:4–10

Chapter 5
Regulation of Transgenerational Toxicity by Transgenerational Communication of Germline Insulin Signals

Abstract Insulin signaling plays a central role in controlling pollutant toxicity in *C. elegans*. We here first discussed the role of transgenerational activation of germline insulin ligands, insulin receptor, and its downstream kinase cascade and inhibition in FOXO transcriptional factor DAF-16 in mediating transgenerational pollutant toxicity. Additionally, transgenerational communication between germline insulin ligands in pollutant-exposed nematodes and insulin receptor in the offspring was introduced. Moreover, the already identified targets for DAF-16 in regulating transgenerational pollutant toxicity, and their functions were introduced and discussed.

Keywords Transgenerational toxicology · Transgenerational communication · Insulin signaling · *Caenorhabditis elegans*

5.1 Introduction

Caenorhabditis elegans is not only a classic model animal but also a useful animal model for the toxicological study, such as molecular toxicology, target organ toxicology, and exposure toxicology [1–4]. Largely due to the high sensitivity, knowledge system on toxicology at environmentally relevant concentrations (ERCs) has been established in *C. elegans* [5]. During the control of toxicity of pollutants at ERCs, the insulin signaling has been shown to play the central function in *C. elegans* [5]. In insulin signaling, insulin ligands activate corresponding insulin receptor DAF-2 and its downstream kinase cascade of AGE-1-PDK-1-AKT-1/AKT-2/SGK-1 to inhibit activity and function of FOXO transcriptional factor DAF-16 and its targets, such as mitochondrial Mn-SODs (SOD-2 and SOD-3) and HSP-6 (a marker of mitochondrial unfolded protein response (mt UPR)) [6–8]. Insulin signaling (such as the DAF-16 expression) showed the response to different pollutants (such as nanomaterials) and stresses (such as heat stress) [9–12]. Accompanied with the formation of response, insulin signaling regulated toxicity induction of pollutants and stresses [13–16]. For example, DAF-16 regulated toxicity of polystyrene nanoparticle (PS-NP) by targeting to SOD-3 and HSP-6 [17, 18]. Normally, DAF-2 and

D. Wang, *Transgenerational Toxicology in Caenorhabditis elegans*,
https://doi.org/10.1007/978-981-92-1389-4_5

DAF-16 acted in the intestine to regulate the toxicity induction of pollutants, such as nanoplastics, and stresses, such as microgravity [3, 5, 19, 20]. Different from this, insulin ligands of INS-6, INS-7, and DAF-28 acted in the intestine to regulate toxicity of 6-PPD quinone (6-PPDQ) [21]. Insulin ligands of INS-4, INS-39, and DAF-28 acted in the neurons to regulate toxicity of nanoplastics [22]. Insulin ligands of INS-3, INS-39, and DAF-28 acted in the germline to regulate toxicity of nanoplastics [23].

In this chapter, we focused on introduction and discussion of transgenerational response of insulin signaling to pollutants and stresses. Moreover, we introduced and discussed the role of transgenerational communication of insulin signaling in regulating transgenerational toxicity of pollutants and stresses. Considering the central role of insulin signaling in regulating toxicity of pollutants and stresses in *C. elegans*, the elucidation of underlying mechanism for insulin signaling in regulating transgenerational toxicity of pollutants and stresses provides the important foundation for further determining and understanding of transgenerational communication of other molecular signaling in regulating the transgenerational toxicity of pollutants and stresses.

5.2 Role of Insulin Ligands

In the germline, certain insulin ligands showed the transgenerational activation after exposure to pollutants at the parental generation (P0-G). As an important member of emergent contaminants, environmental occurrence and exposure risk to environmental organisms and humans for nanoplastics have received worldwide attention [24, 25]. At P0-G in *C. elegans*, exposure to PS-NP, the examined example of nanoplastics, resulted in several aspects of toxicity, including intestinal damage, neurotoxicity, reproductive toxicity, lifespan reduction, and even transgenerational toxicity [26–32]. At the P0-G, germline expressions of *ins-3*, *ins-4*, *ins-17*, *ins-34*, *ins-39*, and *daf-28* encoding insulin ligands were activated by 20 nm PS-NP with a concentration-dependent property (Fig. 5.1) [23]. Moreover, after exposure to 20 nm PS-NP (1 μg/L), transgenerational activation of germline *ins-3*, *ins-39*, and *daf-28* was induced till to the F2-G (Fig. 5.1) [23], which suggested the formation of transgenerational response of germline insulin ligands to pollutants. This also implied that higher concentrations of pollutants would cause transgenerational activation of more germline insulin ligands.

Using DCL569 as a tool for germline RNA of genes, it was shown that RNAi of *ins-3*, *ins-39*, and *daf-28* in the germline caused formation of resistance of nematodes to transgenerational neurotoxicity and transgenerational reproductive toxicity of PS-NP (Fig. 5.2) [23]. These data indicated that transgenerational activation of insulin ligands mediated induction of transgenerational toxicity of pollutants.

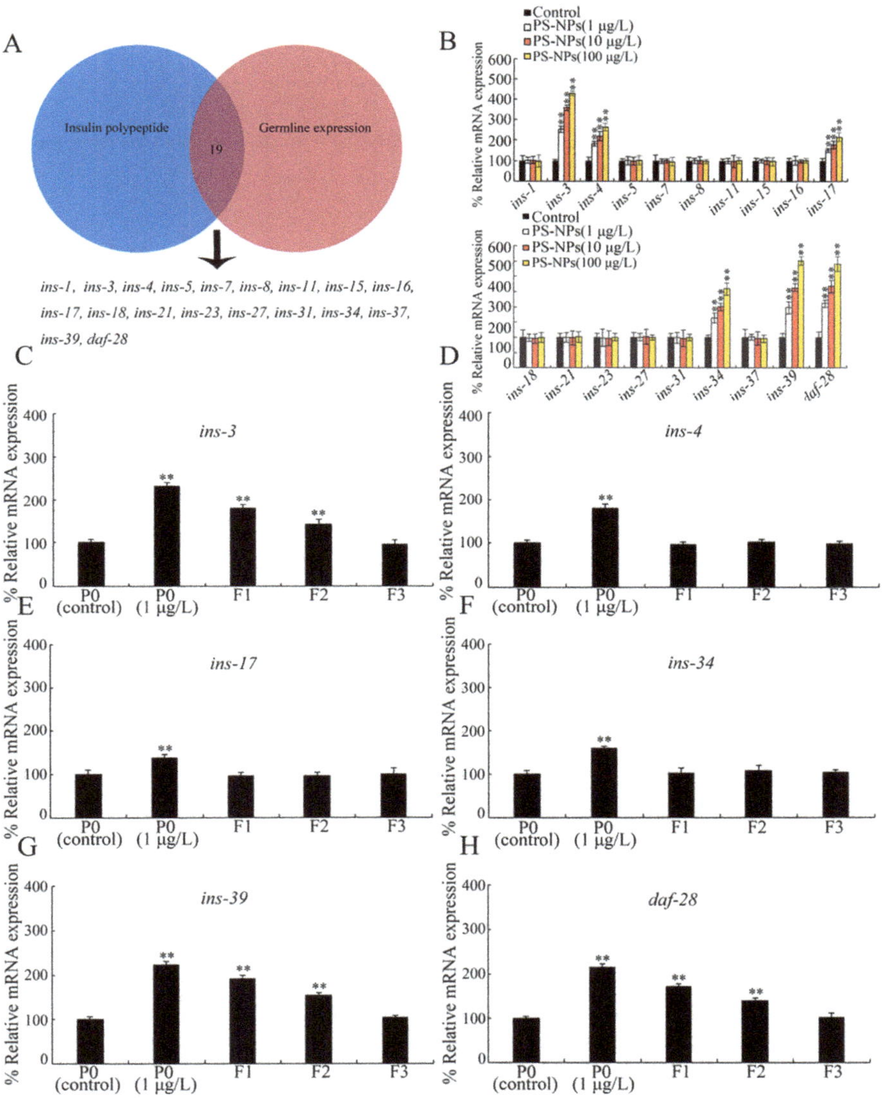

Fig. 5.1 Identification of potential germline insulin peptides in controlling transgenerational PS-NP toxicity [23]. (**a**) Information of insulin ligands expressed in the germline. (**b**) Effect of PS-NP (1–100 μg/L) on expression of germline insulin at P0-G. (**c–h**) Transgenerational expression of *ins-3*, *ins-4*, *ins-17*, *ins-34*, *ins-39*, and *daf-28* with 1 μg/L PS-NP exposure. PS-NP, Polystyrene nanoparticle. $^{**}P < 0.01$ vs control or P0 (control). Control, without PS-NP exposure

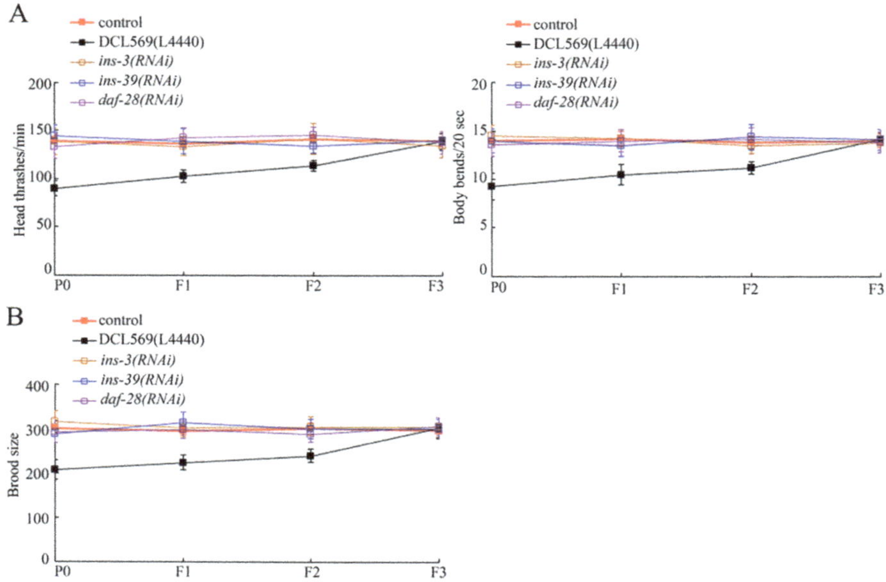

Fig. 5.2 Effect of germline RNAi of *ins-3*, *ins-39*, and *daf-28* on transgenerational PS-NP toxicity in inhibiting locomotion (**a**) and brood size (**b**) [23]. 1 μg/L was used as the exposure concentration. Curve of DCL569(L4440) indicated a difference significantly ($P < 0.01$) in comparison of control. Curves of *ins-3(RNAi)*, *ins-39(RNAi)*, and *daf-28(RNAi)* indicated a difference significantly ($P < 0.01$) compared to DCL569(L4440). PS-NP, Polystyrene nanoparticle. Control, without PS-NP exposure

5.3 Role of Insulin Signaling Pathway

Accompanied with transgenerational activation of insulin ligands, exposure to pollutants can meanwhile induce transgenerational response of insulin signaling pathway to mediate the transgenerational toxicity induction. Further with PS-NP as example of pollutants, at the P0-G, PS-NP exposure in the range of μg/L increased *daf-2* expression, as well as *age-1* and *akt-1* expressions, and decreased *daf-16* expression (Fig. 5.3) [23]. Moreover, transgenerational activation of DAF-2, AGE-1, and AKT-1 and transgenerational inhibition in DAF-16 were induced by 1 μg/L PS-NP till to the F2-G (Fig. 5.3) [23], which suggested formation of transgenerational activation of insulin receptor and its downstream kinase cascade and transgenerational inhibition in FOXO transcriptional factor by pollutants.

Using the RNAi technique, RNAi of *daf-2*, *age-1*, and *akt-1* resulted in formation of resistance of nematodes to transgenerational neurotoxicity and transgenerational reproductive toxicity of PS-NP (Fig. 5.4) [23]. In contrast, RNAi of *daf-16* led to formation of susceptibility of nematodes to transgenerational neurotoxicity and transgenerational reproductive toxicity of PS-NP (Fig. 5.4) [23]. Therefore, transgenerational activation of insulin receptor and its downstream kinase cascade and

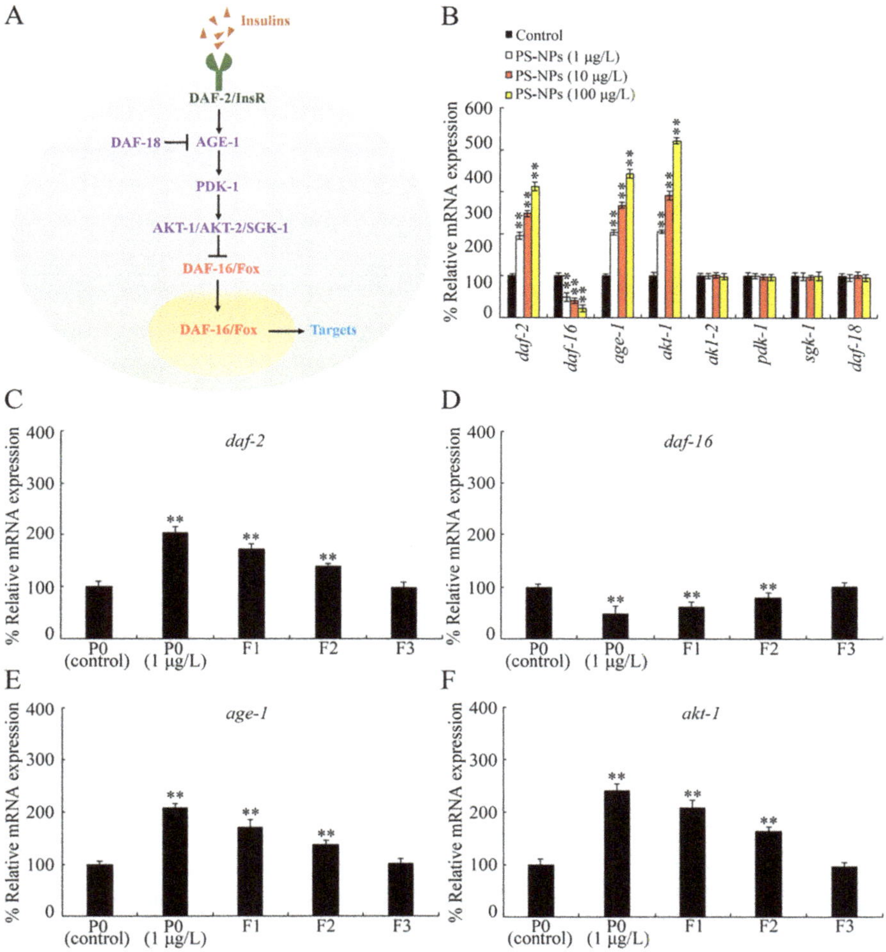

Fig. 5.3 Identification of potential cascade of INS signal transduction in controlling transgenerational PS-NP toxicity [23]. (**a**) Signaling cascade in the insulin signaling pathway. (**b**) The expression alteration of genes encoding signaling cascade in the insulin signaling pathway with PS-NP (1–100 μg/L) exposure at P0-G. (**c–f**) Transgenerational expression of *daf-2*, *daf-16*, *age-1*, and *akt-1* with 1 μg/L PS-NP treatment. PS-NP, Polystyrene nanoparticle. $^{**}P < 0.01$ vs control or P0 (control). Control, without PS-NP exposure

transgenerational inhibition in DAF-16 also mediated induction of transgenerational toxicity of pollutants.

Genetic interaction analysis demonstrated that DAF-2, AGE-1, and AKT-1 function upstream of DAF-16 to regulate transgenerational pollutant toxicity. The *daf-2*-, *age-1*-, and *akt-1* RNAi–induced transgenerational resistance to PS-NP neurotoxicity was suppressed by *daf-16* RNAi (Fig. 5.5) [23]. Similarly, *daf-16* RNAi inhibited transgenerational resistance to PS-NP reproductive toxicity caused

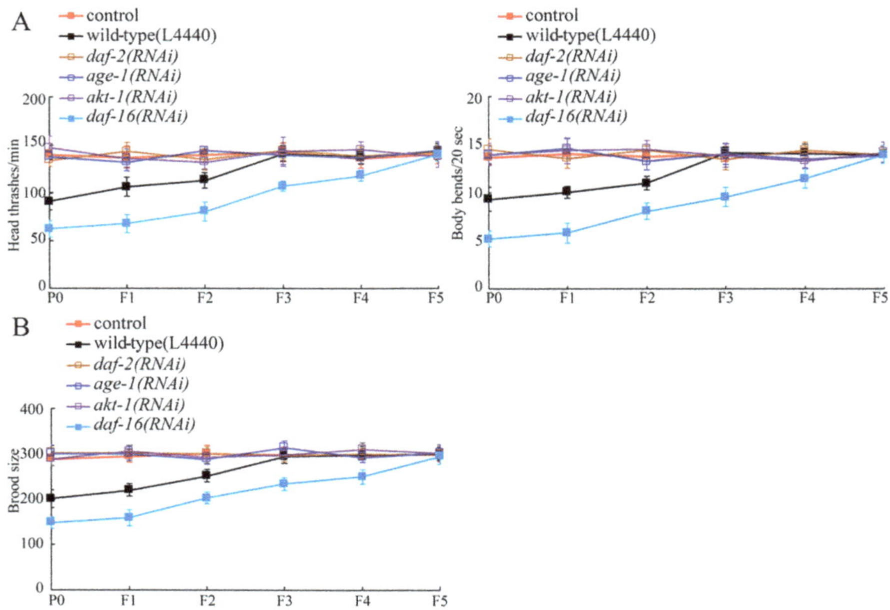

Fig. 5.4 Effect of RNAi of *daf-2*, *age-1*, *akt-1*, and *daf-16* on transgenerational PS-NP toxicity in inhibiting locomotion (**a**) and in decreasing brood size (**b**) [23]. 1 μg/L was used as the exposure concentration. Curve of wild-type (L4440) indicated difference significantly ($P < 0.01$) in comparison to control. Curves of *daf-2(RNAi)*, *age-1 (RNAi)*, *akt-1(RNAi)*, and *daf-16(RNAi)* indicated a significant difference ($P < 0.01$) in contrast to wild-type (L4440). PS-NP, Polystyrene nanoparticle. Control, without PS-NP exposure

by *daf-2*, *age-1*, and *akt-1* RNAi (Fig. 5.5) [23]. These observations provide direct functional evidence for genetic interaction between DAF-16 and DAF-2 or its downstream kinase cascade during the control of transgenerational pollutant toxicity. This also implied that the induced transcriptional activation of insulin receptor by pollutants at environmentally relevant concentrations caused transgenerational inhibition in DAF-16 by activating limited number of members in the downstream kinase cascade.

5.4 Transgenerational Communication Between Insulin Ligands and Its Receptor in Regulating Transgenerational Toxicity of Pollutants

Certain evidence has been raised to suggest the existence of transgenerational communication between germline insulin ligands and its receptor in regulating transgenerational pollutant toxicity in *C. elegans*. To confirm this, insulin ligands were overexpressed in the germline at the P0-G, and after that the *daf-2* RNAi was

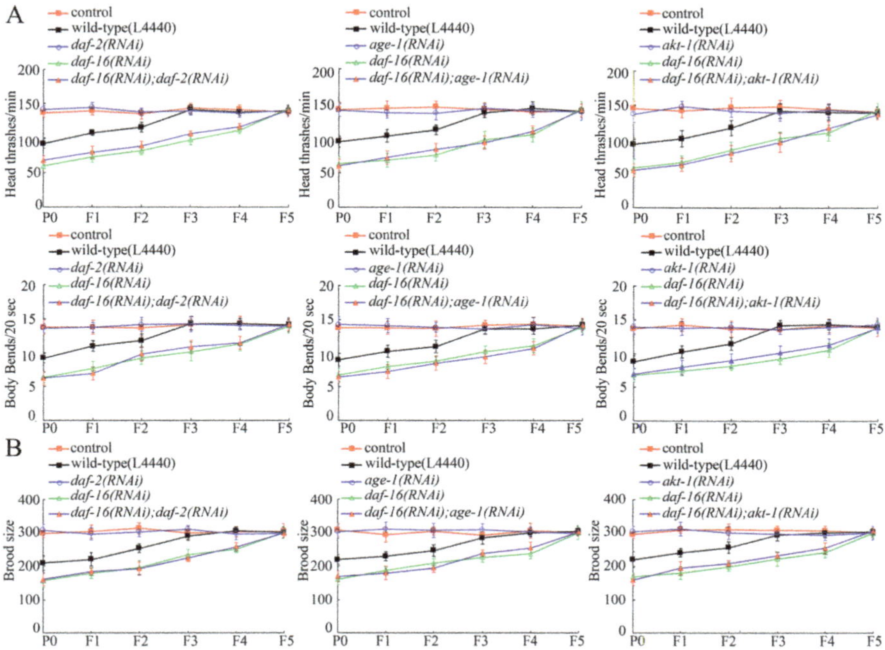

Fig. 5.5 DAF-16 functioned downstream of AKT-1, AGE-1, and DAF-2 to control transgenerational PS-NP toxicity in inhibiting locomotion (**a**) and in reducing brood size (**b**) [23]. 1 μg/L was used as the exposure concentration. Curves of wild-type (L4440) indicated difference significantly ($P < 0.01$) in comparison of control. Curves of *daf-2(RNAi)*, *age-1(RNAi)*, *akt-1(RNAi)*, and *daf-16(RNAi)* indicated a significant difference ($P < 0.01$) in comparison to wild-type (L4440). Curves of *daf-16(RNAi);daf-2(RNAi)* indicated a significant difference ($P < 0.01$) in comparison to *daf-2(RNAi)*. Curves of *daf-16(RNAi);age-1(RNAi)* indicated a significant difference ($P < 0.01$) in comparison to *age-1(RNAi)*. Curves of *daf-16(RNAi);akt-1(RNAi)* indicated a significant difference ($P < 0.01$) in comparison to *akt-1(RNAi)*. PS-NP, Polystyrene nanoparticle. Control, without PS-NP exposure

performed at the F1-G. Germline overexpression of *ins-3*, *ins-39*, and *daf-28* resulted in transgenerational susceptibility to neurotoxicity and reproductive toxicity of PS-NP (Fig. 5.6) [23]. Moreover, susceptibility to neurotoxicity and reproductive toxicity of PS-NP induced by germline overexpression of insulin ligands was inhibited by *daf-2* RNAi at the F1-G (Fig. 5.6) [23]. That is, after *daf-2* RNAi at the F1-G, the nematodes overexpressing germline insulin ligands at the P0-G showed resistance to neurotoxicity and reproductive toxicity of PS-NP (Fig. 5.6) [23]. Therefore, germline insulin ligands mediate transgenerational pollutant toxicity by activating corresponding receptor in the offspring of pollutant-exposed nematodes.

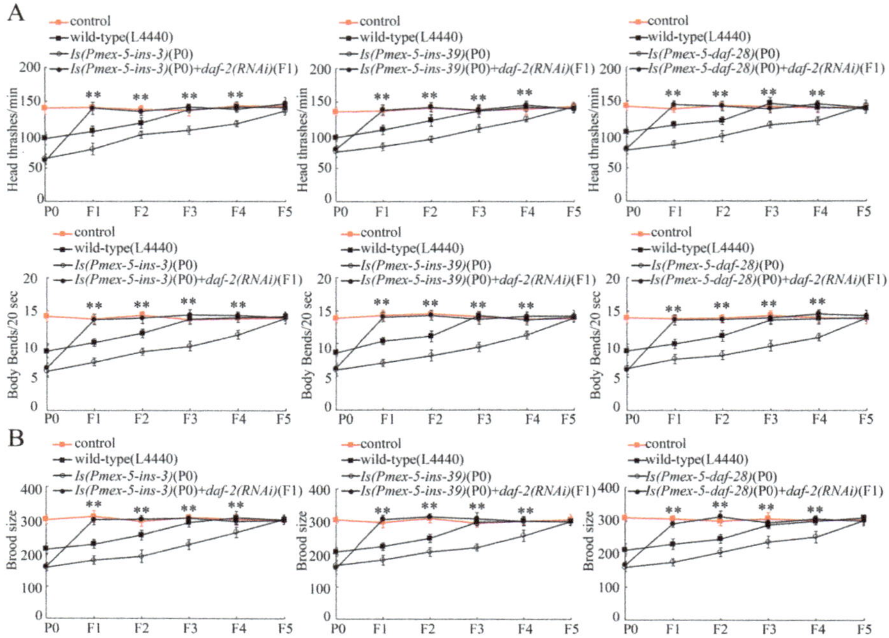

Fig. 5.6 Insulin ligands transgenerationally affected the function of DAF-2 to control transgenerational PS-NP toxicity in decreasing locomotion (**a**) and in reducing brood size (**b**) [23]. After the PS-NP exposure in *Is(Pmex-5-ins-3)*, *Is(Pmex-5-ins-39)*, and *Is(Pmex-5-daf-28)* nematodes (P0), the RNAi of *daf-2* was performed at F1-G. Curves of wild-type (L4440) indicated a significant difference ($P < 0.01$) in comparison to control. Curves of *Is(Pmex-5-ins-3)*, *Is(Pmex-5-ins-39)*, and *Is(Pmex-5-daf-28)* indicated a significant difference ($P < 0.01$) in comparison to wild-type (L4440). 1 μg/L was used as the exposure concentration. $^{**}P < 0.01$ showing significance difference for *Is(Pmex-5-ins-3)*(P0) + *daf-2(RNAi)*(F1) vs *Is(Pmex-5-ins-3)*, *Is(Pmex-5-ins-39)*(P0) + *daf-2(RNAi)*(F1) vs *Is(Pmex-5-ins-39)*, and *Is(Pmex-5-daf-28)*(P0) + *daf-2(RNAi)*(F1) vs *Is(Pmex-5-daf-28)*. PS-NP, Polystyrene nanoparticle. Control, without PS-NP exposure

5.5 Targets of Transcriptional Factor DAF-16

5.5.1 SOD-3 and HSP-6

SOD-3 and HSP-6, two mitochondrial proteins, were first identified as targets of DAF-16 in regulating transgenerational pollutant toxicity. SOD-3 and HSP-6 participated in control of multiple toxicities induced by pollutants [33–38]. With PS-NP as the example of pollutants, PS-NP exposure induced the formation of transgenerational inhibition in *sod-3* and *hsp-6* expressions (Fig. 5.7) [23].

Fig. 5.7 (continued) ($P < 0.01$) in comparison of *Is(Pges-1-daf-16)*. Curves of *Is(Pges-1-daf-16);sod-3(RNAi)* indicated a significant difference ($P < 0.01$) compared to *Is(Pges-1-daf-16)*. Control, without PS-NP exposure. (**d**) A diagram showing mechanism of insulin signaling in controlling transgenerational PS-NP toxicity in nematodes. PS-NP, Polystyrene nanoparticle

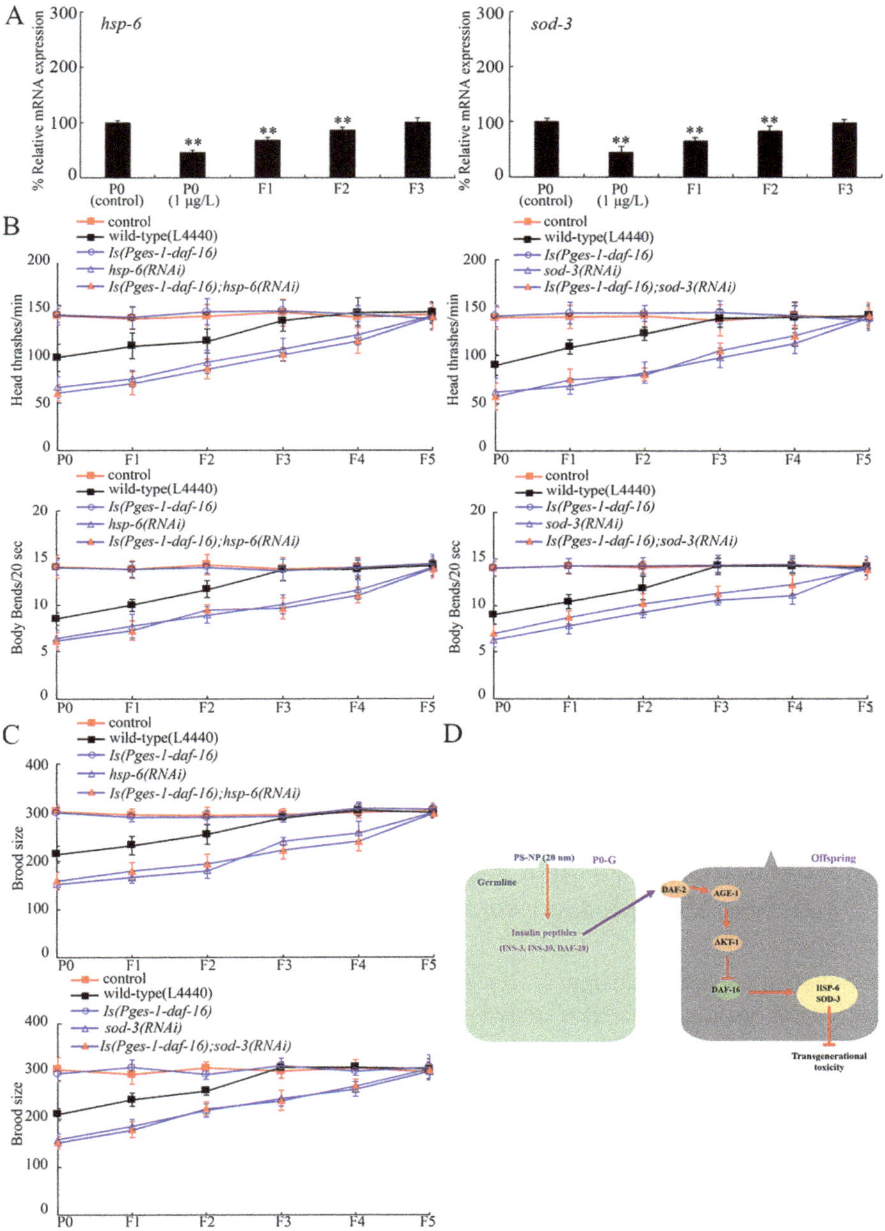

Fig. 5.7 Identification of candidate targets of DAF-16 in controlling transgenerational PS-NP toxicity [23]. (**a**) Transgenerational expression of *hsp-6* and *sod-3* with 1 μg/L PS-NP exposure. $^{**}P < 0.01$ vs P0 (control). Control, without PS-NP exposure. (**b**, **c**) Genetic interaction between DAF-16 and HSP-6 or SOD-3 to control transgenerational PS-NP toxicity in inhibiting locomotion and in reducing brood size. 1 μg/L was used as the exposure concentration. Curves of wild-type (L4440) indicated a significant difference ($P < 0.01$) compared to control. Curves of *Is(Pges-1-daf-16)*, *hsp-6(RNAi)*, and *sod-3(RNAi)* indicated a significant difference ($P < 0.01$) in comparison of wild-type (L4440). Curves of *Is(Pges-1-daf-16);hsp-6(RNAi)* indicated a significant difference

Overexpression of DAF-16 caused transgenerational resistance to neurotoxicity and reproductive toxicity of PS-NP (Fig. 5.7) [23]. More importantly, the transgenerational resistance to neurotoxicity and reproductive toxicity of PS-NP induced by DAF-16 overexpression was suppressed by RNAi of *sod-3* and *hsp-6* (Fig. 5.7) [23]. That is, after RNAi of *sod-3* and *hsp-6*, nematodes overexpressing DAF-16 showed transgenerational susceptibility to neurotoxicity and reproductive toxicity of PS-NP (Fig. 5.7) [23].

5.5.2 Hedgehog Ligands

Some hedgehog ligands were also identified as downstream targets for DAF-16 in regulating transgenerational pollutant toxicity. The hedgehog signals are required for control of toxicity induced by pollutants and stresses [39–41]. With PS-NP as the example of pollutants, expressions of *wrt-3*, *grl-15*, *grl-16*, *qua-1*, and *wrt-1* encoding hedgehog ligands were increased by RNAi of *daf-16* in PS-NP-exposed nematodes (Fig. 5.8) [42]. Expressions of these hedgehog ligand genes were increased by PS-NP exposure at P0-G (Fig. 5.8) [42, 43]. Additionally, exposure to 1 μg/L PS-NPs induced transgenerational activation of *wrt-3*, *grl-15*, *grl-16*, *qua-1*, and *wrt-1* till to the F2-G (Fig. 5.8) [42, 43]. Meanwhile, germline RNAi of these hedgehog ligand genes resulted in transgenerational resistance to neurotoxicity and reproductive toxicity of PS-NPs [42, 43]. Thus, transgenerational inhibition in DAF-16 mediated transgenerational pollutant toxicity by causing activation of hedgehog ligands.

5.6 Formation of Positive Feedback Regulation Mechanism Across Multiple Generations

In pollutant-exposed nematodes, a feedback loop can be formed between DAF-16 and insulin ligands. In PS-NP-exposed nematodes, RNAi of *daf-16* further increased expressions of *ins-3*, *ins-39*, and *daf-28* (Fig. 5.9) [33]. That is, on the one hand, transgenerational activation of germline insulin ligands mediated transgenerational pollutant toxicity by activating DAF-2 and its downstream kinase cascade and inhibiting DAF-16 and some of its targets (Fig. 5.9) [33]. On the other hand, transgenerational inhibition in DAF-16 caused by PS-NPs exposure further resulted in activation of insulin ligands to amply the induction of transgenerational pollutant toxicity (Fig. 5.9) [33].

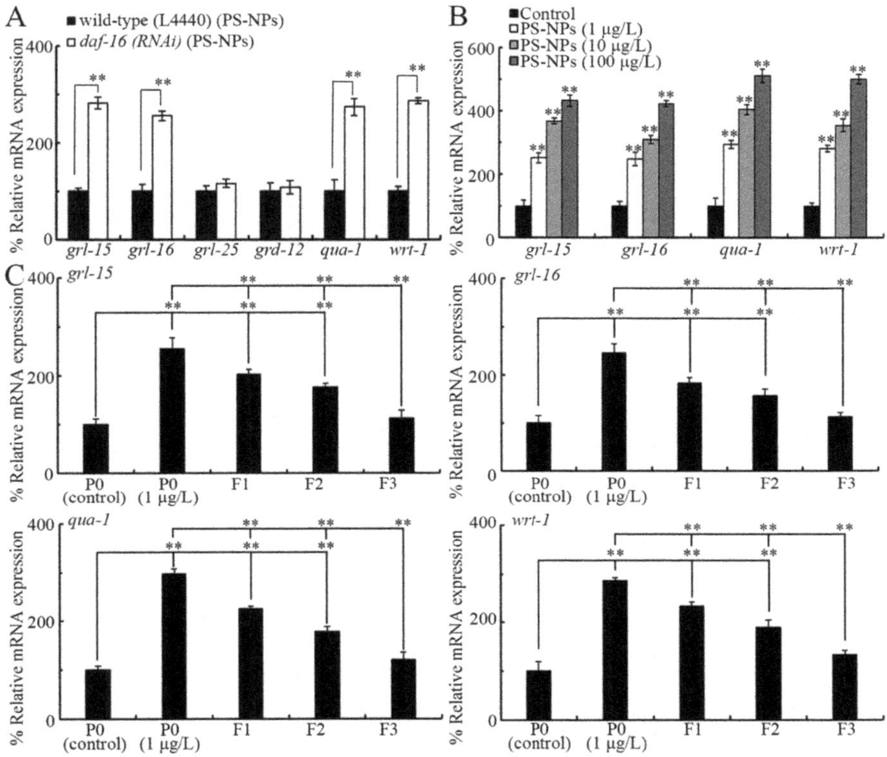

Fig. 5.8 Transgenerational expression of germline hedgehog ligands after PS-NP exposure [42]. (**a**) Effect of RNAi of *daf-16* on expressions of *grl-15*, *grl-16*, *grl-25*, *grd-12*, *qua-1*, and *wrt-1* in PS-NP-exposed nematodes. PS-NP exposure concentration was 1 μg/L. $^{**}P < 0.01$. (**b**) Effect of 1–100 μg/L PS-NP exposure on expressions of germline *grl-15*, *grl-16*, *qua-1*, and *wrt-1*. $^{**}P < 0.01$ vs Control. (**c**) Transgenerational expressions of germline *grl-15*, *grl-16*, *qua-1*, and *wrt-1* after PS-NP exposure. PS-NP exposure concentration was 1 μg/L. $^{**}P < 0.01$. PS-NP, Polystyrene nanoparticle. A total of 30 intact gonads were used for the qRT-PCR assay

5.7 Perspectives

In this chapter, we discussed the role of transgenerational activation of insulin ligands and receptor and inhibition in DAF-16 in mediating induction of transgenerational pollutant toxicity. Considering the central role of insulin signaling in regulating pollutant toxicity at P0-G as described previously [5], it is supposed that insulin signaling may also play the central role in regulating transgenerational pollutant toxicity.

So far, certain evidence has been raised to indicate the role of transgenerational communication between germline insulin ligands in pollutant-exposed nematodes and insulin receptor in the offspring in driving transgenerational toxicity formation. Meanwhile, some evidence also demonstrated the involvement of alteration in

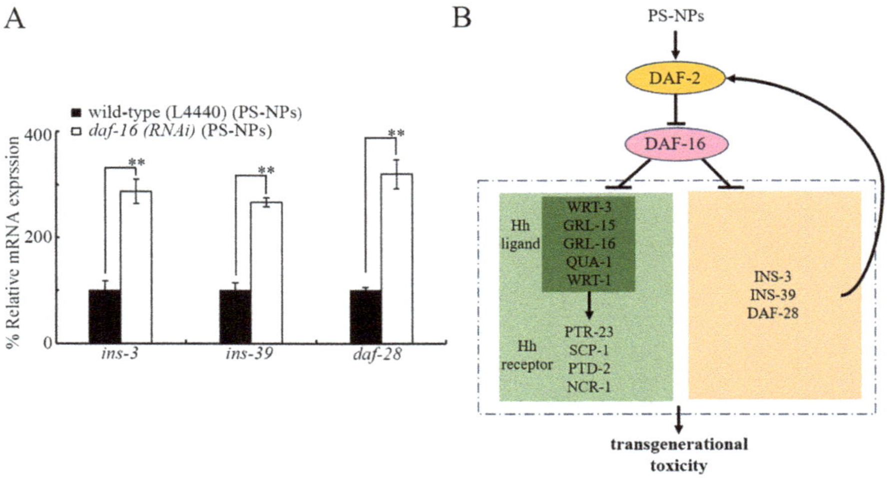

Fig. 5.9 Effect of RNAi of *daf-16* on expression of insulin peptide genes in PS-NP exposed nematodes [42]. (**a**) Effect of RNAi of *daf-16* on expression of *ins-3*, *ins-39*, and *daf-28* in PS-NP-exposed nematodes. PS-NP exposure concentration was 1 μg/L. $^{**}P < 0.01$. (**b**) A diagram showing molecular basis for DAF-16 and its targets in regulating transgenerational toxicity of PS-NP in nematodes. PS-NP, Polystyrene nanoparticle

transgenerational epigenetic signals in mediating transgenerational pollutant toxicity [44–49]. Therefore, transgenerational epigenetic regulation and transgenerational signaling communication may act simultaneously to control transgenerational pollutant toxicity. For example, during the control of transgenerational PS-NP toxicity, WRT-3 was also identified as target of MET-2, a histone methyltransferase [42].

For the role of insulin signaling in regulating transgenerational pollutant toxicity, some molecular mechanisms were identified and raised that amplify the transgenerational pollutant toxicity. One underlying mechanism is the formation of feedback loop between DAF-16 and insulin ligands. Another underlying mechanism is that transgenerational inhibition in DAF-16 could further cause activation in hedgehog ligands to activate another form of transgenerational signaling communication.

References

1. Wang D-Y (2018) Nanotoxicology in *Caenorhabditis elegans*. Springer Nature Singapore Pte Ltd., Singapore
2. Wang D-Y (2019) Molecular toxicology in *Caenorhabditis elegans*. Springer Nature Singapore Pte Ltd., Singapore
3. Wang D-Y (2019) Target organ toxicology in *Caenorhabditis elegans*. Springer Nature Singapore Pte Ltd., Singapore
4. Wang D-Y (2020) Exposure toxicology in *Caenorhabditis elegans*. Springer Nature Singapore Pte Ltd, Singapore

5. Wang D-Y (2022) Toxicology at environmentally relevant concentrations in *Caenorhabditis elegans*. Springer Nature Singapore Pte Ltd., Singapore
6. Zhu R, Chin-Sang ID (2024) *C. elegans* insulin-like peptides. Mol Cell Endocrinol 585:112173
7. Mukhopadhyay A, Oh SW, Tissenbaum HA (2006) Worming pathways to and from DAF-16/FOXO. Exp Gerontol 41:928–934
8. Finch CE, Ruvkun G (2001) The genetics of aging. Annu Rev Genomics Hum Genet 2:435–462
9. Liu Z-J, Li Y-H, Wang D-Y (2024) Exposure to 6-PPD quinone disrupts glucose metabolism associated with lifespan reduction by affecting insulin and AMPK signals in *Caenorhabditis elegans*. Chemosphere 363:142975
10. Yang R-L, Rui Q, Kong Y, Zhang N, Li Y, Wang X-Y, Tao J, Tian P-Y, Ma Y, Wei J-R, Li G-J, Wang D-Y (2016) Metallothioneins act downstream of insulin signaling to regulate toxicity of outdoor fine particulate matter ($PM_{2.5}$) during spring festival in Beijing in nematode *Caenorhabditis elegans*. Toxicol Res 5:1097–1105
11. Savory FR, Sait SM, Hope IA (2011) DAF-16 and delta9 desaturase genes promote cold tolerance in long-lived *Caenorhabditis elegans age-1* mutants. PLoS One 6:e24550
12. Ensaka N, Sakamoto K (2020) alpha-Pinene odor exposure enhances heat stress tolerance through *Daf-16* in *Caenorhabditis elegans*. Biochem Biophys Res Commun 528:726–731
13. Wang Y-X, Hu G-C, Wang D-Y (2025) 6-PPD quinone reduces lifespan by activating a feedback loop between cholesterol transformation related signal and insulin signaling in *C. elegans*. J Environ Sci. https://doi.org/10.1016/j.jes.2025.10.044
14. Wu J-W, Shen S-H, Wang D-Y (2024) 6-PPD quinone at environmentally relevant concentrations induces immunosenescence by causing immunosuppression during the aging process. Chemosphere 368:143719
15. Zhuang Z-H, Li M, Liu H, Luo L-B, Gu W-D, Wu Q-L, Wang D-Y (2016) Function of RSKS-1-AAK-2-DAF-16 signaling cascade in enhancing toxicity of multi-walled carbon nanotubes can be suppressed by *mir-259* activation in *Caenorhabditis elegans*. Sci Rep 6:32409
16. Zhao Y-L, Yang R-L, Rui Q, Wang D-Y (2016) Intestinal insulin signaling encodes two different molecular mechanisms for the shortened longevity induced by graphene oxide in *Caenorhabditis elegans*. Sci Rep 6:24024
17. Hua X, Wang D-Y (2023) Exposure to 6-PPD quinone at environmentally relevant concentrations inhibits both lifespan and healthspan in *C. elegans*. Environ Sci Technol 57:19295–19303
18. Liu H-L, Wang D-Y (2021) Intestinal mitochondrial unfolded protein response induced by nanoplastic particles in *Caenorhabditis elegans*. Chemosphere 267:128917
19. Shao H-M, Han Z-Y, Krasteva N, Wang D-Y (2019) Identification of signaling cascade in the insulin signaling pathway in response to nanopolystyrene particles. Nanotoxicology 13:174–188
20. Kong Y, Liu H-L, Li W-J, Wang D-Y (2019) Intestine-specific activity of insulin signaling pathway in response to microgravity stress in *Caenorhabditis elegans*. Biochem Biophys Res Commun 517:278–284
21. Hua X, Wang D-Y (2025) 6-PPD quinone at environmentally relevant concentrations activates feedback response of electron transport chain to mediate damage on mitochondrial function and longevity in *Caenorhabditis elegans*. Environ Chem Ecotoxicol 7:2356–2365
22. Qu M, Li D, Qiu Y-X, Wang D-Y (2020) Neuronal ERK MAPK signaling in response to low-dose nanopolystyrene exposure by suppressing insulin peptide expression in *Caenorhabditis elegans*. Sci Total Environ 724:138378
23. Liu H-L, Zhao Y-L, Hua X, Wang D-Y (2022) Induction of transgenerational toxicity is associated with the activated germline insulin signals in nematodes exposed to nanoplastic at predicted environmental concentrations. Ecotoxicol Environ Saf 243:114022
24. Hua X, Wang D-Y (2023) Tire-rubber related pollutant 6-PPD quinone: a review of its transformation, environmental distribution, bioavailability, and toxicity. J Hazard Mater 459:132265
25. Wan X, Liang G-Y, Wang D-Y (2024) Potential human health risk of the emerging environmental contaminant 6-PPD quinone. Sci Total Environ 949:175057
26. Qu M, Wang D-Y (2020) Toxicity comparison between pristine and sulfonate modified nanopolystyrene particles in affecting locomotion behavior, sensory perception, and neuronal development in *Caenorhabditis elegans*. Sci Total Environ 703:134817

27. Qiu Y-X, Luo L-B, Yang Y-H, Kong Y, Li Y-H, Wang D-Y (2020) Potential toxicity of nanopolystyrene on lifespan and aging process of nematode *Caenorhabditis elegans*. Sci Total Environ 705:135918
28. Qiu Y-X, Liu Y-Q, Li Y-H, Li G-J, Wang D-Y (2020) Effect of chronic exposure to nanopolystyrene on nematode *Caenorhabditis elegans*. Chemosphere 256:127172
29. Qu M, Kong Y, Yuan Y-J, Wang D-Y (2019) Neuronal damage induced by nanopolystyrene particles in nematode *Caenorhabditis elegans*. Environ Sci Nano 6:2591–2601
30. Qu M, Qiu Y-X, Kong Y, Wang D-Y (2019) Amino modification enhances reproductive toxicity of nanopolystyrene on gonad development and reproductive capacity in nematode *Caenorhabditis elegans*. Environ Pollut 254:112978
31. Qu M, Xu K-N, Li Y-H, Wong G, Wang D-Y (2018) Using *acs-22* mutant *Caenorhabditis elegans* to detect the toxicity of nanopolystyrene particles. Sci Total Environ 643:119–126
32. Zhao L, Qu M, Wong G, Wang D-Y (2017) Transgenerational toxicity of nanopolystyrene particles in the range of μg/L in nematode *Caenorhabditis elegans*. Environ Sci Nano 4:2356–2366
33. Li Y-X, Wang W, Wu Q-L, Li Y-P, Tang M, Ye B-P, Wang D-Y (2012) Molecular control of TiO_2-NPs toxicity formation at predicted environmental relevant concentrations by Mn-SODs proteins. PLoS One 7:e44688
34. Wu Q-L, Zhao Y-L, Li Y-P, Wang D-Y (2014) Susceptible genes regulate the adverse effects of TiO_2-NPs at predicted environmental relevant concentrations on nematode *Caenorhabditis elegans*. Nanomedicine: Nanotechnol. Biol Med 10:1263–1271
35. Rui Q, Zhao Y-L, Wu Q-L, Tang M, Wang D-Y (2013) Biosafety assessment of titanium dioxide nanoparticles in acutely exposed nematode *Caenorhabditis elegans* with mutations of genes required for oxidative stress or stress response. Chemosphere 93:2289–2296
36. Xu R-R, Hua X, Rui Q, Wang D-Y (2022) Polystyrene nanoparticles caused dynamic alteration in mitochondrial unfolded protein response from parents to the offspring in *C. elegans*. Chemosphere 308:136154
37. Hua X, Wang D-Y (2024) 6-PPD quinone at environmentally relevant concentrations induced damage on longevity in *C. elegans*: mechanistic insight from inhibition in mitochondrial UPR response. Sci Total Environ 954:176275
38. Hua X, Wang D-Y (2025) Environmentally relevant concentration of 6-PPD quinone inhibits two types of mitophagy to cause mitochondrial dysfunction and lifespan reduction in *Caenorhabditis elegans*. Environ Sci: Processes Impacts 27:1928–1940
39. Shao Y-T, Li Y-H, Wang D-Y (2024) Polylactic acid microplastics cause transgenerational reproductive toxicity associated with activation of insulin and hedgehog ligands in *C. elegans*. Sci Total Environ 942:173746
40. Zhou L, Jiang L, Li L, Ma C, Xia P, Ding W, Liu Y (2024) A germline-to-soma signal triggers an age-related decline of mitochondrial stress response. Nat Commun 15:8723
41. Zárate-Potes A, Ali I, Ribeiro Camacho M, Brownless H, Benedetto A (2022) Meta-analysis of *Caenorhabditis elegans* transcriptomics implicates hedgehog-like signaling in host-microbe interactions. Front Microbiol 13:853629
42. Wu J-W, Shao Y-T, Hua X, Wang D-Y (2024) Activated hedgehog and insulin ligands by decreased transcriptional factor DAF-16 mediate transgenerational nanoplastic toxicity in *Caenorhabditis elegans*. J Hazard Mater 480:135909
43. Zhang L, Wang S-T, Zhao Y, Bi K, Wang D-Y (2022) Increase in germline methyltransferases governing methylation of histone H3K9 is associated with transgenerational nanoplastic toxicity in *Caenorhabditis elegans*. Environ Sci Nano 9:265–274
44. Wu J-W, Shao Y-T, Hua X, Li Y-H, Wang D-Y (2025) Photo-aged polylactic acid microplastics causes severe transgenerational decline in reproductive capacity in *C. elegans*: insight into activation of DNA damage checkpoints affected by multiple germline histone methyltransferases. Environ Pollut 382:126697
45. Hua X, Zhang L, Wang D-Y (2024) Nanoplastic at environmentally relevant concentrations activates germline *mir-240-rab-5* signaling cascade to affect secreted ligands associated with transgenerational toxicity induction in *C. Elegans*. Environ Sci Nano 11:3615

46. Hua X, Wang D-Y (2024) Transgenerational response of germline histone acetyltransferases and deacetylases to nanoplastics at predicted environmental doses in *Caenorhabditis elegans*. Sci Total Environ 952:175903
47. Zhao Y-L, Chen J-Y, Wang R, Pu X-X, Wang D-Y (2023) A review of transgenerational and multigenerational toxicology in the *in vivo* model animal *Caenorhabditis elegans*. J Appl Toxicol 43:122–145
48. Hua X, Zhao Y, Yuan Y-J, Zhang L, Bian Q, Wang D-Y (2022) Nanoplastics cause transgenerational toxicity through inhibiting germline microRNA *mir-38* in *C. elegans*. J Hazard Mater 437:129302
49. Zhao Y-Y, Xu R-R, Hua X, Rui Q, Wang D-Y (2022) Multi-walled carbon nanotubes induce transgenerational toxicity associated with activation of germline long non-coding RNA *linc-7* in *C. elegans*. Chemosphere 301:134687

Chapter 6
Regulation of Transgenerational Toxicity by Transgenerational Communication of Germline Wnt Signals

Abstract Wnt signaling pathway is conserved among different organisms. In *Caenorhabditis elegans*, transgenerational activation of Wnt ligands and transgenerational inhibition in Wnt receptors mediate induction of transgenerational pollutants toxicity. After pollutant exposure at parental generation, activated Wnt ligands have the potential to transgenerationally target to their receptors in the offspring to cause transgenerational toxicity. Transgenerational inhibition in Wnt receptors induced by pollutant exposure can lead to further transgenerational inhibition in transcriptional factor β-catenin/BAR-1 by affecting signaling cascade of Dishevelled proteins-APC complex components. Transgenerational BAR-1 inhibition results in suppression in mitochondrial unfolded protein response to mediate transgenerational pollutant toxicity by inhibiting DVE-1 expression.

Keywords Transgenerational toxicology · Transgenerational communication · Wnt signaling · *Caenorhabditis elegans*

6.1 Introduction

Largely based on established system of exposure toxicology and high sensitivity to environmental exposures, *Caenorhabditis elegans* has been widely used for toxicity assessment of pollutants and stresses at low doses, such as at environmentally relevant concentrations (ERCs) for pollutants [1–3]. Meanwhile, as a classic model animal, *C. elegans* allows the successful establishment of knowledge systems of molecular toxicology and target organ toxicology in this animal model [4, 5].

Wnt signaling pathway is one of the well-described signaling pathways in *C. elegans*, such as its organization and functions. After the release, Wnt ligands containing the LIN-44, EGL-20, MOM-2, CWN-1, and CWN-2 can target to their receptors including the MIG-1, MOM-5, CFZ-2, and LIN-17 [6, 7]. The Wnt receptors activate transcriptional factor β-catenin/BAR-1 by activating Dishevelled proteins containing MIG-5, DSH-1, and DSH-2 and inhibiting APC complex containing components of KIN-19, GSK-3, APR-1, and PRY-1 [6, 7].

D. Wang, *Transgenerational Toxicology in Caenorhabditis elegans*,
https://doi.org/10.1007/978-981-92-1389-4_6

In *C. elegans*, Wnt signaling participates in the control of various biological processes. During the developmental control, Wnt signaling was required for the cell fate, abundance of neurotransmitter receptor, neuronal lineage organization, neuroblast migration, vulval development, male development, and developmental timing [8–15]. Moreover, Wnt signaling was involved in control of innate immunity and pathogen infection, such as bacterial infection [16–19]. Additionally, response of Wnt signaling to simulated microgravity could be observed at parental generation (P0-G), and Wnt signaling could function in intestinal cells to control toxicity of simulated microgravity by affecting certain targets [20, 21]. More recently, response of Wnt signaling to pollutants of graphene oxide (GO), 6-PPD quinone (6-PPDQ), and polystyrene nanoparticle (PS-NP) was detected at P0-G, and Wnt signaling played important function in modulating toxicity of these pollutants at different aspects [22–26].

In this chapter, we first introduced roles of Wnt ligands and Wnt signaling pathway in response to pollutant exposure and their function in regulating transgenerational pollutant toxicity. After that, the evidence for transgenerational communication between Wnt ligands and their receptor in regulating transgenerational pollutant toxicity was introduced. Moreover, potential targets of transcriptional factor BAR-1 in regulating transgenerational pollutant toxicity were discussed.

6.2 Role of Wnt Ligands

In *C. elegans*, certain Wnt ligand was identified to exhibit transgenerational response to pollutant exposure. With PS-NP as the example of pollutants, exposure to PS-NP caused several aspects of toxicity at P0-G, such as neurotoxicity, reproductive inhibition, and intestinal damage [27–35]. Using intestinal, neuronal, and reproductive endpoints, it was found that transgenerational PS-NP toxicity was further induced by PS-NP exposure at P0-G [36–39]. Among Wnt ligands, exposure to PS-NPs in the range of μg/L activated LIN-44 expression, although PS-NP expression did not affect CWN-1, CWN-2, EGL20, and MOM-2 expressions at P0-G (Fig. 6.1) [40]. Moreover, PS-NP exposure induced transgenerational activation of LIN-44, and activation of LIN-44 by 1 μg/L could be detected till to the F2-G (Fig. 6.1) [40]. Germline RNAi of *lin-44* caused resistance to transgenerational neurotoxicity and transgenerational reproductive toxicity of PS-NPs (Fig. 6.1) [40]. Therefore, transgenerational activation of germline Wnt ligands, such as LIN-44, mediated formation of transgenerational pollutant toxicity.

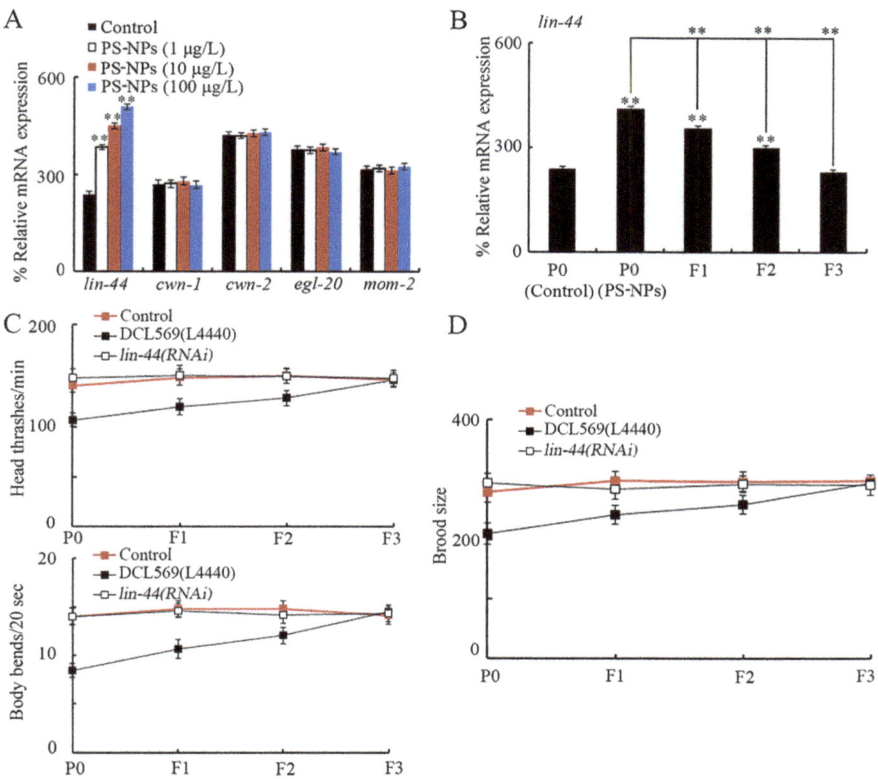

Fig. 6.1 Wnt ligand of LIN-44 in the germline was involved in transgenerational PS-NP toxicity [40]. (**a**) Expressions of *lin-44*, *cwn-1*, *cwn-2*, *egl-20*, and *mom-2* at P0-G. $^{**}P < 0.01$ vs Control. (**b**) Transgenerational expression of germline *lin-44* after PS-NP exposure. Exposure concentration of PS-NP was 1 μg/L. A total of 30 intact gonads were used for the qRT-PCR assay. $^{**}P < 0.01$ vs Control (if not specially indicated). (**c**) Effect of germline RNAi knockdown of *lin-44* on transgenerational PS-NP toxicity in inhibiting locomotion behavior. Exposure concentration of PS-NP was 1 μg/L. L4440, empty vector. Curves of DCL569(L4440) ($P < 0.01$) showed significant difference compared to control. Curves of *lin-44(RNAi)* ($P < 0.01$) showed significant difference compared to DCL569(L4440). (**d**) Effect of germline RNAi knockdown of *lin-44* on transgenerational PS-NP toxicity in reducing brood size. Exposure concentration of PS-NP was 1 μg/L. L4440, empty vector. The curve of DCL569(L4440) ($P < 0.01$) showed a significant difference compared to control. The curve of *lin-44(RNAi)* ($P < 0.01$) showed a significant difference compared to DCL569(L4440). PS-NP, Polystyrene nanoparticle

6.3 Role of Wnt Signaling Pathway

6.3.1 Wnt Receptor

Accompanied with transgenerational activation of Wnt ligands, transgenerational inhibition in Wnt receptors was induced by pollutants. With PS-NPs as the pollutant example, among Wnt receptors, PS-NP exposure inhibited MIG-1 and CFZ-2

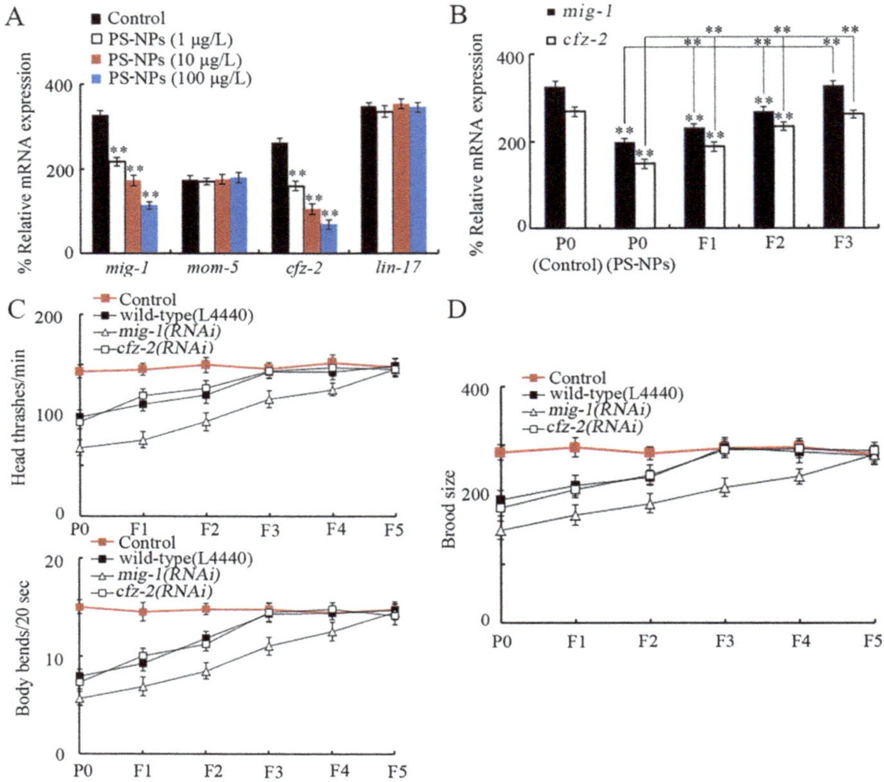

Fig. 6.2 Wnt receptor of MIG-1 was involved in controlling transgenerational PS-NP toxicity [40]. (**a**) Effect of PS-NP exposure on expressions of *mig-1*, *mom-5*, *cfz-2*, and *lin-17* at P0-G. $^{**}P < 0.01$ vs Control. (**b**) Transgenerational expression of *cfz-2* and *mig-1* after PS-NP exposure. Exposure concentration of PS-NP was 1 μg/L. $^{**}P < 0.01$ vs Control (if not specially indicated). (**c**) Effect of RNAi knockdown of *cfz-2* or *mig-1* on transgenerational PS-NP toxicity in inhibiting locomotion behavior. Exposure concentration of PS-NP was 1 μg/L. L4440, empty vector. Curves of wild-type (L4440) ($P < 0.01$) showed a significant difference compared to control. Curves of *mig-1(RNAi)* ($P < 0.01$) showed a significant difference compared to wild-type (L4440). Curves of *cfz-2(RNAi)* (head thrash, $P = 0.978$; body bend, $P = 0.891$) did not show significant difference compared to wild-type (L4440). (**d**) Effect of RNAi knockdown of *cfz-2* or *mig-1* on transgenerational PS-NP toxicity in reducing brood size. Exposure concentration of PS-NP was 1 μg/L. L4440, empty vector. Curve of wild-type (L4440) ($P < 0.01$) showed a significant difference compared to control. Curve of *mig-1(RNAi)* ($P < 0.01$) showed a significant difference compared to wild-type (L4440). Curve of *cfz-2(RNAi)* ($P = 0.804$) did not show significant difference compared to wild-type (L4440). PS-NP, Polystyrene nanoparticle

expressions, although PS-NP exposure did not influence MOM-5 and LIN-17 expressions at P0-G (Fig. 6.2) [40]. PS-NPs exposure further resulted in transgenerational inhibition in MIG-1 and CFZ-2 expressions, and inhibition in MIG-1 and CFZ-2 expressions by 1 μg/L PS-NPs could be observed till to the F2-G (Fig. 6.2) [40]. *mig-1* RNAi induced susceptibility to transgenerational neurotoxicity and transgenerational reproductive toxicity of PS-NPs (Fig. 6.2) [40]. Nevertheless, *cfz-2* RNAi did not influence transgenerational neurotoxicity and transgenerational reproductive toxicity of PS-NPs (Fig. 6.2) [40]. Thus, transgenerational activation of LIN-44 and transgenerational inhibition in MIG-1 acted together to mediate induction of transgenerational pollutant toxicity.

6.3.2 Dishevelled Proteins

Pollutant exposure could lead to transgenerational inhibition in Dishevelled proteins. With PS-NP as the pollutant example, among members of Dishevelled proteins, although MIG-5 expression was not changed, DSH-1 and DSH-2 expressions were inhibited by PS-NPs in the range of μg/L at P0-G (Fig. 6.3) [40]. Exposure to PS-NPs also induced transgenerational inhibition in DSH-1 and DSH-2 expressions, and this transgenerational inhibition could be detected till to the F2-G after exposure to 1 μg/L PS-NP (Fig. 6.3) [40]. Meanwhile, susceptibility to transgenerational neurotoxicity and transgenerational reproductive toxicity of PS-NPs was caused by RNAi of *dsh-1* and *dsh-2* (Fig. 6.3) [40]. More importantly, after pollutant exposure, *dsh-1* and *dsh-2* expressions were further decreased by *mig-1* RNAi (Fig. 6.3) [40]. Therefore, transgenerational inhibition in Wnt receptor mediated induction of transgenerational pollutant toxicity by causing the further transgenerational inhibition in Dishevelled proteins.

6.3.3 APC Complex

Exposure to pollutants also resulted in transgenerational activation of components of APC complex. With PS-NPs as the pollutant example, among members of APC complex, although GSK-3, KIN-19, and PRY-1 expressions were not changed, APR-1 was activated by PS-NPs in the range of μg/L at P0-G (Fig. 6.4) [40]. Exposure to PS-NPs induced transgenerational activation of APR-1, and this transgenerational APR-1 activation was found till to the F2-G (Fig. 6.4) [40]. The resistance to transgenerational neurotoxicity and transgenerational reproductive toxicity of PS-NPs could be caused by *apr-1* RNAi (Fig. 6.4) [40]. Moreover, after pollutant exposure, *apr-1* expression was further activated by RNAi of *dsh-1* and *dsh-2* (Fig. 6.4) [40]. That is, transgenerational inhibition in Dishevelled proteins further mediated induction of transgenerational pollutant toxicity by causing the following transgenerational activation of components in APC components.

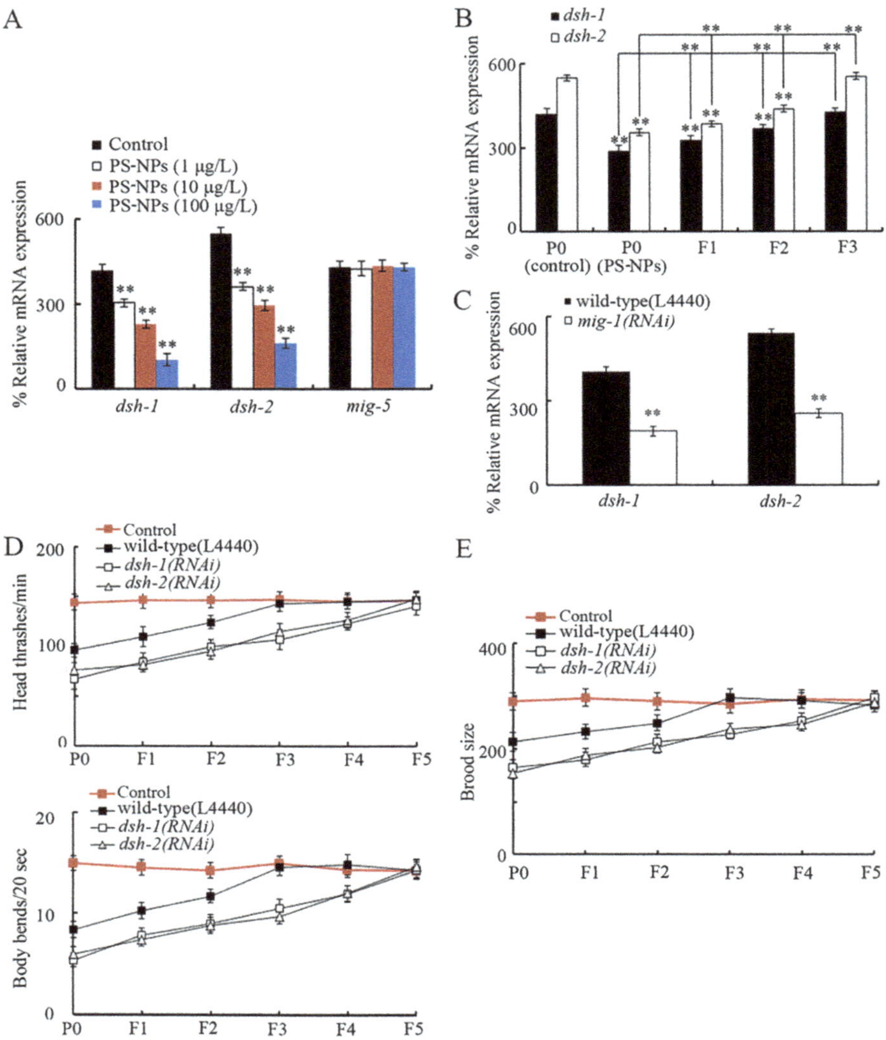

Fig. 6.3 Dishevelled proteins were involved in controlling transgenerational PS-NP toxicity [40]. (**a**) Effect of PS-NP exposure on expressions of *dsh-1*, *dsh-2*, and *mig-5* at P0-G. $^{**}P < 0.01$ vs Control. (**b**) Transgenerational expression of *dsh-1* and *dsh-2* after PS-NP exposure. Exposure concentration of PS-NP was 1 μg/L. $^{**}P < 0.01$ vs P0(control) (if not specially indicated). (**c**) Effect of RNAi knockdown of *mig-1* on *dsh-1* or *dsh-2* expressions in PS-NP exposed nematodes. L4440, empty vector. Exposure concentration of PS-NP was 1 μg/L. $^{**}P < 0.01$ vs wild-type (L4440). (**d**) Effect of RNAi knockdown of *dsh-1* or *dsh-2* on transgenerational PS-NP toxicity in inhibiting locomotion behavior. L4440, empty vector. Exposure concentration of PS-NP was 1 μg/L. Curves of wild-type (L4440) ($P < 0.01$) showed a significant difference compared to control. Curves of *dsh-1(RNAi)* ($P < 0.01$) or *dsh-2(RNAi)* ($P < 0.01$) showed a significant difference compared to wild-type (L4440). (**e**) Effect of RNAi knockdown of *dsh-1* or *dsh-2* on transgenerational PS-NP toxicity in reducing brood size. L4440, empty vector. Exposure concentration of PS-NP was 1 μg/L. Curve of wild-type (L4440) ($P < 0.01$) showed a significant difference compared to control. Curve of *dsh-1(RNAi)* ($P < 0.01$) or *dsh-2(RNAi)* ($P < 0.01$) showed a significant difference compared to wild-type (L4440). PS-NP, Polystyrene nanoparticle

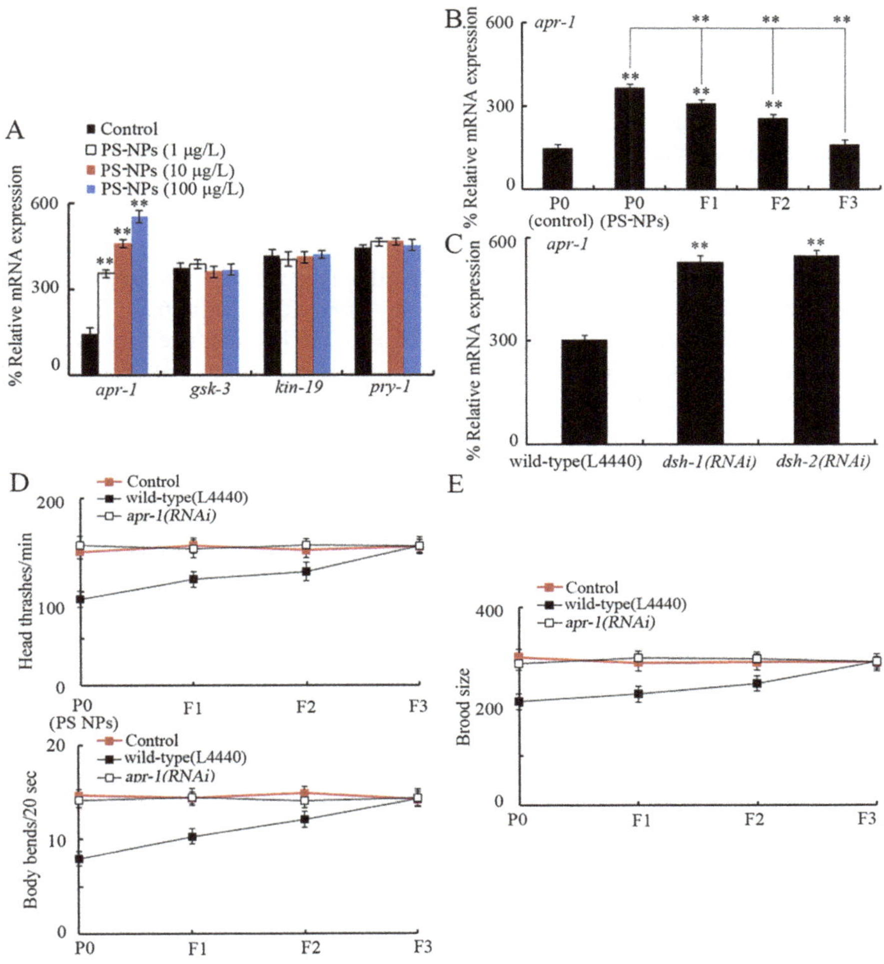

Fig. 6.4 APC complex component APR-1 was involved in the control of transgenerational PS-NP toxicity [40]. (**a**) Effect of PS-NP exposure on expressions of *apr-1*, *gsk-3*, *kin-19*, and *pry-1* at P0-G. $^{**}P < 0.01$ vs Control. (**b**) Transgenerational expression of *apr-1* after PS-NP exposure. Exposure concentration of PS-NP was 1 μg/L. $^{**}P < 0.01$ vs P0(control) (if not specially indicated). (**c**) Effect of RNAi knockdown of *dsh-1* or *dsh-2* on *apr-1* expression in PS-NP exposed nematodes. L4440, empty vector. Exposure concentration of PS-NP was 1 μg/L. $^{**}P < 0.01$ vs wild-type (L4440). (**d**) Effect of RNAi knockdown of *apr-1* on transgenerational PS-NP toxicity in inhibiting locomotion behavior. L4440, empty vector. Exposure concentration of PS-NP was 1 μg/L. Curves of wild-type (L4440) ($P < 0.01$) showed a significant difference compared to control. Curves of *apr-1(RNAi)* ($P < 0.01$) showed a significant difference compared to wild-type (L4440). (**e**) Effect of RNAi knockdown of *apr-1* on transgenerational PS-NP toxicity in reducing brood size. L4440, empty vector. Exposure concentration of PS-NP was 1 μg/L. Curve of wild-type (L4440) ($P < 0.01$) showed a significant difference compared to control. Curve of *apr-1(RNAi)* ($P < 0.01$) showed a significant difference compared to wild-type (L4440). PS-NP, Polystyrene nanoparticle

6.3.4 Transcriptional Factor BAR-1

In *C. elegans*, transgenerational inhibition in transcriptional factor β-catenin/BAR-1 mediates transgenerational pollutant toxicity induction. The *bar-1* expression could be decreased by pollutants, such as PS-NPs in the range of μg/L (Fig. 6.5) [40]. In addition, transgenerational inhibition in BAR-1 was induced by PS-NPs (1 μg/L), which was observed till to the F2-G (Fig. 6.5) [40]. Accompanied with transgenerational inhibition in BAR-1, RNAi of *bar-1* resulted in susceptibility to transgenerational neurotoxicity and transgenerational reproductive toxicity of PS-NPs (Fig. 6.5) [40]. Considering that BAR-1 was activated by *apr-1* RNAi (Fig. 6.5) [40], transgenerational inhibition in Wnt receptors, such as MIG-1, mediated induction of transgenerational pollutants toxicity by causing transgenerational inhibition in BAR-1 through affecting signaling cascade of Dishevelled proteins-APC complex components across multiple generations.

6.3.5 Other Upregulators for BAR-1

Besides the APR-1, some other upregulators were identified recently for BAR-1 in regulating the transgenerational pollutant toxicity. In *C. elegans*, PS-NPs induced the transgenerational activation of FGF receptor EGL-15, which mediated formation of transgenerational PS-NP toxicity [41]. EGL-15 functions in both the intestine and the neurons to regulate transgenerational pollutants toxicity [41]. In the intestine, BAR-1 was identified as target of EGL-15 to control transgenerational pollutant toxicity. Two lines of evidence supported this. Firstly, after pollutant exposure, BAR-1 was activated by intestinal RNAi of *egl-15* [41]. Secondly, resistance to transgenerational reproductive toxicity induced by intestinal RNAi of *egl-15* could be suppressed by RNAi of *bar-1* [41]. That is, transgenerational activation of intestinal EGL-15 mediated transgenerational pollutant toxicity by causing transgenerational inhibition in BAR-1 expression.

6.4 Transgenerational Communication Between Wnt Ligands and Their Receptors in Regulating Transgenerational Toxicity of Pollutants

Certain evidence was provided to indicate the potential transgenerational communication between Wnt ligands and their receptors in regulating transgenerational pollutant toxicity. With the PS-NP as the example of pollutants, *lin-44* RNAi was performed at P0-G after PS-NP exposure, and *mig-1* RNAi was performed at F1-G [40]. After pollutant exposure, MIG-1 at F1-G could be activated by germline RNAi of *lin-44* at P0-G (Fig. 6.6) [40]. The resistance to neurotoxicity and reproductive

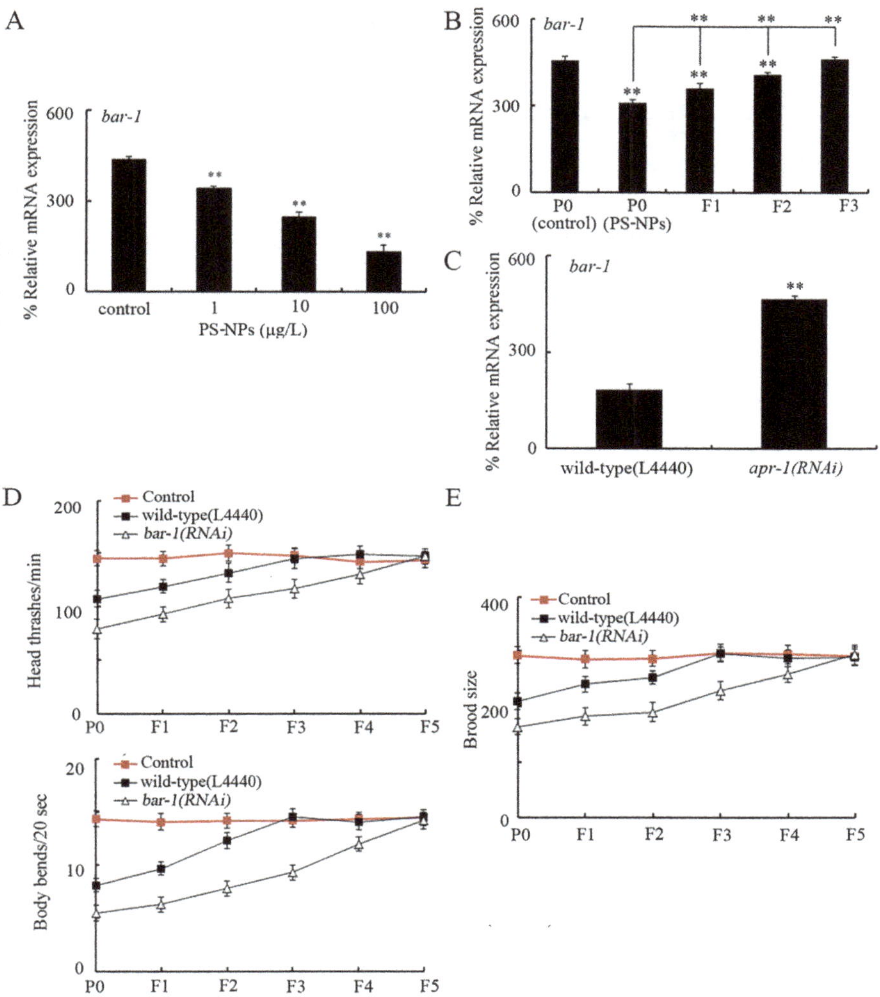

Fig. 6.5 Function of BAR-1/β-catenin in controlling transgenerational PS-NP toxicity [40]. (**a**) Effect of PS-NP exposure on expression of *bar-1* at P0-G. $^{**}P < 0.01$ vs Control. (**b**) Transgenerational expression of *bar-1* after PS-NP exposure. Exposure concentration of PS-NP was 1 μg/L. $^{**}P < 0.01$ vs P0(control) (if not specially indicated). (**c**) Effect of RNAi of *apr-1* on *bar-1* expression in PS-NP exposed nematodes. L4440, empty vector. Exposure concentration of PS-NP was 1 μg/L. $^{**}P < 0.01$ vs wild-type (L4440). (**d**) Effect of RNAi knockdown of *bar-1* on transgenerational PS-NP toxicity in inhibiting locomotion behavior. L4440, empty vector. Exposure concentration of PS-NP was 1 μg/L. Curves of wild-type (L4440) ($P < 0.01$) showed a significant difference compared to control. Curves of *bar-1(RNAi)* ($P < 0.01$) showed a significant difference compared to wild-type (L4440). (**e**) Effect of RNAi knockdown of *bar-1* on transgenerational PS-NP toxicity in reducing brood size. L4440, empty vector. Exposure concentration of PS-NP was 1 μg/L. Curve of wild-type (L4440) ($P < 0.01$) showed significant difference compared to control. Curve of *bar-1(RNAi)* ($P < 0.01$) showed significant difference compared to wild-type (L4440). PS-NP, Polystyrene nanoparticle

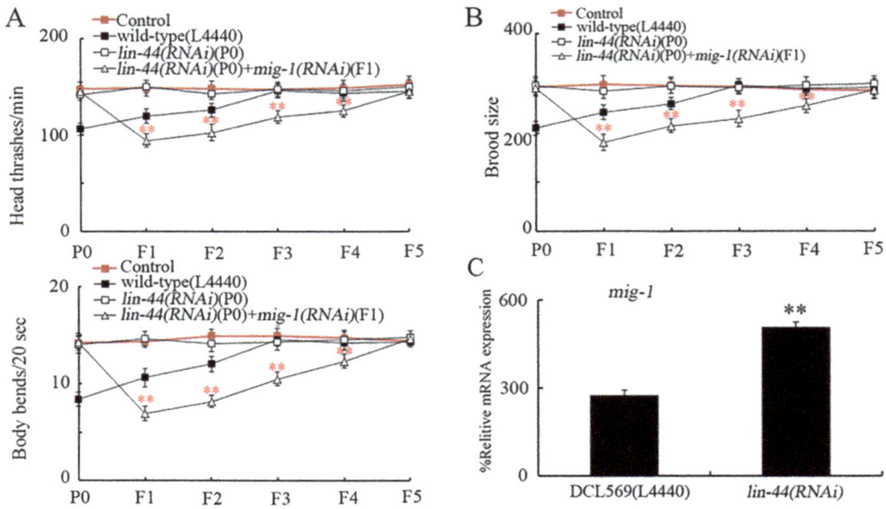

Fig. 6.6 Wnt ligand LIN-44 transgenerationally affected the function of Wnt receptor MIG-1 to modulate PS-NP toxicity [40]. (**a**) Genetic interaction between LIN-44 and MIG-1 in regulating transgenerational PS-NP toxicity in decreasing locomotion behavior. After the PS-NPs exposure in *lin-44(RNAi)* nematodes (P0), the RNAi knockdown of *mig-1* was performed in their F1 generation. $^{**}P < 0.01$ vs *lin-44 (RNAi)*(P0). The curves of wild-type (L4440) ($P < 0.01$) showed significant difference compared to control. Curves of *lin-44(RNAi)*(P0) ($P < 0.01$) showed a significant difference compared to wild-type (L4440). (**b**) Genetic interaction between LIN-44 and MIG-1 in regulating transgenerational PS-NP toxicity in reducing brood size. After the PS-NPs exposure in *lin-44(RNAi)* nematodes (P0), the RNAi knockdown of *mig-1* was performed in their F1 generation. $^{**}P < 0.01$ vs *lin-44(RNAi)*(P0). Curve of wild-type (L4440) ($P < 0.01$) showed a significant difference compared to control. The curve of *lin-44(RNAi)*(P0) ($P < 0.01$) showed a significant difference compared to wild-type (L4440). (**c**) Effect of germline RNAi knockdown of *lin-44* on *mig-1* expression in PS-NP exposed nematodes. The RNAi of *lin-44* was performed at P0 generation, and the *mig-1* expression was analyzed in F1 generation. $^{**}P < 0.01$ vs Control. L4440, empty vector. PS-NP exposure concentration was 1 μg/L. PS-NP, Polystyrene nanoparticle

toxicity of PS-NPs induced by *lin-44* RNAi could be suppressed by *mig-1* RNAi at the F1-G, and the *lin-44(RNAi)* animals showed susceptibility to PS-NP toxicity after RNAi of *mig-1* at the F1-G (Fig. 6.6) [40]. These observations suggested that activation of Wnt ligands may transgenerationally inhibit their receptors in the offspring to mediate transgenerational pollutant toxicity.

6.5 Targets of Transcriptional Factor BAR-1

In *C. elegans*, transgenerational inhibition in BAR-1 mediated transgenerational pollutant toxicity by resulting in transgenerational suppression in mitochondrial unfolded protein response (mt UPR). HSP-6 is a marker of mt UPR, and activation of mt UPR provides a protective strategy for animals against damage from

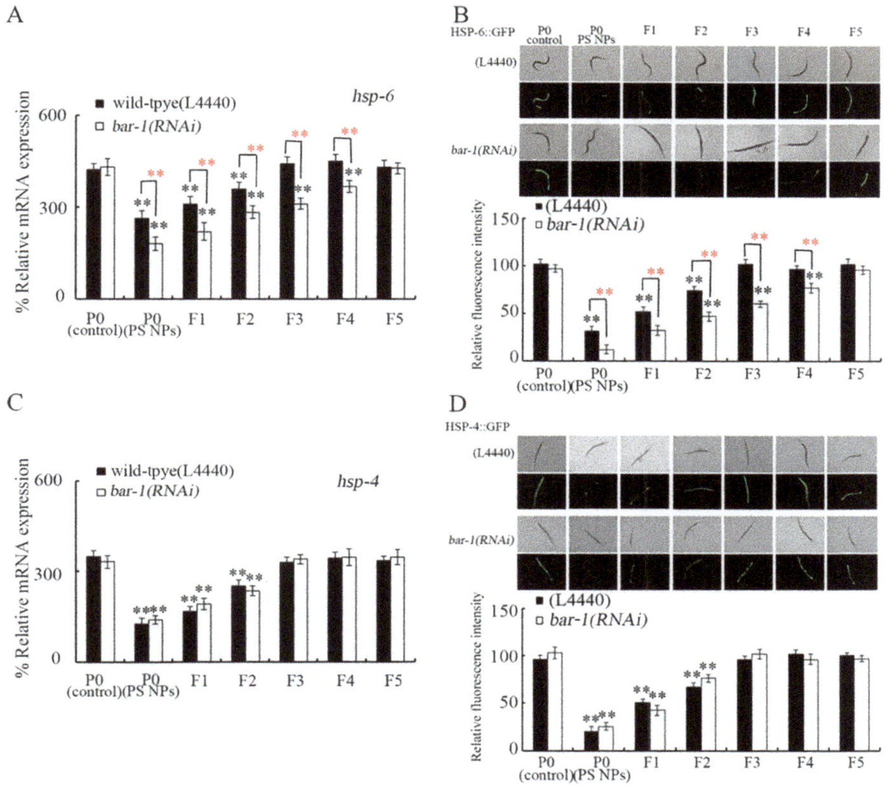

Fig. 6.7 Transgenerational mt UPR and ER UPR responses in wild-type and *bar-1(RNAi)* nematodes after PS-NP exposure [40]. (**a**) Transgenerational mt UPR response as reflected by *hsp-6* expression in wild-type and *bar-1(RNAi)* nematodes after PS-NP exposure. (**b**) Transgenerational mt UPR response as reflected by HSP-6::GFP expression in wild-type and *bar-1(RNAi)* nematodes after PS-NP exposure. (**c**) Transgenerational ER UPR response as reflected by *hsp-4* expression in wild-type and *bar-1(RNAi)* nematodes after PS-NP exposure. (**d**) Transgenerational ER UPR response as reflected by HSP-4::GFP expression in wild-type and *bar-1(RNAi)* nematodes after PS-NP exposure. L4440, empty vector. Exposure concentration of PS-NP was 1 μg/L. $^{**}P < 0.01$ vs P0 (control) (if not specially indicated). PS-NP, Polystyrene nanoparticle

pollutants [42]. Suppression in mt UPR contributes to induction of pollutants toxicity by causing mitochondrial damage [43, 44]. RNAi of *bar-1* strengthened transgenerational suppression in mt UPR induced by PS-NPs (Fig. 6.7) [40]. HSP-4 is a marker of endoplasmic reticulum UPR (ER UPR), another form of protective response against damage of pollutants [45]. Nevertheless, transgenerational suppression in ER UPR caused by PS-NPs was not affected by RNAi of *bar-1* (Fig. 6.7) [40]. Therefore, transgenerational inhibition in BAR-1 mediated transgenerational pollutant toxicity by resulting in transgenerational suppression in mt UPR.

In *C. elegans*, activation of three transcriptional factors (ATFS-1, DVE-1, and UBL-5) mediates induction of mt UPR [4]. However, after pollutant exposure,

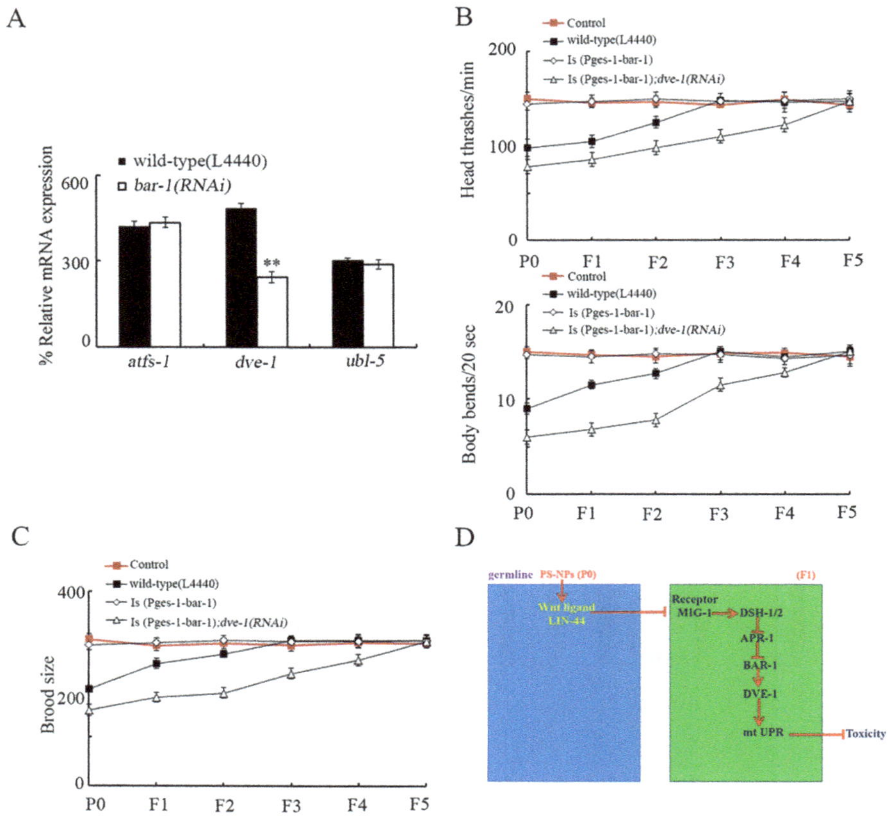

Fig. 6.8 BAR-1 acted upstream of DVE-1 to control transgenerational PS-NP toxicity [40]. (**a**) Effect of RNAi of *bar-1* on *ubl-5*, *dve-1*, and *atfs-1* expressions in PS-NP exposed nematodes. L4440, empty vector. Exposure concentration of PS-NP was 1 μg/L. $^{**}P < 0.01$ vs wild-type(L4440). (**b**) Genetic interaction between BAR-1 and DVE-1 in controlling transgenerational PS-NP toxicity to inhibit locomotion. L4440, empty vector. Exposure concentration of PS-NP was 1 μg/L. The curves of wild-type (L4440) ($P < 0.01$) showed a significant difference compared to control. Curves of *Is(Pges-1-bar-1)* ($P < 0.01$) showed a significant difference compared to wild-type (L4440). Curves of *Is(Pges-1-bar-1);dve-1(RNAi)* ($P < 0.01$) showed a significant difference compared to *Is(Pges-1-bar-1)*. (**c**) Genetic interaction between BAR-1 and DVE-1 in controlling transgenerational PS-NP toxicity to reduce brood size. L4440, empty vector. Exposure concentration of PS-NP was 1 μg/L. Curve of wild-type (L4440) ($P < 0.01$) showed a significant difference compared to control. Curve of *Is(Pges-1-bar-1)* ($P < 0.01$) showed a significant difference compared to wild-type (L4440). Curve of *Is(Pges-1-bar-1);dve-1(RNAi)* ($P < 0.01$) showed a significant difference compared to *Is(Pges-1-bar-1)*. L4440, empty vector. $^{**}P < 0.01$. (**d**) A diagram showing molecular basis for germline Wnt signaling in regulating transgenerational PS-NP toxicity in nematodes. PS-NP, Polystyrene nanoparticle

bar-1 RNAi only decreased *dve-1* expression and did not influence *atfs-1* and *ubl-5* expressions (Fig. 6.8) [40]. Moreover, resistance to transgenerational neurotoxicity and transgenerational reproductive toxicity of PS-NPs caused by intestinal BAR-1 overexpression could be inhibited by *dve-1* RNAi (Fig. 6.8) [40]. That is, transgenerational inhibition in BAR-1 caused transgenerational suppression in mt UPR to mediate the transgenerational pollutant toxicity by inhibiting DVE-1 expression in the nucleus.

6.6 Perspectives

In this chapter, transgenerational activation of Wnt ligands and transgenerational inhibition in Wnt receptors, as well as the following transgenerational responses of Dishevelled proteins and APC complex components, mediated induction of transgenerational pollutant toxicity. We here used the nanoplastics as the example of pollutants. Considering the limited toxicity of nanoplastics, it is supposed that exposure to other pollutants may cause transgenerational responses of more Wnt ligands, Wnt receptors, Dishevelled proteins, and APC complex components.

During control of transgenerational pollutants toxicity, only DVE-1 was identified as target of BAR-1 in *C. elegans* [40]. Considering its function as a transcriptional factor, normally, BAR-1 will have multiple downstream targets and affect their expressions and functions. Therefore, more downstream targets for BAR-1 are needed to be identified in controlling transgenerational pollutants toxicity.

In the Wnt signaling pathway, APC-1 acted upstream and inhibited expression of BAR-1 to regulate the transgenerational pollutant toxicity [40]. Meanwhile, FGF receptor EGL-15 could also act as upregulator of BAR-1 to control transgenerational pollutant toxicity [41]. Besides these, nuclear hormone receptor NHR-12 could function upstream of germline Wnt ligand LIN-44 to affect Wnt receptor MIG-1 so as to regulate transgenerational PS-NP toxicity [46]. Additionally, some histone acetyltransferases and histone deacetylases acted upstream of germline Wnt ligand LIN-44 to modulate the transgenerational PS-NP toxicity [47]. Therefore, the Wnt signaling can coordinate with different molecular signals to regulate transgenerational pollutant toxicity. In addition, during control of transgenerational pollutant toxicity, the other molecular signals may affect different sites in the Wnt signaling pathway.

References

1. Wang D-Y (2018) Nanotoxicology in *Caenorhabditis elegans*. Springer Nature Singapore Pte Ltd, Singapore
2. Wang D-Y (2020) Exposure toxicology in *Caenorhabditis elegans*. Springer Nature Singapore Pte Ltd, Singapore
3. Wang D-Y (2022) Toxicology at environmentally relevant concentrations in *Caenorhabditis elegans*. Springer Nature Singapore Pte Ltd., Singapore
4. Wang D-Y (2019) Molecular toxicology in *Caenorhabditis elegans*. Springer Nature Singapore Pte Ltd., Singapore

5. Wang D-Y (2019) Target organ toxicology in *Caenorhabditis elegans*. Springer Nature Singapore Pte Ltd, Singapore
6. Eisenmann DM (2005) Wnt signaling. WormBook. https://doi.org/10.1895/wormbook.1.7.1
7. Korswagen HC, Herman MA, Clevers HC (2000) Distinct beta-catenins mediate adhesion and signalling functions in *C. elegans*. Nature 406:527–532
8. Kidd AR 3rd, Miskowski JA, Siegfried KR, Sawa H, Kimble J (2005) A beta-catenin identified by functional rather than sequence criteria and its role in Wnt/MAPK signaling. Cell 121:761–772
9. Dreier L, Burbea M, Kaplan JM (2005) LIN-23-mediated degradation of beta-catenin regulates the abundance of GLR-1 glutamate receptors in the ventral nerve cord of *C. elegans*. Neuron 46:51–64
10. Korswagen HC, Coudreuse DY, Betist MC, van de Water S, Zivkovic D, Clevers HC (2002) The Axin-like protein PRY-1 is a negative regulator of a canonical Wnt pathway in *C. elegans*. Genes Dev 16:1291–1302
11. Eisenmann DM, Maloof JN, Simske JS, Kenyon C, Kim SK (1998) The beta-catenin homolog BAR-1 and LET-60 Ras coordinately regulate the Hox gene *lin-39* during *Caenorhabditis elegans* vulval development. Development 125:3667–3680
12. van der Bent ML, Sterken MG, Volkers RJ, Riksen JA, Schmid T, Hajnal A, Kammenga JE, Snoek LB (2014) Loss-of-function of beta-catenin bar-1 slows development and activates the Wnt pathway in *Caenorhabditis elegans*. Sci Rep 4:4926
13. Yu H, Seah A, Herman MA, Ferguson EL, Horvitz HR, Sternberg PW (2009) Wnt and EGF pathways act together to induce *C. elegans* male hook development. Dev Biol 327:419–432
14. Zheng M, Messerschmidt D, Jungblut B, Sommer RJ (2005) Conservation and diversification of Wnt signaling function during the evolution of nematode vulva development. Nat Genet 37:300–304
15. Maloof JN, Whangbo J, Harris JM, Jongeward GD, Kenyon C (1999) A Wnt signaling pathway controls hox gene expression and neuroblast migration in *C. elegans*. Development 126:37–49
16. Zhang L, Wang Y, Wang D (2023) Paeoniflorin increases the survival of *Pseudomonas aeruginosa* infected *Caenorhabditis elegans* at the immunosuppression stage by activating PMK-1, BAR-1, and EGL-1 signals. Arch Pharm Res 46:616–628
17. Sang Y, Ren J, Aballay A (2022) The transcription factor HLH-26 controls probiotic-mediated protection against intestinal infection through up-regulation of the Wnt/BAR-1 pathway. PLoS Biol 20:e3001581
18. Wang Y, Zhang L, Yuan X, Wang D (2023) Treatment with paeoniflorin increases lifespan of *Pseudomonas aeruginosa* infected *Caenorhabditis elegans* by inhibiting bacterial accumulation in intestinal lumen and biofilm formation. Front Pharmacol 14:1114219
19. Irazoqui JE, Ng A, Xavier RJ, Ausubel FM (2008) Role for beta-catenin and HOX transcription factors in *Caenorhabditis elegans* and mammalian host epithelial-pathogen interactions. Proc Natl Acad Sci USA 105:17469–17474
20. Zhao Y, Dong S, Kong Y, Rui Q, Wang D (2020) Molecular basis of intestinal canonical Wnt/beta-catenin BAR-1 in response to simulated microgravity in *Caenorhabditis elegans*. Biochem Biophys Res Commun 522:198–204
21. Rui Q, Dong S, Jiang W, Wang D (2019) Response of canonical Wnt/beta-catenin signaling pathway in the intestine to microgravity stress in *Caenorhabditis elegans*. Ecotoxicol Environ Saf 186:109782
22. Wu J, Shen S, Wang D (2024) 6-PPD quinone at environmentally relevant concentrations induces immunosenescence by causing immunosuppression during the aging process. Chemosphere 368:143719
23. Hua X, Wang D (2025) An environmentally relevant concentration of 6-PPD quinone inhibits two types of mitophagy to cause mitochondrial dysfunction and lifespan reduction in *Caenorhabditis elegans*. Environ Sci Process Impacts 27:1928–1940
24. Wang S, Zhang R, Wang D (2021) Induction of protective response to polystyrene nanoparticles associated with methylation regulation in *Caenorhabditis elegans*. Chemosphere 271:129589

25. Liu P, Shao H, Kong Y, Wang D (2020) Effect of graphene oxide exposure on intestinal Wnt signaling in nematode *Caenorhabditis elegans*. J Environ Sci 88:200–208
26. Shao Y-T, Kong Y, Wang D-Y (2020) Response of intestinal signaling communication between nucleus and peroxisome to nanopolystyrene at predicted environmental concentration. Environ Sci Nano 7:250–261
27. Qu M, Kong Y, Yuan Y-J, Wang D-Y (2019) Neuronal damage induced by nanopolystyrene particles in nematode *Caenorhabditis elegans*. Environ Sci Nano 6:2591–2601
28. Qu M, Qiu Y-X, Kong Y, Wang D-Y (2019) Amino modification enhances reproductive toxicity of nanopolystyrene on gonad development and reproductive capacity in nematode *Caenorhabditis elegans*. Environ Pollut 254:112978
29. Qu M, Nida A, Kong Y, Du H-H, Xiao G-S, Wang D-Y (2019) Nanopolystyrene at predicted environmental concentration enhances microcystin-LR toxicity by inducing intestinal damage in *Caenorhabditis elegans*. Ecotoxicol Environ Saf 183:109568
30. Qu M, Xu K-N, Li Y-H, Wong G, Wang D-Y (2018) Using *acs-22* mutant *Caenorhabditis elegans* to detect the toxicity of nanopolystyrene particles. Sci Total Environ 643:119–126
31. Qiu Y-X, Liu Y-Q, Li Y-H, Li G-J, Wang D-Y (2020) Effect of chronic exposure to nanopolystyrene on nematode *Caenorhabditis elegans*. Chemosphere 256:127172
32. Qiu Y-X, Luo L-B, Yang Y-H, Kong Y, Li Y-H, Wang D-Y (2020) Potential toxicity of nanopolystyrene on lifespan and aging process of nematode *Caenorhabditis elegans*. Sci Total Environ 705:135918
33. Qu M, Wang D-Y (2020) Toxicity comparison between pristine and sulfonate modified nanopolystyrene particles in affecting locomotion behavior, sensory perception, and neuronal development in *Caenorhabditis elegans*. Sci Total Environ 703:134817
34. Li D, Deng Y-J, Wang S-T, Du H-H, Xiao G-S, Wang D-Y (2020) Assessment of nanopolystyrene toxicity under fungal infection condition in *Caenorhabditis elegans*. Ecotoxicol Environ Saf 197:110625
35. Hua X, Wang D-Y (2022) Cellular uptake, transport, and organelle response after exposure to microplastics and nanoplastics: current knowledge and perspectives for environmental and health risks. Rev Environ Contam Toxicol 260:12
36. Zhao L, Qu M, Wong G, Wang D-Y (2017) Transgenerational toxicity of nanopolystyrene particles in the range of μg/L in nematode *Caenorhabditis elegans*. Environ Sci Nano 4:2356–2366
37. Liu H-L, Tian L-J, Wang S-T, Wang D-Y (2021) Size-dependent transgenerational toxicity induced by nanoplastics in nematode *Caenorhabditis elegans*. Sci Total Environ 790:148217
38. Sun L-M, Liao K, Wang D-Y (2021) Comparison of transgenerational reproductive toxicity induced by pristine and amino modified nanoplastics in *Caenorhabditis elegans*. Sci Total Environ 768:144362
39. Xu R-R, Hua X, Rui Q, Wang D-Y (2022) Polystyrene nanoparticles caused dynamic alteration in mitochondrial unfolded protein response from parents to the offspring in *C. elegans*. Chemosphere 308:136154
40. Xu R-R, Hua X, Rui Q, Wang D-Y (2022) Alteration in Wnt signaling mediates induction of transgenerational toxicity of polystyrene nanoplastics in *C. elegans*. NanoImpact 28:100425
41. Hua X, Cao C, Zhang L, Wang D-Y (2023) Activation of FGF signal in germline mediates transgenerational toxicity of polystyrene nanoparticles at predicted environmental concentrations in *Caenorhabditis elegans*. J Hazard Mater 451:131174
42. Liu H-L, Wang D-Y (2021) Intestinal mitochondrial unfolded protein response induced by nanoplastic particles in *Caenorhabditis elegans*. Chemosphere 267:128917
43. Hua X, Wang D-Y (2024) 6-PPD quinone at environmentally relevant concentrations induced damage on longevity in *C. elegans*: mechanistic insight from inhibition in mitochondrial UPR response. Sci Total Environ 954:176275
44. Wang Y-X, Hu G-C, Wang D-Y (2025) Increased S-adenosyl methionine strengthens the suppression in mitochondrial unfolded protein response induced by 6-PPD quinone at environmentally relevant concentrations in *Caenorhabditis elegans*. Environ Pollut 386:127231

45. Qu M, Liu Y-Q, Xu K-N, Wang D-Y (2019) Activation of p38 MAPK signaling-mediated endoplasmic reticulum unfolded protein response by nanopolystyrene particles. Adv Biosyst 3:1800325
46. Liu Z-Y, Wang Y-X, Bian Q, Wang D-Y (2024) Transgenerational response of germline nuclear hormone receptor genes to nanoplastics at predicted environmental doses in *Caenorhabditis elegans*. Toxics 12:420
47. Hua X, Wang D-Y (2024) transgenerational response of germline histone acetyltransferases and deacetylases to nanoplastics at predicted environmental doses in *Caenorhabditis elegans*. Sci Total Environ 952:175903

Chapter 7
Regulation of Transgenerational Toxicity by Transgenerational Communication of Germline Hedgehog Signals

Abstract Hedgehog signaling pathway is an important signaling pathway to control biological processes, including stress response. In *Caenorhabditis elegans*, transgenerational activation of germline Hedgehog ligands together with transgenerational activation of Hedgehog receptors mediate the induction of transgenerational pollutant toxicity. Certain evidence further suggested the potential role of activated germline Hedgehog ligands to transgenerationally target to and activate corresponding receptors in the offspring after pollutants exposure.

Keywords Transgenerational toxicology · Transgenerational communication · Hedgehog signaling · *Caenorhabditis elegans*

7.1 Introduction

Caenorhabditis elegans is a classic model animal with tiny size of body and well-described genetic and genomic backgrounds. Moreover, it has been shown that *C. elegans* is highly sensitive to stresses and pollutants. These backgrounds allow the establishment of knowledge system of exposure toxicology in *C. elegans* [1]. In this animal model, the increasing number of studies have been carried out surrounding the toxicology at environmentally relevant concentrations (ERCs) of pollutants at the whole animal level [2], such as the toxicology of nanomaterials at ERCs [3].

During the study of target organ toxicology, it has been shown that molecular signals acted in different organs or tissues (such as in the germline) to regulate toxicity of stresses and pollutants [4]. At the parental generation (P0-G), a series of molecular signaling pathways have been identified to be involved in regulating pollutant toxicity [5]. Among these molecular signals, the Hedgehog ligands were initially raised to be required for developmental control, such as the control of cuticle development, sensory cilium development, and axon guidance [6–8]. Besides the Hedgehog ligands, their receptors were also involved in controlling the biological processes, including the developmental process [9, 10]. Moreover, the Hedgehog

D. Wang, *Transgenerational Toxicology in Caenorhabditis elegans*,
https://doi.org/10.1007/978-981-92-1389-4_7

signaling participated in the control of stress response, such as innate immunity, mitochondrial stress, and reproductive aging [11–13].

In *C. elegans*, among Hedgehog ligands, WRT-3, GRL-15, GRL-16, GRL-25, GRD-12, QUA-1, and WRT-1 can be expressed in the germline [14]. The corresponding Hedgehog receptors contain PTC-1, PTC-3, PTR-2, PTR-3, PTR-9, PTR-13, PTR-16, PTR-21, PTR-23, PTR-24, SCP-1, PTD-2, and NCR-1 [14]. In this chapter, we first introduced the roles of Hedgehog ligands and their receptors in regulating transgenerational toxicity of pollutants. Moreover, the potential transgenerational communication between Hedgehog ligands and their receptors during the control of transgenerational pollutants toxicity was discussed.

7.2 Role of Hedgehog Ligands

7.2.1 Transgenerational Response of Hedgehog Ligands to Pollutants

In *C. elegans*, some of the germline Hedgehog ligands were found to show transgenerational response to pollutants. The polystyrene nanoparticle (PS-NP), a member of nanoplastics, was used as the example of pollutants. At the P0-G, exposure to PS-NPs could result in several aspects of toxicity on nematodes, such as neuronal damage, reproductive inhibition, intestinal toxicity, and lifespan reduction [15–23]. WRT-3 is the first identified Hedgehog ligand to exhibit transgenerational response in the germline to pollutant exposure. After PS-NP exposure at P0-G, transgenerational activation of germline WRT-3 was induced, and this germline WRT-3 activation lasted till to F2-G after exposure to 1 μg/L PS-NPs at P0-G (Fig. 7.1) [24]. After that, several other germline Hedgehog ligands were found to also show the transgenerational response to pollutants exposure. After PS-NP exposure at P0-G, transgenerational activation of germline GRL-15, GRL-16, QUA-1, and WRT-1 was caused, and the transgenerational activation of these four germline Hedgehog ligands could be observed till to F2-G at P0-G (Fig. 7.2) [14]. That is, pollutant exposure at P0-G potentially induces the transgenerational activation of Hedgehog ligands in the germline.

Fig. 7.1 (continued) means ± SD. (**d**) Effect of germline RNAi knockdown of *pat-12* or *wrt-3* on transgenerational PS-NP toxicity that reduces brood size. Control, DCL569 nematodes without PS-NP exposure. Bars represent means ± SD. PS-NPs, Polystyrene nanoparticles. L4440, empty vector. The exposure concentration of PS-NPs was 1 μg L^{-1}

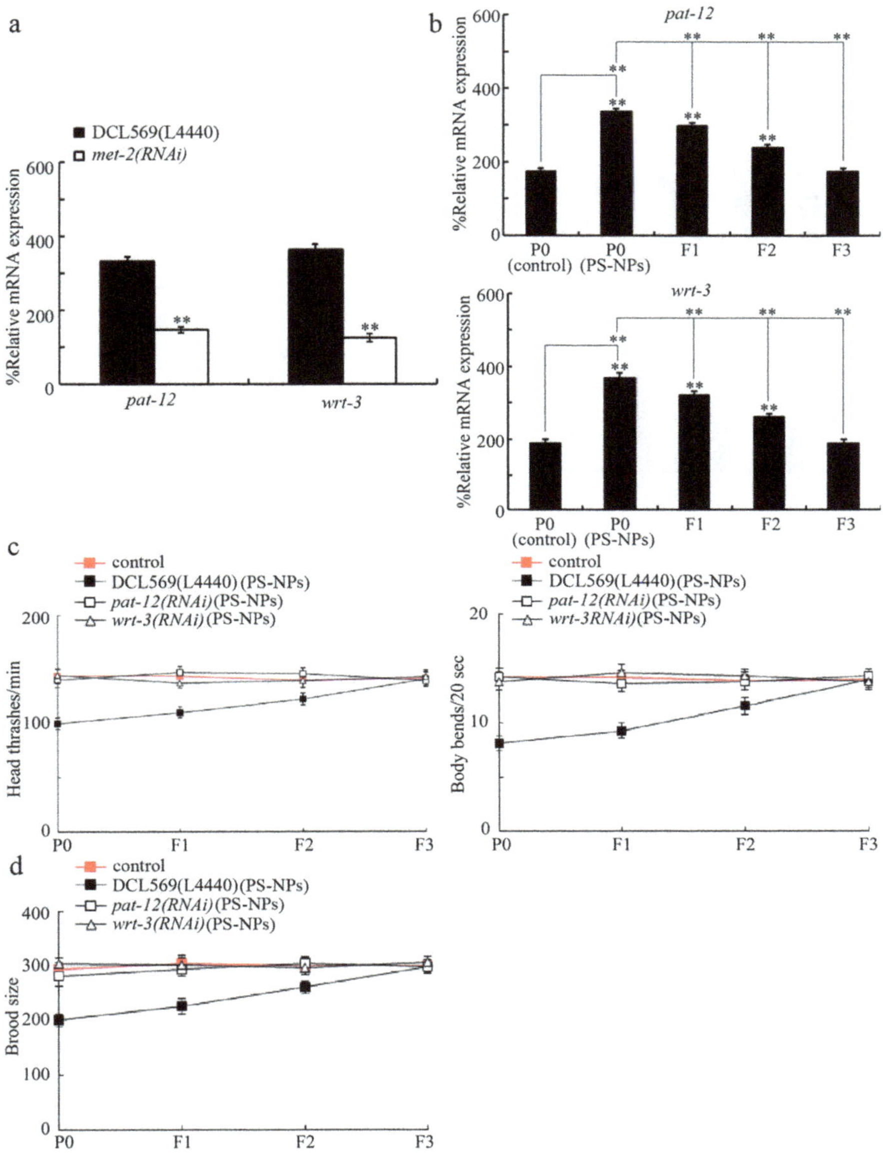

Fig. 7.1 Identification of PAT-12 and WRT-3 as potential targets of germline MET-2 in regulating transgenerational PS-NP toxicity [24]. (**a**) Effect of germline RNAi knockdown of *met-2* on expressions of *pat-12* and *wrt-3* in PS-NP-exposed DCL569 nematodes. Bars represent means ± SD. $^{**}P < 0.01$ vs. DCL569(L4440). (b) Transgenerational expressions of germline PAT-12 and WRT-3 after PS-NP exposure. A total of 30 intact gonads were used for the qRT-PCR assay. Bars represent means ± SD. $^{**}P < 0.01$ vs. P0 (control) (if not specially indicated). (**c**) Effect of germline RNAi knockdown of *pat-12* or *wrt-3* on transgenerational PS-NP toxicity that decreases locomotion behavior. Control, DCL569 nematodes without PSNP exposure. Bars represent

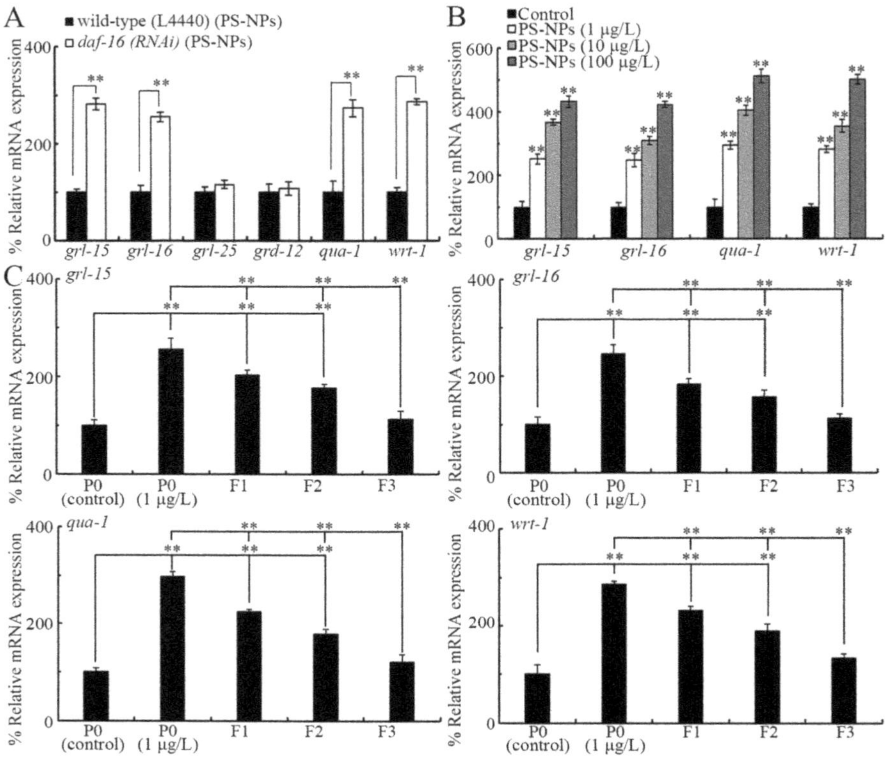

Fig. 7.2 Transgenerational expression of germline hedgehog ligands after PS-NP exposure [14]. (**a**) Effect of RNAi of *daf-16* on expressions of *grl-15*, *grl-16*, *grl-25*, *grd-12*, *qua-1*, and *wrt-1* in PS-NP-exposed nematodes. PS-NP exposure concentration was 1 μg/L. $^{**}P < 0.01$. (**b**) Effect of 1–100 μg/L PS-NP exposure on expressions of germline *grl-15*, *grl-16*, *qua-1*, and *wrt-1*. $^{**}P < 0.01$ vs Control. (**c**) Transgenerational expressions of germline *grl-15*, *grl-16*, *qua-1*, and *wrt-1* after PS-NP exposure. PS-NP exposure concentration was 1 μg/L. $^{**}P < 0.01$. A total of 30 intact gonads were used for the qRT-PCR assay

7.2.2 Function of Hedgehog Ligands in Regulating Transgenerational Toxicity of Pollutants

Functional analysis has indicated the role of Hedgehog ligands in mediating induction of transgenerational pollutants toxicity. Germline RNAi of *wrt-3* caused resistance to transgenerational neurotoxicity and transgenerational reproductive toxicity of PS-NPs (Fig. 7.1) [24]. Similarly, transgenerational neurotoxicity and transgenerational reproductive toxicity of PS-NPs were also suppressed by germline RNAi of *grl-15*, *grl-16*, *qua-1*, and *wrt-1* (Fig. 7.3) [14]. Therefore, transgenerational activation of Hedgehog ligands could mediate formation of transgenerational toxicity of pollutants after the exposure at the P0-G.

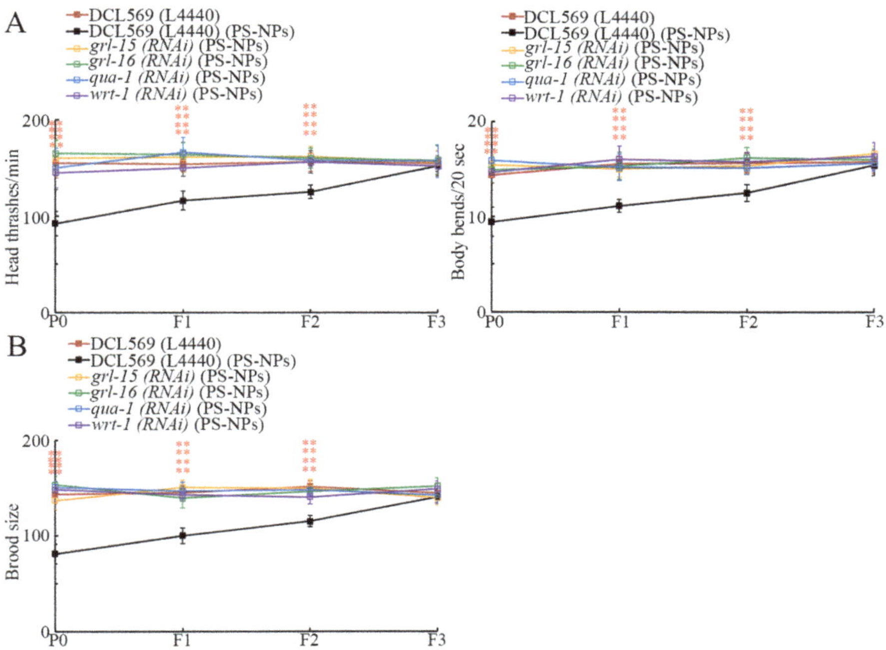

Fig. 7.3 Effect of germline RNAi of *grl-15*, *grl-16*, *qua-1*, and *wrt-1* on transgenerational toxicity of PS-NP on locomotion behavior (**a**) and brood size (**b**) [14]. PS-NP exposure concentration was 1 μg/L. $^{**}P < 0.01$ vs DCL560(L4440)(PS-NPs)

7.2.3 Upregulators for Hedgehog Ligands in Regulating Transgenerational Toxicity of Pollutants

1. DAF-16

   DAF-16 is a FOXO transcriptional factor in insulin signaling pathway and regulates toxicity of stresses and pollutants by affecting expressions of its downstream targeted genes [25–32]. In Chap. 5, we have introduced the role of insulin signaling pathway in regulating transgenerational pollutants toxicity. In PS-NP-exposed nematodes, germline expressions of *grl-15*, *grl-16*, *qua-1*, and *wrt-1* were increased by RNAi of *daf-16* (Fig. 7.2) [14]. This suggested that *grl-15*, *grl-16*, *qua-1*, and *wrt-1* can act as the downstream targets of *daf-16*, and insulin signaling regulates the transgenerational pollutants toxicity by inhibiting germline Hedgehog ligands.

2. MET-2

   MET-2 is a histone methyltransferase required for the control of toxicity of different pollutants in nematodes [33–39]. Exposure to PS-NPs caused the transgenerational activation of MET-2, and germline RNAi of *met-2* resulted in the resistance to transgenerational neurotoxicity and transgenerational reproductive toxicity of PS-NPs [24]. In PS-NP-exposed nematodes, germline RNAi of *met-2* decreased *wrt-3* expression (Fig. 7.1) [24]. These observations suggested that, in the germline, MET-2 regulates the transgenerational pollutants toxicity by activating the downstream Hedgehog ligands.

3. *mir-38*

   In *C. elegans*, some microRNAs (miRNAs) were identified to function in different tissues to regulate the toxicity of stresses and pollutants [40–59]. Among these miRNAs, *mir-38* could act in the germline to regulate transgenerational pollutant toxicity. Exposure to PS-NPs at P0-G caused transgenerational inhibition in expression of germline *mir-38*, and germline overexpression of *mir-38* resulted in resistance to transgenerational neurotoxicity and transgenerational reproductive toxicity of PS-NPs [60]. Moreover, germline *wrt-3* expression could be reduced by germline overexpression of *mir-38* in PS-NP-exposed nematodes [60]. Therefore, in the germline, *mir-38* can further regulate transgenerational pollutant toxicity by inhibiting the downstream Hedgehog ligands.

7.3 Role of Hedgehog Receptors

7.3.1 Transgenerational Response of Hedgehog Receptors to Pollutants

In *C. elegans*, exposure to pollutants can cause the transgenerational activation of Hedgehog receptors. The PS-NP was used as an example of pollutants. Among the examined 10 Hedgehog receptor genes, expressions of *ptr-23*, *scp-1*, *ptd-2*, and *ncr-1* were increased by 1–100 μg/L PS-NPs at P0-G (Fig. 7.4) [14]. Moreover, transgenerational activation of Hedgehog receptors was induced by PS-NPs, and activation of these four Hedgehog receptors could be detected till to the F2-G after exposure to 1 μg/L PS-NPs at P0-G (Fig. 7.4) [14].

7.3.2 Function of Hedgehog Receptors in Regulating Transgenerational Toxicity of Pollutants

Using RNAi technique, it has been shown that RNAi of Hedgehog receptor genes conferred a resistance to transgenerational pollutant toxicity. With PS-NP as an example of pollutants, RNAi of *ptd-2*, *ptr-23*, *scp-1*, and *ncr-1* could cause the resistance to transgenerational neurotoxicity and transgenerational reproductive toxicity of PS-NPs (Fig. 7.5) [14]. Therefore, activation of Hedgehog receptors across multiple generations mediated induction of transgenerational pollutant toxicity.

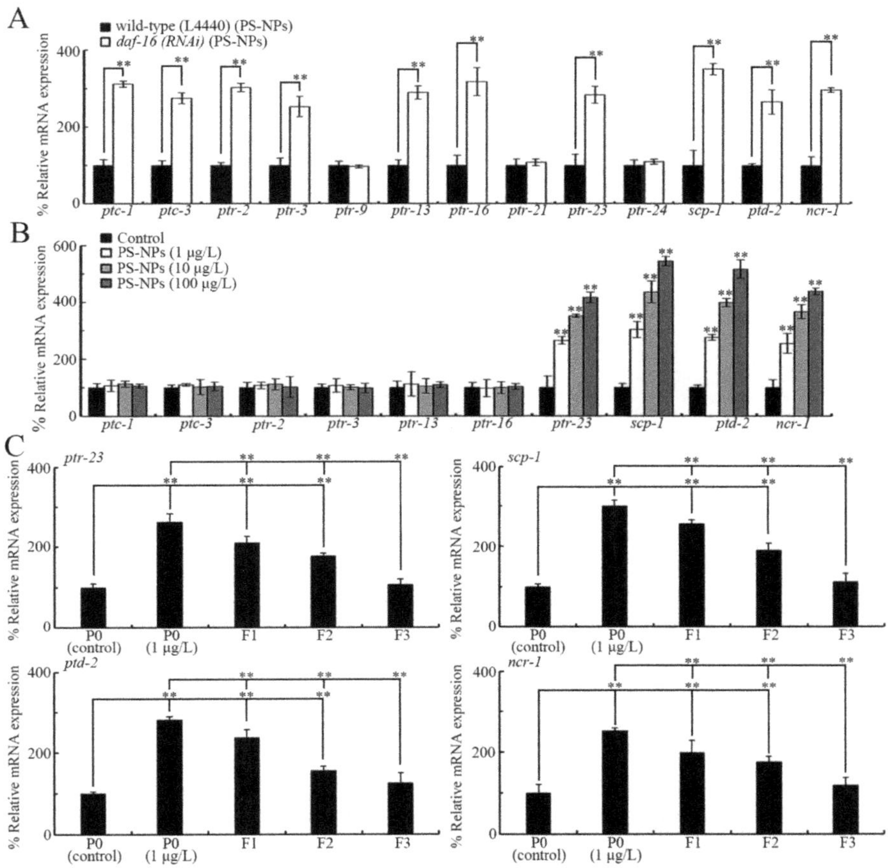

Fig. 7.4 Transgenerational expression of hedgehog receptors after PS-NP exposure [14]. (A) Effect of RNAi of *daf-16* on expressions of hedgehog receptor genes in PS-NP-exposed nematodes. PS-NP exposure concentration was 1 μg/L. $^{**}P < 0.01$. (B) Effect of 1–100 μg/L PS-NP exposure on expressions of hedgehog receptor genes. $^{**}P < 0.01$ vs Control. (C) Transgenerational expressions of *ptr-23*, *scp-1*, *ptd-2*, and *ncr-1* after PS-NP exposure. PS-NP exposure concentration was 1 μg/L. $^{**}P < 0.01$

7.4 Transgenerational Communication Between Hedgehog Ligands and Their Receptors in Regulating Transgenerational Toxicity of Pollutants

Certain evidence was raised to show the potential transgenerational communication between Hedgehog ligands and their receptors in regulating transgenerational pollutants toxicity. PS-NP was used as an example of pollutants. RNAi of Hedgehog ligands was performed in PS-NP-exposed nematodes at the P0-G, and the expression of corresponding Hedgehog receptor genes was examined at the F1-G [14].

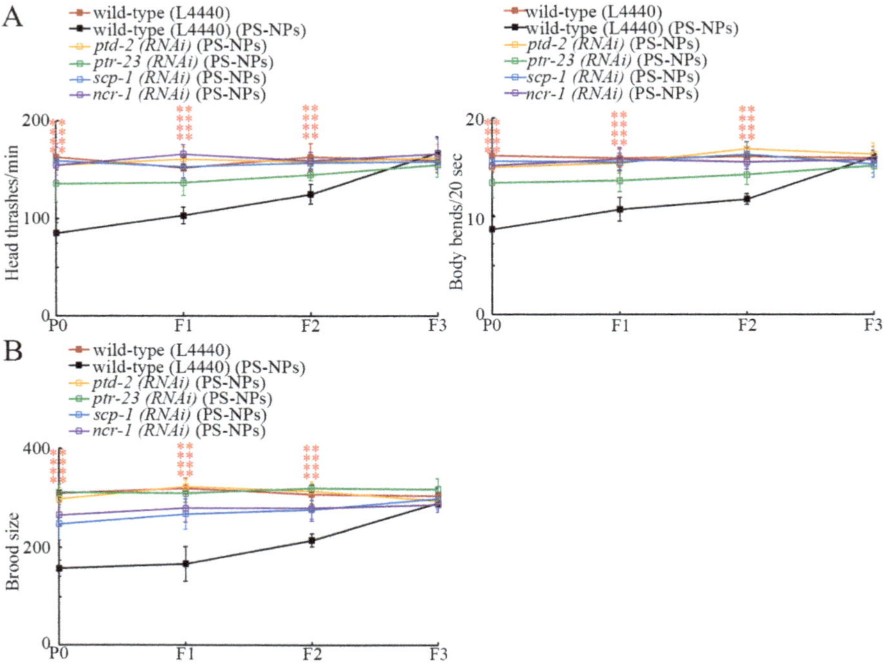

Fig. 7.5 Effect of RNAi of *ptd-2*, *ptr-23*, *scp-1*, and *ncr-1* on transgenerational toxicity of PS-NP on locomotion behavior (**a**) and brood size (**b**) [14]. PS-NP exposure concentration was 1 μg/L. $^{**}P < 0.01$ vs wild-type (L4440)(PS-NPs)

After PS-NP exposure, germline RNAi of *wrt-3* at the P0-G caused inhibition in expressions of *ptr-23*, *scp-1*, *ptd-2*, and *ncr-1* at the F1-G (Fig. 7.6) [14]. Similarly, after PS-NP exposure, expressions of *ptr-23*, *scp-1*, *ptd-2*, and *ncr-1* at the F1-G were also inhibited by germline RNAi of *grl-15*, *grl-16*, *qua-1*, and *wrt-1* at the P0-G (Fig. 7.6) [14]. That is, transgenerational activation of Hedgehog ligands mediates induction of transgenerational pollutants toxicity by targeting to and activating corresponding Hedgehog receptors in the offspring.

7.5 Perspectives

So far, five Hedgehog ligands (WRT-3, GRL-15, GRL-16, QUA-1, and WRT-1) were identified to be required for controlling transgenerational pollutant toxicity. Nevertheless, the other Hedgehog ligands involved in controlling transgenerational pollutants toxicity still need to be further identified. One reason is that different pollutants with different exposure durations and exposure concentrations will cause different transgenerational response of germline Hedgehog ligands. The other

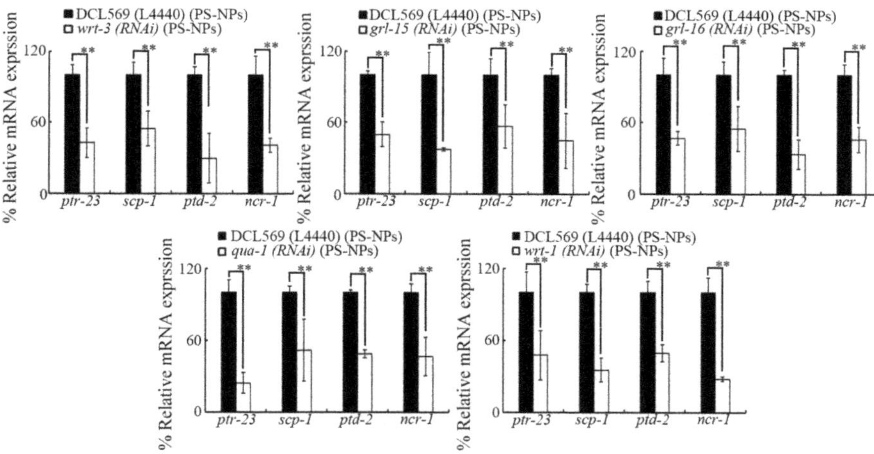

Fig. 7.6 Effect of germline RNAi of *wrt-3*, *grl-15*, *grl-16*, *qua-1*, and *wrt-1* on expressions of *ptd-2*, *ptr-23*, *scp-1*, and *ncr-1* at F1-G of PS-NP-exposed nematodes [14]. PS-NP exposure concentration was 1 μg/L. $^{**}P < 0.01$

reason is that the already identified five Hedgehog ligands were based on identification of downstream targets of *daf-16*, *met-2*, and *mir-38*. That is, even in PS-NP-exposed nematodes, some other germline Hedgehog ligands may still exist to be involved in regulating the transgenerational toxicity. For the same reasons, the other Hedgehog receptors involved in regulating transgenerational pollutants toxicity also need to be further identified.

In this chapter, we discussed the upregulators for Hedgehog ligands during control of transgenerational pollutant toxicity. One of the upregulators is the DAF-16 in the insulin signaling pathway. This suggested that transgenerational response of insulin signaling can function sequentially together with transgenerational response of Hedgehog signaling to regulate the transgenerational pollutants toxicity. Besides the DAF-16, MET-2 and *mir-38* were also identified as the upregulators for Hedgehogs to control transgenerational pollutant toxicity. That is, transgenerational response of epigenetic signals can also coordinate with Hedgehog signaling in controlling transgenerational pollutant toxicity.

During the control of transgenerational pollutant toxicity, although several Hedgehog receptors have been identified, their downstream targeted genes or signaling cascades remain unclear. That is, in the offspring of pollutant exposed nematodes, more efforts are needed to be performed to elucidate the mechanism for how Hedgehog receptors control the transgenerational pollutant toxicity. Actually, during the control of biological processes, some genes mediate the functions of Hedgehog signaling [61–63]. In addition, we cannot exclude such a possibility that the response of Hedgehog receptors in the offspring of pollutant-exposed nematodes may further activate or inhibit other important molecular ligands.

References

1. Wang D-Y (2020) Exposure toxicology in *Caenorhabditis elegans*. Springer Nature Singapore Pte Ltd, Singapore
2. Wang D-Y (2022) Toxicology at environmentally relevant concentrations in *Caenorhabditis elegans*. Springer Nature Singapore Pte Ltd, Singapore
3. Wang D-Y (2018) Nanotoxicology in *Caenorhabditis elegans*. Springer Nature Singapore Pte Ltd., Singapore
4. Wang D-Y (2019) Target organ toxicology in *Caenorhabditis elegans*. Springer Nature Singapore Pte Ltd, Singapore
5. Wang D-Y (2019) Molecular toxicology in *Caenorhabditis elegans*. Springer Nature Singapore Pte Ltd., Singapore
6. Serra ND, Darwin CB, Sundaram MV (2024) *Caenorhabditis elegans* hedgehog-related proteins are tissue- and substructure-specific components of the cuticle and precuticle. Genetics 227:iyae081
7. Moussaif M, Sze JY (2009) Intraflagellar transport/hedgehog-related signaling components couple sensory cilium morphology and serotonin biosynthesis in *Caenorhabditis elegans*. J Neurosci 29:4065–4075
8. Riveiro AR, Mariani L, Malmberg E, Amendola PG, Peltonen J, Wong G, Salcini AE (2017) JMJD-1.2/PHF8 controls axon guidance by regulating hedgehog-like signaling. Development 144:856–865
9. Emans SW, Yerevanian A, Ahsan FM, Rotti JF, Zhou Y, Cedillo L, Soukas AA (2023) GRD-1/PTR-11, the *C. elegans* hedgehog/patched-like morphogen-receptor pair, modulates developmental rate. Development 150:dev201974
10. Wang Q, Fu R, Li G, Xiong S, Zhu Y, Zhang H (2023) Hedgehog receptors exert immune-surveillance roles in the epidermis across species. Cell Rep 42:112929
11. Zárate-Potes A, Ali I, Ribeiro Camacho M, Brownless H, Benedetto A (2022) Meta-analysis of *Caenorhabditis elegans* transcriptomics implicates hedgehog-like signaling in host-microbe interactions. Front Microbiol 13:853629
12. Templeman NM, Cota V, Keyes W, Kaletsky R, Murphy CT (2020) CREB non-autonomously controls reproductive aging through hedgehog/patched signaling. Dev Cell 54:92–105
13. Zhou L, Jiang L, Li L, Ma C, Xia P, Ding W, Liu Y (2024) A germline-to-soma signal triggers an age-related decline of mitochondrial stress response. Nat Commun 15:8723
14. Wu J-W, Shao Y-T, Hua X, Wang D-Y (2024) Activated hedgehog and insulin ligands by decreased transcriptional factor DAF-16 mediate transgenerational nanoplastic toxicity in *Caenorhabditis elegans*. J Hazard Mater 480:135909
15. Qu M, Kong Y, Yuan Y-J, Wang D-Y (2019) Neuronal damage induced by nanopolystyrene particles in nematode *Caenorhabditis elegans*. Environ Sci Nano 6:2591–2601
16. Qu M, Qiu Y-X, Kong Y, Wang D-Y (2019) Amino modification enhances reproductive toxicity of nanopolystyrene on gonad development and reproductive capacity in nematode *Caenorhabditis elegans*. Environ Pollut 254:112978
17. Qu M, Nida A, Kong Y, Du H-H, Xiao G-S, Wang D-Y (2019) Nanopolystyrene at predicted environmental concentration enhances microcystin-LR toxicity by inducing intestinal damage in *Caenorhabditis elegans*. Ecotoxicol Environ Saf 183:109568
18. Qu M, Xu K-N, Li Y-H, Wong G, Wang D-Y (2018) Using *acs-22* mutant *Caenorhabditis elegans* to detect the toxicity of nanopolystyrene particles. Sci Total Environ 643:119–126
19. Qiu Y-X, Liu Y-Q, Li Y-H, Li G-J, Wang D-Y (2020) Effect of chronic exposure to nanopolystyrene on nematode *Caenorhabditis elegans*. Chemosphere 256:127172
20. Qiu Y-X, Luo L-B, Yang Y-H, Kong Y, Li Y-H, Wang D-Y (2020) Potential toxicity of nanopolystyrene on lifespan and aging process of nematode *Caenorhabditis elegans*. Sci Total Environ 705:135918
21. Qu M, Wang D-Y (2020) Toxicity comparison between pristine and sulfonate modified nanopolystyrene particles in affecting locomotion behavior, sensory perception, and neuronal development in *Caenorhabditis elegans*. Sci Total Environ 703:134817

22. Li D, Deng Y-J, Wang S-T, Du H-H, Xiao G-S, Wang D-Y (2020) Assessment of nanopolystyrene toxicity under fungal infection condition in *Caenorhabditis elegans*. Ecotoxicol Environ Saf 197:110625
23. Hua X, Wang D-Y (2022) Cellular uptake, transport, and organelle response after exposure to microplastics and nanoplastics: current knowledge and perspectives for environmental and health risks. Rev Environ Contam Toxicol 260:12
24. Zhang L, Wang S-T, Zhao Y, Bi K, Wang D-Y (2022) Increase in germline methyltransferases governing methylation of histone H3K9 is associated with transgenerational nanoplastic toxicity in *Caenorhabditis elegans*. Environ Sci Nano 9:265–274
25. Shao Y-T, Li Y-H, Wang D-Y (2024) Polylactic acid microplastics cause transgenerational reproductive toxicity associated with activation of insulin and hedgehog ligands in *C. elegans*. Sci Total Environ 942:173746
26. Liu Z-J, Li Y-H, Wang D-Y (2024) Exposure to 6-PPD quinone disrupts glucose metabolism associated with lifespan reduction by affecting insulin and AMPK signals in *Caenorhabditis elegans*. Chemosphere 363:142975
27. Hua X, Wang D-Y (2023) Exposure to 6-PPD quinone at environmentally relevant concentrations inhibits both lifespan and healthspan in *C. elegans*. Environ Sci Technol 57:19295–19303
28. Liu T-W, Zhuang Z-H, Wang D-Y (2023) Paeoniflorin mitigates high glucose-induced lifespan reduction by inhibiting insulin signaling in *Caenorhabditis elegans*. Front Pharmacol 14:1202379
29. Liu H-L, Zhao Y-L, Hua X, Wang D-Y (2022) Induction of transgenerational toxicity is associated with the activated germline insulin signals in nematodes exposed to nanoplastic at predicted environmental concentrations. Ecotoxicol Environ Saf 243:114022
30. Shao H-M, Han Z-Y, Krasteva N, Wang D-Y (2019) Identification of signaling cascade in the insulin signaling pathway in response to nanopolystyrene particles. Nanotoxicology 13:174–188
31. Kong Y, Liu H-L, Li W-J, Wang D-Y (2019) Intestine-specific activity of insulin signaling pathway in response to microgravity stress in *Caenorhabditis elegans*. Biochem Biophys Res Commun 517:278–284
32. Zhao Y-L, Yang R-L, Rui Q, Wang D-Y (2016) Intestinal insulin signaling encodes two different molecular mechanisms for the shortened longevity induced by graphene oxide in *Caenorhabditis elegans*. Sci Rep 6:24024
33. Wu J, Shao Y, Hua X, Wang Y, Wang D (2024) Nanoplastic at environmentally relevant concentrations induces toxicity across multiple generations associated with inhibition in germline G protein-coupled receptor CED-1 in *Caenorhabditis elegans*. Chemosphere 364:143011
34. Hua X, Wang D (2024) 6-PPD quinone at environmentally relevant concentrations induced damage on longevity in *C. elegans*: mechanistic insight from inhibition in mitochondrial UPR response. Sci Total Environ 954:176275
35. Wang S, Zhang R, Wang D (2021) Induction of protective response to polystyrene nanoparticles associated with methylation regulation in *Caenorhabditis elegans*. Chemosphere 271:129589
36. Wu J, Shao Y, Hua X, Li Y, Wang D (2025) Photo-aged polylactic acid microplastics causes severe transgenerational decline in reproductive capacity in *C. elegans*: insight into activation of DNA damage checkpoints affected by multiple germline histone methyltransferases. Environ Pollut 382:126697
37. Yu CW, Luk TC, Liao VH (2021) Long-term nanoplastics exposure results in multi and transgenerational reproduction decline associated with germline toxicity and epigenetic regulation in *Caenorhabditis elegans*. J Hazard Mater 412:125173
38. Chen H, Gu Y, Jiang Y, Yu J, Chen C, Shi C, Li H (2023) Photoaged polystyrene nanoplastics result in transgenerational reproductive toxicity associated with the methylation of histone H3K4 and H3K9 in *Caenorhabditis elegans*. Environ Sci Technol 57:19341–19351
39. Yang J, Chatterjee N, Kim Y, Roh JY, Kwon JH, Park MS, Choi J (2018) Histone methylation-associated transgenerational inheritance of reproductive defects in *Caenorhabditis elegans* exposed to crude oil under various exposure scenarios. Chemosphere 200:358–365
40. Zhao Y-L, Wu Q-L, Wang D-Y (2015) A microRNAs-mRNAs network involved in the control of graphene oxide toxicity in *Caenorhabditis elegans*. RSC Adv 5:92394–92405

41. Zhao Y-L, Wu Q-L, Li Y-P, Nouara A, Jia R-H, Wang D-Y (2014) *In vivo* translocation and toxicity of multi-walled carbon nanotubes are regulated by microRNAs. Nanoscale 6:4275–4284
42. Wu Q-L, Zhao Y-L, Zhao G, Wang D-Y (2014) microRNAs control of *in vivo* toxicity from graphene oxide in *Caenorhabditis elegans*. Nanomedicine: Nanotechnol. Biol Med 10:1401–1410
43. Yang R-L, Ren M-X, Rui Q, Wang D-Y (2016) A *mir-231*-regulated protection mechanism against the toxicity of graphene oxide in nematode *Caenorhabditis elegans*. Sci Rep 6:32214
44. Zhuang Z-H, Li M, Liu H, Luo L-B, Gu W-D, Wu Q-L, Wang D-Y (2016) Function of RSKS-1-AAK-2-DAF-16 signaling cascade in enhancing toxicity of multi-walled carbon nanotubes can be suppressed by *mir-259* activation in *Caenorhabditis elegans*. Sci Rep 6:32409
45. Zhao Y-L, Yang J-N, Wang D-Y (2016) A microRNA-mediated insulin signaling pathway regulates the toxicity of multi-walled carbon nanotubes in nematode *Caenorhabditis elegans*. Sci Rep 6:23234
46. Zhi L-T, Yu Y-L, Jiang Z-X, Wang D-Y (2017) *mir-355* functions as an important link between p38 MAPK signaling and insulin signaling in the regulation of innate immunity. Sci Rep 7:14560
47. Zhi L-T, Yu Y-L, Li X-Y, Wang D-Y (2017) Molecular control of innate immune response to *Pseudomonas aeruginosa* infection by intestinal *let-7* in *Caenorhabditis elegans*. PLoS Pathog 13:e1006152
48. Wu Q-L, Han X-X, Wang D, Zhao F, Wang D-Y (2017) Coal combustion related fine particulate matter ($PM_{2.5}$) induces toxicity in *Caenorhabditis elegans* by dysregulating microRNA expression. Toxicol Res 6:432–441
49. Zhao L, Wan H-X, Liu Q-Z, Wang D-Y (2017) Multi-walled carbon nanotubes-induced alterations in microRNA *let-7* and its targets activate a protection mechanism by conferring a developmental timing control. Part Fibre Toxicol 14:27
50. Xiao G-S, Zhi L-T, Ding X-C, Rui Q, Wang D-Y (2017) Value of *mir-247* in warning graphene oxide toxicity in nematode *Caenorhabditis elegans*. RSC Adv 7:52694–52701
51. Zhao Y-L, Jin L, Wang Y, Kong Y, Wang D-Y (2019) Prolonged exposure to multi-walled carbon nanotubes dysregulates intestinal *mir-35* and its direct target MAB-3 in nematode *Caenorhabditis elegans*. Sci Rep 9:12144
52. Zhao L, Dong S-S, Zhao Y-L, Shao H-M, Krasteva N, Wu Q-L, Wang D-Y (2019) Dysregulation of *let-7* by PEG modified graphene oxide in nematodes with deficit in epidermal barrier. Ecotoxicol Environ Saf 169:1–7
53. Qu M, Luo L-B, Yang Y-H, Kong Y, Wang D-Y (2019) Nanopolystyrene-induced microRNAs response in *Caenorhabditis elegans* after long-term and lose-dose exposure. Sci Total Environ 697:134131
54. Sun L-M, Li W-J, Li D, Wang D-Y (2020) microRNAs involved in the control of toxicity on locomotion behavior induced by simulated microgravity stress in *Caenorhabditis elegans*. Sci Rep 10:17510
55. Yang Y-H, Wu Q-L, Wang D-Y (2020) Epigenetic response to nanopolystyrene in germline of nematode *Caenorhabditis elegans*. Ecotoxicol Environ Saf 206:111404
56. Wang S-T, Liu H-L, Zhao Y-Y, Rui Q, Wang D-Y (2020) Dysregulated *mir-354* enhanced the protective response to nanopolystyrene by affecting the activity of TGF-β signaling pathway in nematode *Caenorhabditis elegans*. NanoImpact 20:100256
57. Qiu Y-X, Liu Y-Q, Li Y-H, Wang D-Y (2020) Intestinal *mir-794* responds to nanopolystyrene by linking insulin and p38 MAPK signaling pathways in nematode *Caenorhabditis elegans*. Ecotoxicol Environ Saf 201:110857
58. Li D, Yuan Y-J, Wang D-Y (2020) Regulation of response to nanopolystyrene by intestinal microRNA *mir-35* in nematode *Caenorhabditis elegans*. Sci Total Environ 736:139677
59. Liu H-L, Zhao Y-Y, Bi Q, Rui Q, Wang D-Y (2021) Dysregulated *mir-76* mediated a protective response to nanopolystyrene by modulating heme homeostasis related molecular signaling in nematode *Caenorhabditis elegans*. Ecotoxicol Environ Saf 212:112018

60. Hua X, Zhao Y, Yuan Y-J, Zhang L, Bian Q, Wang D-Y (2022) Nanoplastics cause transgenerational toxicity through inhibiting germline microRNA *mir-38* in *C. elegans*. J Hazard Mater 437:129302
61. VanDerMolen KR, Newman MA, Breen PC, Gao Y, Huff LA, Dowen RH (2025) Non-cell-autonomous regulation of mTORC2 by hedgehog signaling maintains lipid homeostasis. Cell Rep 44:115191
62. Cadena Del Castillo CE, Hannich JT, Kaech A, Chiyoda H, Brewer J, Fukuyama M, Færgeman NJ, Riezman H, Spang A (2021) Patched regulates lipid homeostasis by controlling cellular cholesterol levels. Nat Commun 12:4898
63. Lin CJ, Wang MC (2017) Microbial metabolites regulate host lipid metabolism through NR5A-hedgehog signalling. Nat Cell Biol 19:550–557

Chapter 8
Regulation of Transgenerational Toxicity by Transgenerational Communication of Germline FGF Signals

Abstract Fibroblast growth factor (FGF) signaling pathway is a conserved signaling pathway. The role of FGF ligand EGL-17 in the germline and its several upregulators in regulating transgenerational pollutants toxicity was introduced. Also, the role of FGF receptor EGL-15 and its tissue-specific activity in regulating transgenerational pollutants toxicity were discussed. In intestinal cells and neurons, EGL-15 regulated transgenerational pollutants toxicity by affecting different downstream targets. Moreover, the potential transgenerational communication between germline EGL-17 and its receptor EGL-15 and its association with induction of transgenerational pollutants toxicity were discussed.

Keywords Transgenerational toxicology · Transgenerational communication · FGF signaling · *Caenorhabditis elegans*

8.1 Introduction

Caenorhabditis elegans has high sensitivity to damage induced by stresses and pollutants [1], which makes it to be widely used for toxicity assessment of different stresses (such as microgravity) and pollutants (such as nanomaterials) [1, 2]. Largely due to the role as a classic model animal, *C. elegans* provides a powerful platform for the study of molecular toxicology [3], especially the toxicity of pollutants at environmentally relevant concentrations [4]. At the parental generation (P0-G), different molecular signaling pathways were involved in controlling toxicity of stresses and pollutants [5–20].

Among the molecular signaling pathways in *C. elegans*, the fibroblast growth factor (FGF) signaling pathway is conserved and important [21] and initially received the attention in 1997 [22]. The FGF ligand EGL-17 is required for developmental control of sex myoblast migration, vulval cell specification, neuronal migration, and sex myoblast specification [23–26]. Meanwhile, the FGF receptor EGL-15 is involved in regulating intestinal tube organization, neurite branching, axon position maintenance, muscle differentiation, extension of muscle membrane,

D. Wang, *Transgenerational Toxicology in Caenorhabditis elegans*,
https://doi.org/10.1007/978-981-92-1389-4_8

sex myoblast migration, architecture of nervous system, and excretory canal development [27–34]. Besides these, FGF signaling was further found to participate in control of metabolic homeostasis and oxidative stress response [34, 35].

In *C. elegans*, the established knowledge system of target organ toxicology allows the possibility to determine transgenerational behavior of germline molecular signals and its association with induction of transgenerational toxicity induced by stresses and pollutants [36]. In this chapter, the role of germline FGF ligand and upregulators was first introduced. After that, we introduced the role of FGF receptor and its tissue-specific activity in regulating transgenerational pollutant toxicity. Moreover, the potential transgenerational communication between germline FGF ligand and its receptor and its association with induction of transgenerational pollutants toxicity were discussed.

8.2 Role of FGF Ligand

8.2.1 Transgenerational Response of Germline EGL-17 to Pollutants

EGL-17 is the *C. elegans* FGF ligand. Exposure to pollutants could cause the transgenerational activation of EGL-17. Nanoplastic is an emergent contaminant, and its exposure at the P0-G caused multiple aspects of toxicity in nematodes, such as damage on different organs, lifespan reduction, and disruption in metabolisms [37–50]. Exposure to polystyrene nanoparticle (PS-NP), a member of nanoplastics, also induced transgenerational toxicity on locomotion and reproduction [51–53]. With the PS-NP as an example of pollutants, exposure to 1–100 μg/L PS-NPs activated germline EGL-17 at the P0-G, and this activation could be further detected in the offspring (Fig. 8.1) [54]. After exposure to 1 μg/L PS-NPs at the P0-G, the activation of germline EGL-17 could last till to the F2-G (Fig. 8.1) [54]. After exposure to 10 μg/L PS-NPs at the P0-G, the activation of germline EGL-17 could be observed till to the F3-G [55]. Moreover, after exposure to 10 μg/L amino-modified PS-NPs at the P0-G, the activation of germline EGL-17 could be even detected till to the F4-G [55].

8.2.2 Function of Germline EGL-17 in Regulating Transgenerational Toxicity of Pollutants

Using RNAi technique, it was found that transgenerational pollutant toxicity could be inhibited by germline RNAi of *egl-17*. With the PS-NPs as an example of pollutants, germline RNAi of *egl-17* caused resistance to transgenerational reproductive toxicity and transgenerational neurotoxicity of PS-NPs (Fig. 8.1) [55]. That is, transgenerational activation of germline EGL-17 mediated induction of transgenerational pollutant toxicity.

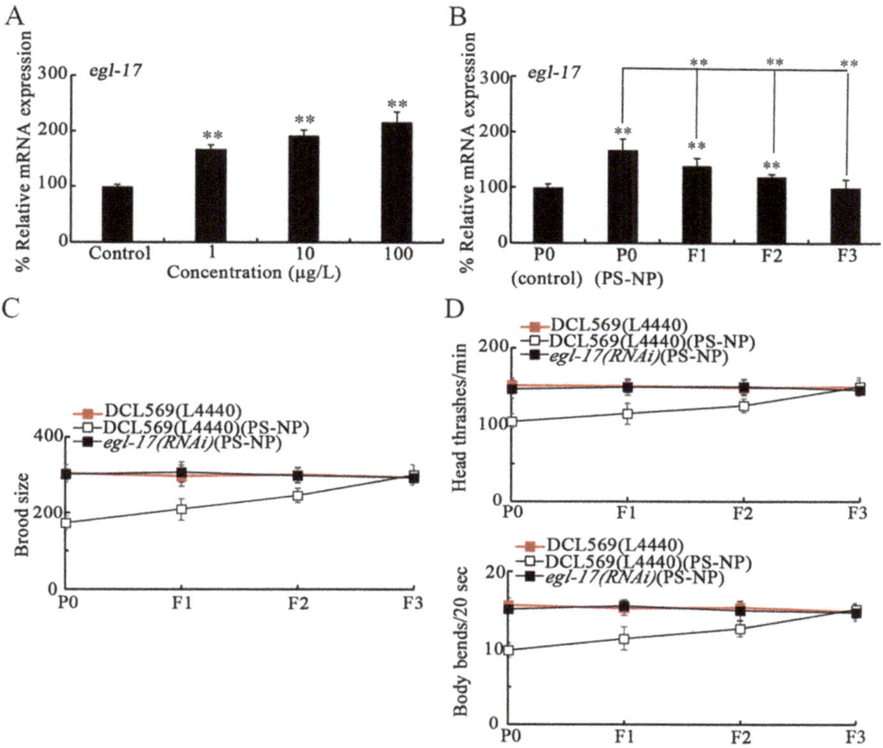

Fig. 8.1 Effect of germline RNAi of *egl-17* on transgenerational PS-NP toxicity [54]. (**a**) Effect of 1–100 μg/L PS-NP exposure on germline *egl-17* expression. (**b**) Transgenerational expression of germline *egl-17* after exposure to 1 μg/L PS-NP. (**c, d**) Effect of germline RNAi of *egl-17* on transgenerational toxicity of PS-NP (1 μg/L) in reducing brood size and locomotion. Curves of DCL569(L4440)(PS-NP) showed a significant difference ($P < 0.01$) compared to DCL569(L4440). After the PS-NP exposure, the curves of *egl-17(RNAi)* showed a significant difference ($P < 0.01$) compared to DCL569(L4440). A total of 30 intact gonads were used for the qRT-PCR assay. $^{**}P < 0.01$ vs control or P0 (control)

8.3 Upregulators for FGF Ligand in Regulating Transgenerational Toxicity of Pollutants

8.3.1 LRP-1

The EGL-17 secretion is controlled by lipoprotein receptors (LRP-1 and LRP-2) and an adaptor protein DAB-1 [56]. Among these three proteins, although germline LRP-2 and DAB-1 activations were not observed, germline activation of LRP-1 was induced by 1–100 μg/L PS-NPs exposure at the P0-G (Fig. 8.2) [54]. More importantly, after exposure to 1 μg/L PS-NPs at the P0-G, the germline activation of LRP-1 lasted till to the F2-G (Fig. 8.2) [54]. Functional analysis indicated that

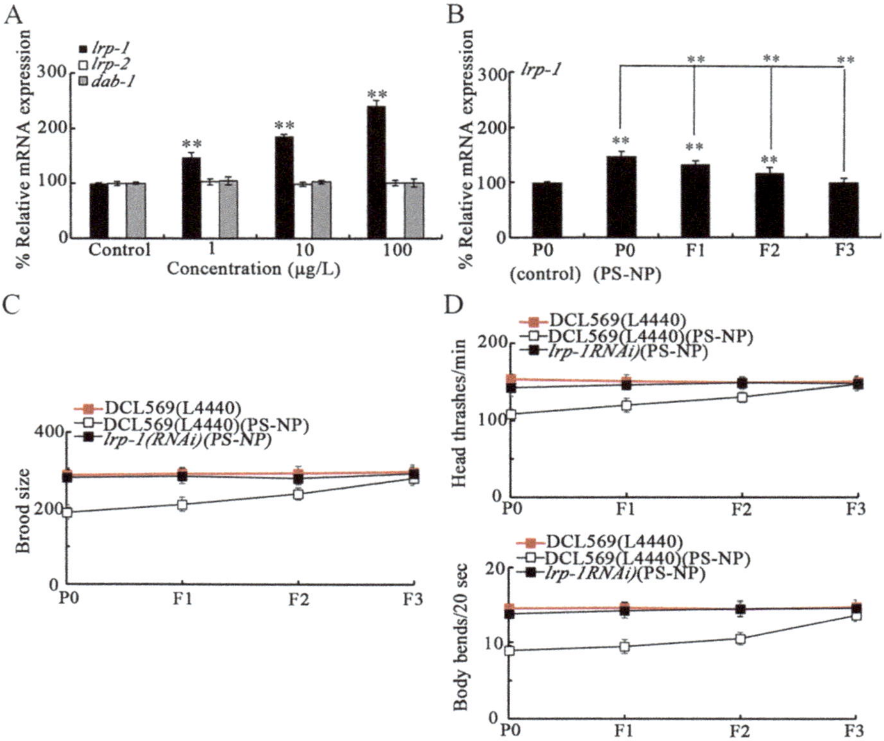

Fig. 8.2 Effect of germline RNAi of *lrp-1* on transgenerational PS-NP toxicity [54]. (**a**) Effect of 1–100 μg/L PS-NP exposure on expressions of germline *lrp-1*, *lrp-2*, and *dab-1*. (**b**) Transgenerational expression of germline *lrp-1* after exposure to 1 μg/L PS-NP. (**c**, **d**) Effect of germline RNAi of *lrp-1* on transgenerational toxicity of PS-NP (1 μg/L) in reducing brood size and locomotion. Curves of DCL569(L4440)(PS-NP) showed a significant difference ($P < 0.01$) compared to DCL569(L4440). After the PS-NP exposure, the curves of *lrp-1(RNAi)* showed a significant difference ($P < 0.01$) compared to DCL569(L4440). A total 30 intact gonads were used for the qRT-PCR assay. $^{**}P < 0.01$ vs control or P0 (control)

germline RNAi of *lrp-1* also resulted in the resistance to transgenerational reproductive toxicity and transgenerational neurotoxicity of PS-NPs (Fig. 8.2) [54]. Therefore, in the germline, LRP-1 acted as an upstream regulator for EGL-17 to enhance its activation so as to mediate the transgenerational pollutant toxicity.

8.3.2 RAB-5

The endosomal Rab-type GTPase RAB-5 is another identified target of EGL-17 in the germline to regulate the transgenerational pollutant toxicity. After the exposure, germline *rab-5* expression was inhibited by PS-NPs at the P0-G, and this inhibition in germline *rab-5* expression could be observed till to the F2-G after exposure to

1 μg/L PS-NPs at the P0-G [57]. Meanwhile, germline RNAi of *rab-5* conferred the susceptibility to transgenerational neurotoxicity and transgenerational reproductive toxicity of PS-NPs, suggesting inhibition in germline RAB-5-mediated induction of transgenerational pollutant toxicity. Moreover, in PS-NP-exposed nematodes, germline RNAi of *rab-5* increased *egl-17* expression [57], suggesting the role of RAB-5 as the upstream regulator of EGL-17 to inhibit its expression during the control of transgenerational pollutant toxicity.

8.3.3 *TAF-1*

TAF-1 is a histone acetyltransferase, and its function reflects the epigenetic regulation mechanism by driving histone acetylation. Epigenetic regulation is an important molecular mechanism regulating the formation of pollutant toxicity, including the transgenerational pollutant toxicity [58–70]. The germline *taf-1* expression was reduced by PS-NP exposure at the P0-G, and the inhibition in germline TAF-1 lasted till to the F4-G after exposure to 100 μg/L PA-NPs [71]. Meanwhile, germline RNAi of *taf-1* caused the susceptibility to transgenerational neurotoxicity and transgenerational reproductive toxicity of PS-NPs and could increase the *egl-17* expression in PS-NP-exposed nematodes [71]. Therefore, in the germline, TAF-1 functions as an upstream negative regulator for EGL-17 to regulate the transgenerational pollutant toxicity.

8.4 Role of FGF Receptor

8.4.1 *Transgenerational Response of EGL-15 to Pollutants*

EGL-15 is corresponding receptor for FGF ligand EGL-17. The transgenerational activation of EGL-15 further mediates the transgenerational pollutant toxicity. The EGL-15 activation could be observed after PS-NP exposure, and the transgenerational EGL-15 activation could be observed till to the F2-G after exposure to 1 μg/L PS-NPs (Fig. 8.3) [54]. That is, pollutant exposure at the P0-G can induce the transgenerational activation of EGL-15.

8.4.2 *Function of EGL-15 in Regulating Transgenerational Toxicity of Pollutants*

Based on RNAi analysis, transgenerational reproductive toxicity and transgenerational neurotoxicity of PS-NPs could be suppressed by RNAi of *egl-15* (Fig. 8.3) [54]. That is, the transgenerational pollutant toxicity could not be formed in

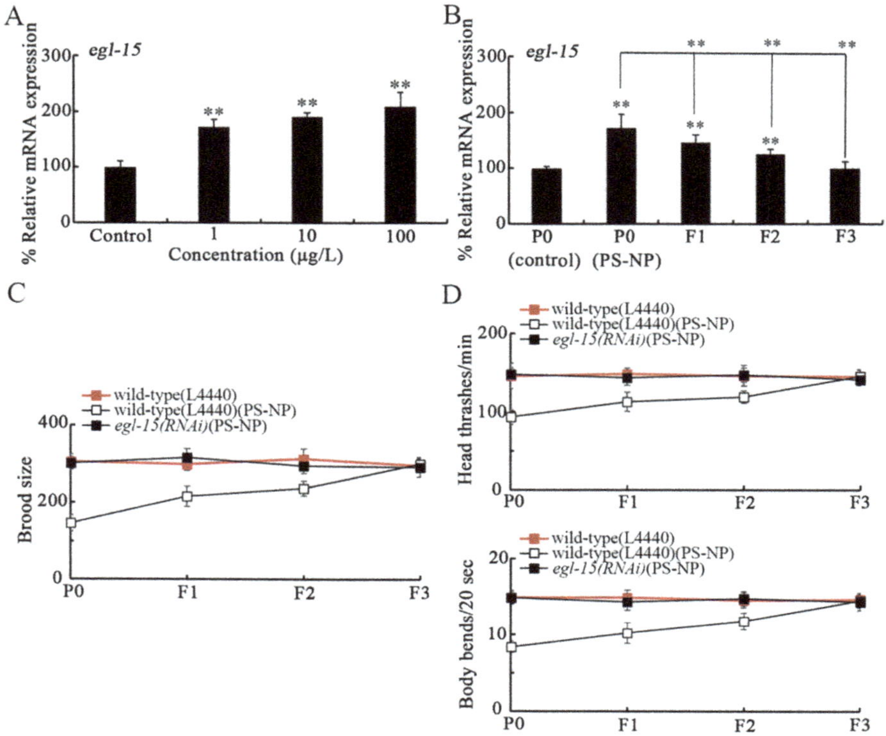

Fig. 8.3 Effect of RNAi of *egl-15* on transgenerational PS-NP toxicity [54]. (**a**) Effect of 1–100 μg/L PS-NP exposure on *egl-15* expression. (**b**) Transgenerational expression of *egl-15* after exposure to 1 μg/L PS-NP. (**c**, **d**) Effect of RNAi of *egl-15* on transgenerational toxicity of PS-NP (1 μg/L) in reducing brood size and locomotion. Curves of wild-type (L4440)(PS-NP) showed a significant difference ($P < 0.01$) compared to wild-type (L4440). After the PS-NP exposure, the curves of *egl-15(RNAi)* showed a significant difference ($P < 0.01$) compared to wild-type (L4440). $^{**}P < 0.01$ vs control or P0 (control)

egl-15(RNAi) nematodes. Therefore, similar to the role of germline EGL-17, transgenerational activation of EGL-15 also mediates induction of transgenerational pollutant toxicity.

8.5 Tissue-Specific Activity of EGL-15 in Regulating Transgenerational Toxicity of Pollutants

In *C. elegans*, EGL-15 functioned in the intestine and the neurons to control the transgenerational pollutant toxicity. EGL-15 can be expressed in several tissues. Nevertheless, transgenerational reproductive toxicity and transgenerational

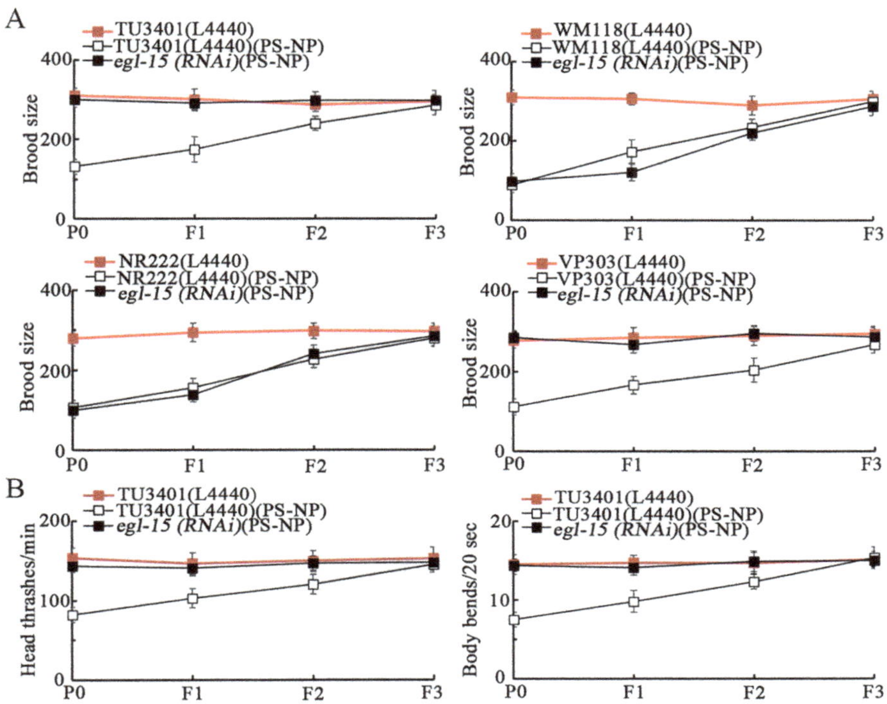

Fig. 8.4 Tissue-specific activity of *egl-15* in controlling transgenerational PS-NP toxicity [54]. (**a**) Tissue-specific activity of *egl-15* in controlling transgenerational PS-NP toxicity in reducing brood size. (**b**) Tissue-specific activity of *egl-15* in controlling transgenerational PS-NP toxicity in decreasing locomotion. Curves of TU3401 (L4440)(PS-NP) showed a significant difference ($P < 0.01$) compared to TU3401(L4440). Curve of WM118(L4440)(PS-NP) showed a significant difference ($P < 0.01$) compared to WM118(L4440). Curve of NR222(L4440)(PS-NP) showed a significant difference ($P < 0.01$) compared to NR222(L4440). Curve of VP303 (L4440)(PS-NP) showed a significant difference ($P < 0.01$) compared to VP303(L4440). After the PS-NP exposure, the curves of *egl-15(RNAi)* showed a significant difference ($P < 0.01$) compared to TU3401(L4440) or VP303(L4440). After PS-NP exposure, the curve of *egl-15(RNAi)* ($P = 0.695$) did not show a significant difference compared to WM118(L4440), and the curve of *egl-15(RNAi)* ($P = 0.705$) did not show a significant difference compared to NR222(L4440)

neurotoxicity of PS-NPs could not be affected by muscle RNAi of *egl-15* and epidermal RNAi of *egl-15* (Fig. 8.4) [54]. In contrast, intestinal RNAi of *egl-15* and neuronal RNAi of *egl-15* resulted in the resistance to transgenerational reproductive toxicity and transgenerational neurotoxicity of PS-NPs (Fig. 8.4) [54]. Therefore, EGL-15 acted in the intestine and the neurons to regulate the transgenerational pollutant toxicity.

8.6 Downstream Targets of Intestinal and Neuronal EGL-15 in Regulating Transgenerational Toxicity of Pollutants

8.6.1 Targets of Intestinal EGL-15 in Regulating Transgenerational Toxicity of Pollutants

In the intestinal cells, DAF-16 (FOXO transcriptional factor in insulin signaling pathway) and BAR-1/β-catenin (transcriptional factor in Wnt signaling pathway) were identified as downstream targets of EGL-15 to regulate the transgenerational pollutant toxicity. DAF-16, BAR-1, ELT-2, and PMK-1 are molecular signals with the function to act in the intestine to regulate the pollutant toxicity [4, 12–13, 16–18, 36, 72–77]. In PS-NP-exposed nematodes, although *elt-2* and *pmk-1* expressions were not changed by intestinal RNAi of *egl-15*, *daf-16* and *bar-1* expressions were increased by intestinal RNAi of *egl-15* (Fig. 8.5) [54]. Additionally, intestinal RNAi of *daf-16* and *bar-1* could lead to the susceptibility to transgenerational reproductive toxicity of PS-NPs (Fig. 8.5) [54]. Therefore, in the intestinal cells, transgenerational activation of EGL-15 mediated transgenerational pollutant toxicity by causing transgenerational inhibition in DAF-16 and BAR-1.

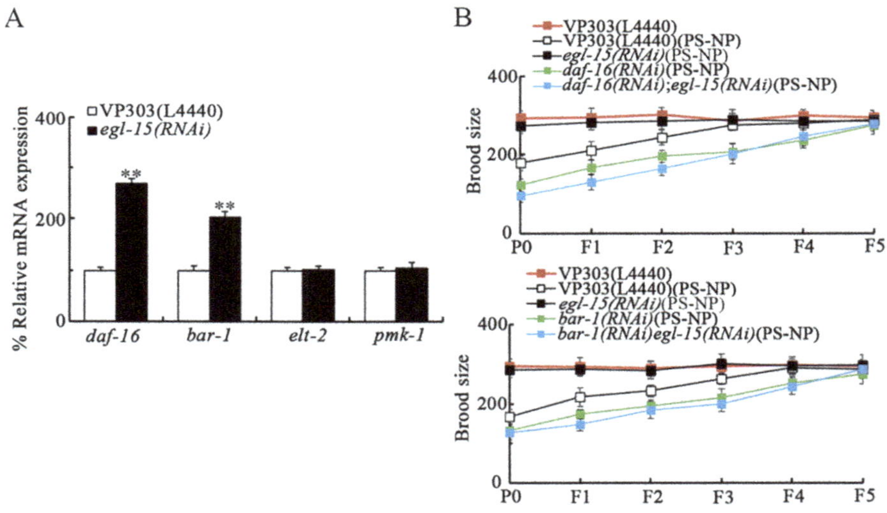

Fig. 8.5 Identification of downstream targets of intestinal EGL-15 in controlling transgenerational PS-NP toxicity [54]. (**a**) Effect of intestinal RNAi of *egl-15* on expressions of *daf-16*, *bar-1*, *elt-2*, and *pmk-1* in PS-NP (1 μg/L)-exposed nematodes. ** $P < 0.01$ vs VP303(L4440). (**b**) Genetic interaction between intestinal EGL-15 and DAF-16 or BAR-1 in controlling transgenerational toxicity of PS-NP (1 μg/L) in reducing brood size. Curve of VP303(L4440)(PS-NP) showed a significant difference ($P < 0.01$) compared to VP303(L4440). After the PS-NP exposure, the curves of *egl-15(RNAi)*, *daf-16(RNAi)*, and *bar-1(RNAi)* showed a significant difference ($P < 0.01$) compared to VP303(L4440), and the curves of *daf-16(RNAi);egl-15(RNAi)* and *bar-1(RNAi) egl-15(RNAi)* showed a significant difference ($P < 0.01$) compared to *egl-15(RNAi)*

8.6.2 Targets of Neuronal EGL-15 in Regulating Transgenerational Toxicity of Pollutants

In the neurons, MPK-1 (ERK MAPK) was identified as a downstream target of EGL-15 to regulate the transgenerational pollutant toxicity. DAF-7, DBL-1, JNK-1, MPK-1, and GLB-10 are molecular signals with the function to act in the neurons to regulate the toxicity of stresses and/or pollutants [14, 15, 62, 66, 78]. In PS-NP-exposed nematodes, although *daf-7*, *dbl-1*, *jnk-1*, and *glb-10* expressions were not changed by neuronal RNAi of *egl-15*, *mpk-1* expression was elevated by neuronal RNAi of *egl-15* (Fig. 8.6) [54]. Moreover, neuronal RNAi of *mpk-1* could result in the susceptibility to transgenerational reproductive toxicity and transgenerational neurotoxicity of PS-NPs (Fig. 8.6) [54]. Therefore, in the neurons, transgenerational activation of EGL-15 mediated the transgenerational pollutant toxicity by causing transgenerational inhibition in MPK-1.

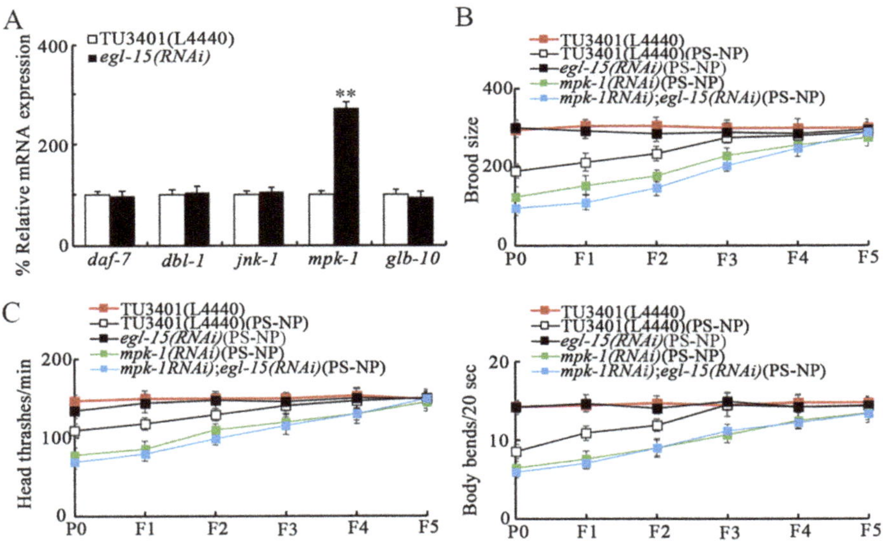

Fig. 8.6 Identification of downstream targets of neuronal EGL-15 in controlling transgenerational PS-NP toxicity [54]. (**a**) Effect of neuronal RNAi of *egl-15* on expressions of *daf-7*, *dbl-1*, *jnk-1*, *mpk-1*, and *glb-10* in PS-NP (1 μg/L)-exposed nematodes. $^{**}P < 0.01$ vs TU3401(L4440). (**b**, **c**) Genetic interaction between intestinal EGL-15 and MPK-1 in controlling transgenerational toxicity of PS-NP (1 μg/L) in reducing brood size and locomotion. Curve of TU3401(L4440)(PS-NP) showed a significant difference ($P < 0.01$) compared to TU3401(L4440). After the PS-NP exposure, the curves of *egl-15(RNAi)* and *mpk-1(RNAi)* showed a significant difference ($P < 0.01$) compared to TU3401(L4440), and the curves of *mpk-1(RNAi);egl-15(RNAi)* showed a significant difference ($P < 0.01$) compared to *egl-15(RNAi)*

8.7 Transgenerational Communication Between FGF Ligand and Its Receptor in Regulating Transgenerational Toxicity of Pollutants

Several lines of evidence have been raised to indicate the potential of transgenerational communication between FGF ligand EGL-17 and its receptor EGL-15 in regulating transgenerational toxicity of pollutants. In *C. elegans*, germline EGL-17 overexpression caused the susceptibility to transgenerational reproductive toxicity and transgenerational neurotoxicity of PS-NPs (Fig. 8.7) [54]. Firstly, after PS-NP exposure, germline EGL-17 overexpression at the P0-G could elevate the *egl-15* expression at the F1-G (Fig. 8.7) [54]. Secondly, the susceptibility to transgenerational reproductive toxicity and transgenerational neurotoxicity of PS-NPs caused by germline EGL-17 overexpression at the P0-G could be inhibited by RNAi of *egl-15* at the F1-G (Fig. 8.7) [54]. These observations suggested the potential transgenerational communication of germline EGL-17 with its receptor EGL-15 in the offspring to mediate the transgenerational pollutant toxicity.

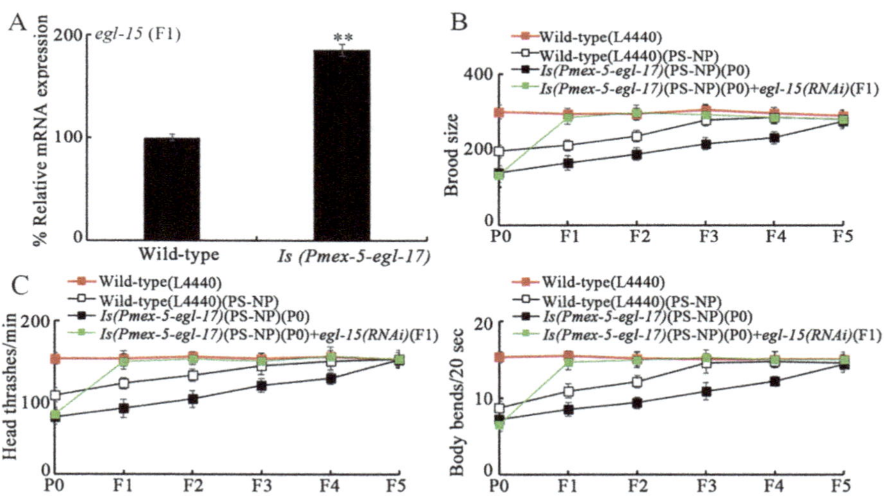

Fig. 8.7 FGF ligand FGF-17 controlled transgenerational PS-NP toxicity by affecting function of FGF receptor FGF-15 transgenerationally [54]. (**a**) Effect of germline overexpression of EGL-17 on *egl-15* expression at F1-G after exposure to 1 μg/L PS-NP at P0-G. (**b, c**) Effect of *egl-15* RNAi on toxicity of exposure to 1 μg/L PS-NP on brood size and locomotion in *Is(Pmex-5-egl-17)* nematodes. After exposure at P0-G for *Is(Pmex-5-egl-17)*, RNAi of *egl-15* was carried out at F1-G. Curves of wild-type (L4440)(PS-NP) showed a significant difference ($P < 0.01$) compared to wild-type (L4440). Curves of *Is(Pmex-5-egl-17)*(PS-NP)(P0) showed a significant difference ($P < 0.01$) compared to wild-type (L4440) (PS-NP). The 1 μg/L was the exposure concentration. $^{**}P < 0.01$ showing significance difference for *Is(Pmex-5-egl-17)*(PS-NP)(P0) + *egl-15(RNAi)*(F1) vs *Is(Pmex-5-egl-17)*(PS-NP)(P0)

8.8 Perspectives

In this chapter, we first discussed the role of transgenerational activation of FGF ligand EGL-17 in the germline to mediate the transgenerational pollutant toxicity. During this process, LRP-1, RAB-5, and TAF-1 were identified as upregulators for EGL-17 in the germline to regulate the transgenerational pollutant toxicity. During the control of biological processes, some other upregulators for EGL-17 exist, such as the protein kinase VRK-1 [79]. Whether these upregulators also participate in the control of transgenerational pollutant toxicity still needs to be further examined.

Accompanied with transgenerational activation of germline EGL-17, transgenerational activation of FGF receptor EGL-15 was also found after pollutant exposure at the P0-G. Evidences were raised for the formation of transgenerational communication between germline EGL-17 and its receptor EGL-15 during the control of transgenerational pollutant toxicity. During the control of transgenerational pollutant toxicity, EGL-15 acted in both intestinal cells and neurons to exert its function, and some downstream targets for EGL-15 in these tissues in controlling transgenerational pollutants toxicity were also identified. Nevertheless, the more detailed underlying mechanisms for EGL-15 in regulating transgenerational pollutant toxicity still need to be further carefully determined. Some clues from the study of life science can be considered and referred to. For example, it was reported that EGL-15 could promote the protein degradation process [80]. In addition, some evidence indicated that different isoforms of *egl-15* are all functional during the control of biological processes, such as sex myoblast migration [81]. Therefore, the exact function of different isoforms of *egl-15* in regulating transgenerational pollutants toxicity also needs further examination.

References

1. Wang D-Y (2020) Exposure toxicology in *Caenorhabditis elegans*. Springer Nature Singapore Pte Ltd, Singapore
2. Wang D-Y (2018) Nanotoxicology in *Caenorhabditis elegans*. Springer Nature Singapore Pte Ltd., Singapore
3. Wang D-Y (2019) Molecular toxicology in *Caenorhabditis elegans*. Springer Nature Singapore Pte Ltd, Singapore
4. Wang D-Y (2022) Toxicology at environmentally relevant concentrations in *Caenorhabditis elegans*. Springer Nature Singapore Pte Ltd., Singapore
5. Liu Z-Y, Bian Q, Wang D-Y (2025) 6-PPD quinone induces response of nuclear hormone receptors in the germline associated with formation of reproductive toxicity in *Caenorhabditis elegans*. J Hazard Mater 495:138815
6. Liu Z-J, Li Y-H, Wang D-Y (2024) Exposure to 6-PPD quinone disrupts glucose metabolism associated with lifespan reduction by affecting insulin and AMPK signals in *Caenorhabditis elegans*. Chemosphere 363:142975
7. Hua X, Wang D-Y (2023) Exposure to 6-PPD quinone at environmentally relevant concentrations inhibits both lifespan and healthspan in *C. elegans*. Environ Sci Technol 57:19295–19303

8. Yang Y-H, Dong W-T, Wu Q-L, Wang D-Y (2021) Induction of protective response associated with expressional alterations in neuronal G protein-coupled receptors in polystyrene nanoparticle exposed *Caenorhabditis elegans*. Chem Res Toxicol 34:1308–1318
9. Liu H-L, Qiu Y-X, Wang D-Y (2021) Alteration in expressions of ion channels in *Caenorhabditis elegans* exposed to polystyrene nanoparticles. Chemosphere 273:129686
10. Wang S-T, Liu H-L, Qu M, Wang D-Y (2021) Response of tyramine and glutamate related signals to nanoplastic exposure in *Caenorhabditis elegans*. Ecotoxicol Environ Saf 217:112239
11. Sun L-M, Li D, Yuan Y-J, Wang D-Y (2021) Intestinal long non-coding RNAs in response to simulated microgravity stress in *Caenorhabditis elegans*. Sci Rep 11:1997
12. Shao H-M, Wang D-Y (2020) Long-term and low-dose exposure to nanopolystyrene induces a protective strategy to maintain functional state of intestine barrier in nematode *Caenorhabditis elegans*. Environ Pollut 258:113649
13. Shao H-M, Kong Y, Wang D-Y (2020) Response of intestinal signaling communication between nucleus and peroxisome to nanopolystyrene at predicted environmental concentration. Environ Sci Nano 7:250–261
14. Qu M, Li D, Qiu Y-X, Wang D-Y (2020) Neuronal ERK MAPK signaling in response to low-dose nanopolystyrene exposure by suppressing insulin peptide expression in *Caenorhabditis elegans*. Sci Total Environ 724:138378
15. Liu H-L, Zhang R-J, Wang D-Y (2020) Response of DBL-1/TGF-β signaling-mediated neuron-intestine communication to nanopolystyrene in nematode *Caenorhabditis elegans*. Sci Total Environ 745:1141047
16. Liu P-D, Shao H-M, Kong Y, Wang D-Y (2020) Effect of graphene oxide exposure on intestinal Wnt signaling in nematode *Caenorhabditis elegans*. J Environ Sci 88:200–208
17. Zhao Y-Y, Dong S-S, Kong Y, Rui Q, Wang D-Y (2020) Molecular basis of intestinal canonical Wnt/β-catenin BAR-1 in response to simulated microgravity in *Caenorhabditis elegans*. Biochem Biophy Res Commun 522:198–204
18. Rui Q, Dong S-S, Jiang W-K, Wang D-Y (2019) Response of canonical Wnt/β-catenin signaling pathway in the intestine to microgravity stress in *Caenorhabditis elegans*. Ecotoxicol Environ Saf 186:109782
19. Chen H, Li H-R, Wang D-Y (2017) Graphene oxide dysregulates Neuroligin/NLG-1-mediated molecular signaling in interneurons in *Caenorhabditis elegans*. Sci Rep 7:41655
20. Ren M-X, Zhao L, Lv X, Wang D-Y (2017) Antimicrobial proteins in the response to graphene oxide in *Caenorhabditis elegans*. Nanotoxicology 11:578–590
21. Borland CZ, Schutzman JL, Stern MJ (2001) Fibroblast growth factor signaling in *Caenorhabditis elegans*. BioEssays 23:1120–1130
22. Burdine RD, Chen EB, Kwok SF, Stern MJ (1997) *egl-17* encodes an invertebrate fibroblast growth factor family member required specifically for sex myoblast migration in *Caenorhabditis elegans*. Proc Natl Acad Sci USA 94:2433–2437
23. Burdine RD, Branda CS, Stern MJ (1998) EGL-17(FGF) expression coordinates the attraction of the migrating sex myoblasts with vulval induction in *C. elegans*. Development 125:1083–1093
24. Cui M, Han M (2003) Cis regulatory requirements for vulval cell-specific expression of the *Caenorhabditis elegans* fibroblast growth factor gene *egl-17*. Dev Biol 257:104–116
25. Fleming TC, Wolf FW, Garriga G (2005) Sensitized genetic backgrounds reveal a role for *C. elegans* FGF EGL-17 as a repellent for migrating CAN neurons. Development 132:4857–4867
26. Photos A, Gutierrez A, Sommer RJ (2006) *sem-4*/spalt and *egl-17*/FGF have a conserved role in sex myoblast specification and migration in *P. pacificus* and *C. elegans*. Dev Biol 293:142–153
27. Hoffmann M, Segbert C, Helbig G, Bossinger O (2010) Intestinal tube formation in *Caenorhabditis elegans* requires *vang-1* and *egl-15* signaling. Dev Biol 339:268–279
28. Díaz-Balzac CA, Lázaro-Peña MI, Ramos-Ortiz GA, Bülow HE (2015) The adhesion molecule KAL-1/anosmin-1 regulates neurite branching through a SAX-7/L1CAM-EGL-15/FGFR receptor complex. Cell Rep 11:1377–1384

29. Bülow HE, Boulin T, Hobert O (2004) Differential functions of the *C. elegans* FGF receptor in axon outgrowth and maintenance of axon position. Neuron 42:367–374
30. Sasson IE, Stern MJ (2004) FGF and PI3 kinase signaling pathways antagonistically modulate sex muscle differentiation in *C. elegans*. Development 131:5381–5392
31. Dixon SJ, Alexander M, Fernandes R, Ricker N, Roy PJ (2006) FGF negatively regulates muscle membrane extension in *Caenorhabditis elegans*. Development 133:1263–1275
32. DeVore DL, Horvitz HR, Stern MJ (1995) An FGF receptor signaling pathway is required for the normal cell migrations of the sex myoblasts in *C. elegans* hermaphrodites. Cell 83:611–620
33. Bénard CY, Boyanov A, Hall DH, Hobert O (2006) DIG-1, a novel giant protein, non-autonomously mediates maintenance of nervous system architecture. Development 133:3329–3340
34. Polanska UM, Edwards E, Fernig DG, Kinnunen TK (2011) The cooperation of FGF receptor and Klotho is involved in excretory canal development and regulation of metabolic homeostasis in *Caenorhabditis elegans*. J Biol Chem 286:5657–5666
35. Kim S, Sieburth D (2019) A receptor tyrosine kinase network regulates neuromuscular function in response to oxidative stress in *Caenorhabditis elegans*. Genetics 211:1283–1295
36. Wang D-Y (2019) Target organ toxicology in *Caenorhabditis elegans*. Springer Nature Singapore Pte Ltd, Singapore
37. Qu M, Kong Y, Yuan Y-J, Wang D-Y (2019) Neuronal damage induced by nanopolystyrene particles in nematode *Caenorhabditis elegans*. Environ Sci Nano 6:2591–2601
38. Qu M, Qiu Y-X, Kong Y, Wang D-Y (2019) Amino modification enhances reproductive toxicity of nanopolystyrene on gonad development and reproductive capacity in nematode *Caenorhabditis elegans*. Environ Pollut 254:112978
39. Qu M, Nida A, Kong Y, Du H-H, Xiao G-S, Wang D-Y (2019) Nanopolystyrene at predicted environmental concentration enhances microcystin-LR toxicity by inducing intestinal damage in *Caenorhabditis elegans*. Ecotoxicol Environ Saf 183:109568
40. Qu M, Xu K-N, Li Y-H, Wong G, Wang D-Y (2018) Using *acs-22* mutant *Caenorhabditis elegans* to detect the toxicity of nanopolystyrene particles. Sci Total Environ 643:119–126
41. Qiu Y-X, Liu Y-Q, Li Y-H, Li G-J, Wang D-Y (2020) Effect of chronic exposure to nanopolystyrene on nematode *Caenorhabditis elegans*. Chemosphere 256:127172
42. Qiu Y-X, Luo L-B, Yang Y-H, Kong Y, Li Y-H, Wang D-Y (2020) Potential toxicity of nanopolystyrene on lifespan and aging process of nematode *Caenorhabditis elegans*. Sci Total Environ 705:135918
43. Qu M, Wang D-Y (2020) Toxicity comparison between pristine and sulfonate modified nanopolystyrene particles in affecting locomotion behavior, sensory perception, and neuronal development in *Caenorhabditis elegans*. Sci Total Environ 703:134817
44. Li D, Deng Y-J, Wang S-T, Du H-H, Xiao G-S, Wang D-Y (2020) Assessment of nanopolystyrene toxicity under fungal infection condition in *Caenorhabditis elegans*. Ecotoxicol Environ Saf 197:110625
45. Hua X, Wang D-Y (2022) Cellular uptake, transport, and organelle response after exposure to microplastics and nanoplastics: current knowledge and perspectives for environmental and health risks. Rev Environ Contam Toxicol 260:12
46. Liu H-L, Shao H-M, Guo Z-J, Wang D-Y (2020) Nanopolystyrene exposure activates a fat metabolism related signaling-mediated protective response in *Caenorhabditis elegans*. NanoImpact 17:100204
47. Dong S-S, Qu M, Rui Q, Wang D-Y (2018) Combinational effect of titanium dioxide nanoparticles and nanopolystyrene particles at environmentally relevant concentrations on nematodes *Caenorhabditis elegans*. Ecotoxicol Environ Saf 161:444–450
48. Zhao Y-Y, Li D, Rui Q, Wang D-Y (2020) Toxicity induction of nanopolystyrene under microgravity stress condition in *Caenorhabditis elegans*. Sci Total Environ 703:135623
49. Liu H-L, Wang D-Y (2021) Intestinal mitochondrial unfolded protein response induced by nanoplastic particles in *Caenorhabditis elegans*. Chemosphere 267:128917

50. Tang M, Ding G, Lu X, Huang Q, Du H, Xiao G, Wang D (2022) Exposure to nanoplastic particles enhances *Acinetobacter* survival, biofilm formation, and serum resistance. Nano 12:4222
51. Zhao L, Qu M, Wong G, Wang D-Y (2017) Transgenerational toxicity of nanopolystyrene particles in the range of μg/L in nematode *Caenorhabditis elegans*. Environ Sci Nano 4:2356–2366
52. Sun L-M, Liao K, Wang D-Y (2021) Comparison of transgenerational reproductive toxicity induced by pristine and amino modified nanoplastics in *Caenorhabditis elegans*. Sci Total Environ 768:144362
53. Liu H-L, Tian L-J, Wang S-T, Wang D-Y (2021) Size-dependent transgenerational toxicity induced by nanoplastics in nematode *Caenorhabditis elegans*. Sci Total Environ 790:148217
54. Hua X, Cao C, Zhang L, Wang D-Y (2023) Activation of FGF signal in germline mediates transgenerational toxicity of polystyrene nanoparticles at predicted environmental concentrations in *Caenorhabditis elegans*. J Hazard Mater 451:131174
55. Song M-X, Ruan Q-L, Wang D-Y (2024) Comparison of transgenerational toxicity on locomotion behavior and development of D-type motor neurons between pristine and amino modified nanoplastics at environmentally relevant doses in *C. elegans*. Toxics 12:555
56. Kamikura DM, Cooper JA (2003) Lipoprotein receptors and a disabled family cytoplasmic adaptor protein regulate EGL-17/FGF export in *C. elegans*. Genes Dev 17:2798–2811
57. Hua X, Zhang L, Wang D-Y (2024) Nanoplastic at environmentally relevant concentrations activates germline *mir-240-rab-5* signaling cascade to affect secreted ligands associated with transgenerational toxicity induction in *C. elegans*. Environ Sci Nano 11:3615
58. Wu J-W, Shao Y-T, Hua X, Li Y-H, Wang D-Y (2025) Photo-aged polylactic acid microplastics causes severe transgenerational decline in reproductive capacity in *C. elegans*: insight into activation of DNA damage checkpoints affected by multiple germline histone methyltransferases. Environ Pollut 382:126697
59. Zhang L, Wang S-T, Zhao Y, Bi K, Wang D-Y (2022) Increase in germline methyltransferases governing methylation of histone H3K9 is associated with transgenerational nanoplastic toxicity in *Caenorhabditis elegans*. Environ Sci Nano 9:265–274
60. Hua X, Zhao Y, Yuan Y-J, Zhang L, Bian Q, Wang D-Y (2022) Nanoplastics cause transgenerational toxicity through inhibiting germline microRNA *mir-38* in *C. elegans*. J Hazard Mater 437:129302
61. Zhao Y-Y, Xu R-R, Hua X, Rui Q, Wang D-Y (2022) Multi-walled carbon nanotubes induce transgenerational toxicity associated with activation of germline long non-coding RNA *linc-7* in *C. elegans*. Chemosphere 301:134687
62. Liu H-L, Zhao Y-Y, Bi K, Rui Q, Wang D-Y (2021) Dysregulated *mir-76* mediated a protective response to nanopolystyrene by modulating heme homeostasis related molecular signaling in nematode *Caenorhabditis elegans*. Ecotoxicol Environ Saf 212:112018
63. Zhao Y-Y, Xu R-R, Chen X, Wang J, Rui Q, Wang D-Y (2021) Induction of protective response to polystyrene nanoparticles associated with dysregulation of intestinal long non-coding RNAs in *Caenorhabditis elegans*. Ecotoxicol Environ Saf 212:111976
64. Li D, Yuan Y-J, Wang D-Y (2020) Regulation of response to nanopolystyrene by intestinal microRNA *mir-35* in nematode *Caenorhabditis elegans*. Sci Total Environ 736:139677
65. Qiu Y-X, Liu Y-Q, Li Y-H, Wang D-Y (2020) Intestinal *mir-794* responds to nanopolystyrene by linking insulin and p38 MAPK signaling pathways in nematode *Caenorhabditis elegans*. Ecotoxicol Environ Saf 201:110857
66. Wang S-T, Liu H-L, Zhao Y-Y, Rui Q, Wang D-Y (2020) Dysregulated *mir-354* enhanced the protective response to nanopolystyrene by affecting the activity of TGF-β signaling pathway in nematode *Caenorhabditis elegans*. NanoImpact 20:100256
67. Yang Y-H, Wu Q-L, Wang D-Y (2020) Epigenetic response to nanopolystyrene in germline of nematode *Caenorhabditis elegans*. Ecotoxicol Environ Saf 206:111404
68. Zhao L, Dong S-S, Zhao Y-L, Shao H-M, Krasteva N, Wu Q-L, Wang D-Y (2019) Dysregulation of *let-7* by PEG modified graphene oxide in nematodes with deficit in epidermal barrier. Ecotoxicol Environ Saf 169:1–7

69. Zhao L, Wan H-X, Liu Q-Z, Wang D-Y (2017) Multi-walled carbon nanotubes-induced alterations in microRNA *let-7* and its targets activate a protection mechanism by conferring a developmental timing control. Part Fibre Toxicol 14:27
70. Zhao Y-L, Wu Q-L, Wang D-Y (2016) An epigenetic signal encoded protection mechanism is activated by graphene oxide to inhibit its induced reproductive toxicity in *Caenorhabditis elegans*. Biomaterials 79:15–24
71. Hua X, Wang D-Y (2024) Transgenerational response of germline histone acetyltransferases and deacetylases to nanoplastics at predicted environmental doses in *Caenorhabditis elegans*. Sci Total Environ 952:175903
72. Shao H-M, Han Z-Y, Krasteva N, Wang D-Y (2019) Identification of signaling cascade in the insulin signaling pathway in response to nanopolystyrene particles. Nanotoxicology 13:174–188
73. Qu M, Liu Y-Q, Xu K-N, Wang D-Y (2019) Activation of p38 MAPK signaling-mediated endoplasmic reticulum unfolded protein response by nanopolystyrene particles. Adv Biosyst 3:1800325
74. Zhao Y-L, Yang J-N, Wang D-Y (2016) A microRNA-mediated insulin signaling pathway regulates the toxicity of multi-walled carbon nanotubes in nematode *Caenorhabditis elegans*. Sci Rep 6:23234
75. Zhao Y-L, Zhi L-T, Wu Q-L, Yu Y-L, Sun Q-Q, Wang D-Y (2016) p38 MAPK-SKN-1/Nrf signaling cascade is required for intestinal barrier against graphene oxide toxicity in *Caenorhabditis elegans*. Nanotoxicology 10:1469–1479
76. Zhao Y-L, Yang R-L, Rui Q, Wang D-Y (2016) Intestinal insulin signaling encodes two different molecular mechanisms for the shortened longevity induced by graphene oxide in *Caenorhabditis elegans*. Sci Rep 6:24024
77. Zhi L-T, Ren M-X, Qu M, Zhang H-Y, Wang D-Y (2016) Wnt ligands differentially regulate toxicity and translocation of graphene oxide through different mechanisms in *Caenorhabditis elegans*. Sci Rep 6:39261
78. Qu M, Li D, Zhao Y-L, Yuan Y-J, Wang D-Y (2020) Exposure to low-dose nanopolystyrene induces the response of neuronal JNK MAPK signaling pathway in nematode *Caenorhabditis elegans*. Environ Sci Euro 32:58
79. Klerkx EP, Alarcón P, Waters K, Reinke V, Sternberg PW, Askjaer P (2009) Protein kinase VRK-1 regulates cell invasion and EGL-17/FGF signaling in *Caenorhabditis elegans*. Dev Biol 335:12–21
80. Szewczyk NJ, Jacobson LA (2003) Activated EGL-15 FGF receptor promotes protein degradation in muscles of *Caenorhabditis elegans*. EMBO J 22:5058–5067
81. Lo TW, Branda CS, Huang P, Sasson IE, Goodman SJ, Stern MJ (2008) Different isoforms of the *C. elegans* FGF receptor are required for attraction and repulsion of the migrating sex myoblasts. Dev Biol 318:268–275

Chapter 9
Regulation of Transgenerational Toxicity by Transgenerational Communication of Germline Notch Signals

Abstract Notch signaling is initially identified to play an important role in the developmental control. In Notch signaling pathway of *Caenorhabditis elegans*, activation of both Notch ligand LAG-2 and Notch receptor GLP-1 mediated transgenerational pollutant toxicity. During this process, TAF-1 and HDA-3 were identified as upregulators for LAG-2 in the germline, and potential transgenerational communication between germline LAG-2 and its receptor GLP-1 may exist. GLP-1 could further act in germline and neurons to regulate the transgenerational pollutant toxicity by affecting certain downstream targeted genes.

Keywords Transgenerational toxicology · Notch ligand · Notch receptor · *Caenorhabditis elegans*

9.1 Introduction

Notch signaling is very conserved and functions to mediate specification of multiple cell fates during the development of organisms [1]. In Notch signaling pathway of *Caenorhabditis elegans*, Notch receptors, GLP-1 and LIN-12, were first identified [2, 3]. During the *C. elegans* development, GLP-1 and LIN-12 are required for the control of mitosis and meiosis, formation of germline tumor, nervous system organization, left-right asymmetry, oocyte growth, and cell fate decision and specification [2–11]. Corresponding to GLP-1 and LIN-22, their ligands were further identified, such as LAG-2 and APX-1 [12, 13]. Similarly, Notch ligands participate in diverse developmental processes, such as vulval cell fate decision, muscle development, ovulation, larval molting, dorsal-ventral polarity, fertility, and left/right asymmetry [14–20]. Besides the developmental control, Notch signaling is also involved in controlling dauer maintenance and toxicity of microgravity stress [21, 22], suggesting its role in regulating the stress response.

D. Wang, *Transgenerational Toxicology in Caenorhabditis elegans*,
https://doi.org/10.1007/978-981-92-1389-4_9

Considering the high sensitivity after environmental exposures, *C. elegans* has been recognized as a widely used animal model for assessing adverse effects of stresses and pollutants, especially at environmentally relevant doses or concentrations [23–25]. The well-described genetic background makes the *C. elegans* suitable for the study of molecular toxicology [26]. The systematical work on target organ toxicology in *C. elegans* provides important foundation for further study on transgenerational toxicology of stresses and pollutants [27]. In this chapter, we first introduced roles of Notch ligands and Notch receptors in regulating transgenerational pollutant toxicity. After that, tissue-specific activity of Notch receptor GLP-1 and the downstream targets of GLP-1 in different tissues to regulate transgenerational pollutants toxicity were introduced. Moreover, involvement of potential transgenerational communication between germline Notch ligand and its receptor in regulating transgenerational pollutants toxicity was discussed.

9.2 Role of Notch Ligand

9.2.1 Transgenerational Response of Germline LAG-2 to Pollutants

In *C. elegans*, after pollutant exposure, certain Notch ligand exhibited the transgenerational activation. Nanoplastic is one of the emergent contaminants receiving the high concern recently [28]. In *C. elegans*, nanoplastic exposure resulted in multiple toxicities, such as neuronal damage, intestinal injury, reproductive inhibition, and metabolic disruption at the parental generation (P0-G) [29–40]. Transgenerational toxicity at different aspects, such as the toxicity on reproduction and locomotion, could be further detected after nanoplastic exposure at the P0-G [41–55]. The nanoplastic toxicity on nematodes can be increased by some chemical modifications, such as amino modification and sulfonate modification [30, 56]. The sulfonate-modified polystyrene nanoparticle (PS-S NP) was used as an example of pollutants. After exposure to 10 μg/L, the transgenerational toxicity of PS-S NP on locomotion and reproductive could be observed till to the F2-G [57].

In *C. elegans*, among the examined three Notch ligands genes, although *apx-1* and *dsl-2* expressions were not changed by PS-S NPs, LAG-2 was activated by 1–100 μg/L PS-S NPs at the P0-G (Fig. 9.1) [57]. Moreover, the 10 μg/L PS-S NP–induced transgenerational activation of LAG-2 could last till to the F2-G (Fig. 9.1) [57]. Besides this, LAG-2 activation could also be induced by pristine and amino-modified PS-NPs at the P0-G [58]. After exposure at the concentration of 10 μg/L, compared to the transgenerational activation of LAG-2 by pristine PS-NPs till to F3-G, transgenerational activation of LAG-2 by amino modified PS-NPs could be observed till to the F4-G [58].

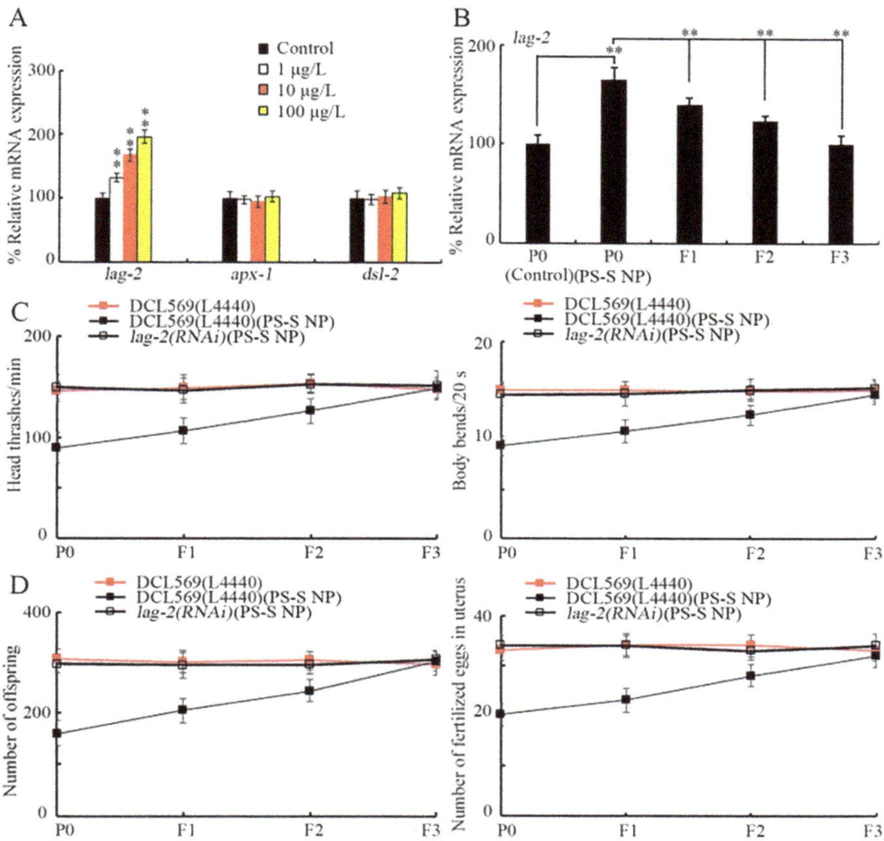

Fig. 9.1 Requirement of germline LAG-2 for controlling the transgenerational PS-S NP toxicity [57]. (**a**) Effect of PS-S NP exposure on expressions of the germlines *lag-2*, *apx-1*, and *dsl-2* at P0-G. $^{**}P < 0.01$ vs. control. (**b**) Germline *lag-2* expressions from P0-G to F3-G after parental exposure to 10 μg/L PS-S NP. $^{**}P < 0.01$. (**c**) Effect of the germline *lag-2* RNAi on transgenerational toxicity of 10 μg/L PS-S NP in inhibiting locomotion. (**d**) Effect of the germline *lag-2* RNAi on transgenerational toxicity of 10 μg/L PS-S NP in reducing reproductive capacity. Curves of DCL569(L4440)(PS-S NP) showed a significant difference ($P < 0.01$) compared with DCL569(L4440). After the PS-S NP exposure, the curves of *lag-2(RNAi)* showed a significant difference ($P < 0.01$) compared with DCL569(L4440)

9.2.2 Function of Germline LAG-2 in Regulating Transgenerational Toxicity of Pollutants

Based on functional analysis, germline RNAi of *lag-2* induced resistance to transgenerational pollutant toxicity. After germline RNAi of *lag-2*, the resistance to transgenerational neurotoxicity and transgenerational reproductive toxicity of PS-S

NPs could be detected (Fig. 9.1) [57]. Therefore, transgenerational activation of germline LAG-2 mediates the induction of transgenerational pollutant toxicity.

9.2.3 Upregulators for Germline LAG-2 in Regulating Transgenerational Toxicity of Pollutants

1. TAF-1

   TAF-1 is a histone acetyltransferase. Exposure to PS-NPs caused the transgenerational inhibition in germline *taf-1* expression [59]. Germline RNAi of *taf-1* led to susceptibility to transgenerational neurotoxicity and transgenerational reproductive toxicity of PS-NPs [59], suggesting transgenerational inhibition of germline TAF-1-mediated transgenerational PS-NP toxicity. Moreover, after PS-NP exposure, germline RNAi of *taf-1* caused activation of germline LAG-2 [59]. That is, inhibition in germline TAF-1 caused transgenerational activation of LAG-2 to mediate the transgenerational pollutant toxicity.

2. HDA-3

   HDA-3 is a histone deacetylase. Transgenerational inhibition in germline *hda-3* expression was also induced by PS-NPs exposure at the P0-G [59]. Transgenerational inhibition of germline HDA-1 also mediated induction of transgenerational PS-NPs toxicity, since susceptibility to transgenerational neurotoxicity and transgenerational reproductive toxicity of PS-NPs was induced by germline RNAi of *hda-3* [59]. In addition, in PS-NP-exposed *C. elegans*, activation of LAG-2 was caused by germline RNAi of *hda-3* [59]. Therefore, transgenerational inhibition in HDA-3 mediates transgenerational pollutants toxicity by causing the following transgenerational activation of LAG-2.

9.3 Role of Notch Receptor

9.3.1 Transgenerational Response of GLP-1 to Pollutants

In *C. elegans*, pollutant exposure also causes transgenerational activation of Notch receptor. With PS-S NPs as example of pollutants, among examined Notch receptor genes, although *lin-12* expression was not altered by PS-S NPs exposure, GLP-1 activation was caused by 1–100 μg/L PS-S NPs at the P0-G (Fig. 9.2) [57]. After exposure at the concentration of 10 μg/L, transgenerational activation of GLP-1 caused by PS-S NPs could be found till to the F2-G (Fig. 9.2) [57].

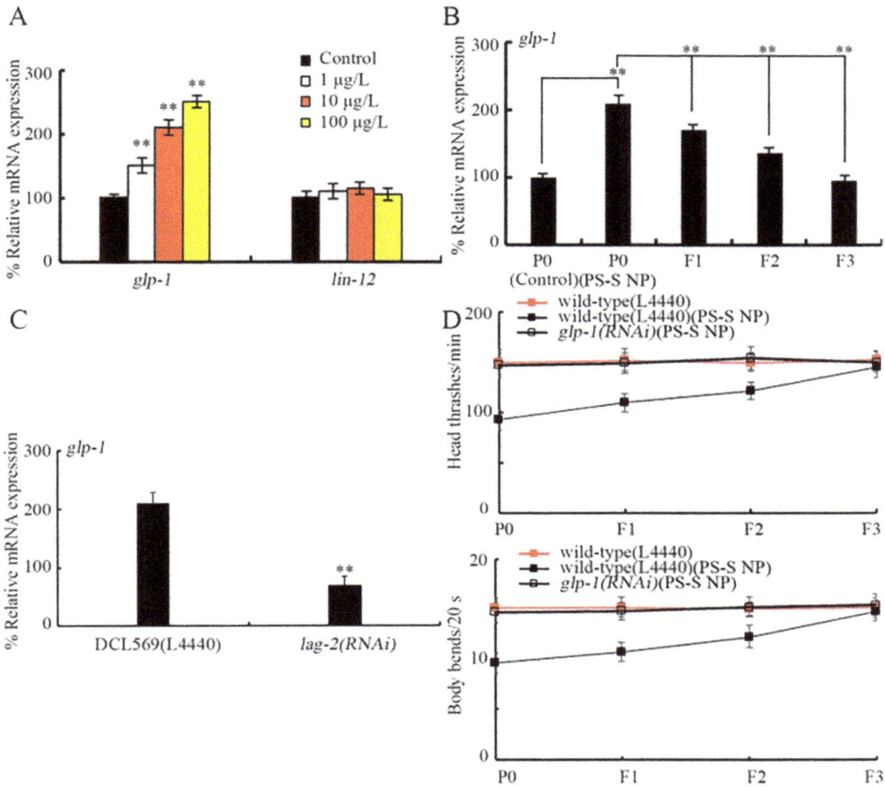

Fig. 9.2 Requirement of Notch receptor GLP-1 for controlling the transgenerational PS-S NP toxicity [57]. (**a**) Effect of PS-S NP exposure on expressions of *glp-1* and *lin-12* at P0-G. $^{**}P < 0.01$ vs control. (**b**) *glp-1* expression from P0-G to F3-G after parental exposure to 10 μg/L PS-S NP. $^{**}P < 0.01$. (**c**) Effect of parental germline RNAi of *lag-2* on *glp-1* expression in the offspring of 10 μg/L PS-S NP–exposed nematodes. The RNAi of *lag-2* was performed at P0-G, and the *glp-1* expression was analyzed at F1-G. (**d**) Effect of *glp-1* RNAi on transgenerational toxicity of 10 μg/L PS-S NP in inhibiting locomotion. Curves of wild-type (L4440)(PS-S NP) showed a significant difference ($P < 0.01$) compared with wild-type (L4440). After the PS-S NP exposure, the curves of *glp-1(RNAi)* showed a significant difference ($P < 0.01$) compared with wild-type (L4440)

9.3.2 *Function of GLP-1 in Regulating Transgenerational Toxicity of Pollutants*

Similar to the observation from functional analysis for germline Notch ligand, RNAi of Notch receptor genes also resulted in the resistance to transgenerational pollutant toxicity. After RNAi of *glp-1*, the resistance to transgenerational neurotoxicity of PS-S NPs could be detected (Fig. 9.2) [57]. Therefore, both transgenerational activation of Notch ligand LAG-2 and transgenerational activation of Notch receptor GLP-1 mediate the induction of transgenerational pollutant toxicity.

9.4 Tissue-Specific Activity of GLP-1 in Regulating Transgenerational Toxicity of Pollutants

GLP-1 can function in both germline and neurons to regulate the transgenerational pollutant toxicity. GLP-1 can be expressed in the germline and neurons. PS-S NP was selected as an example of a pollutant. Transgenerational neurotoxicity of PS-S NPs could be inhibited by germline RNAi of *glp-1* (Fig. 9.3) [57]. Besides this, the transgenerational neurotoxicity of PS-S NPs was also suppressed by neuronal RNAi of *glp-1* (Fig. 9.3) [57].

9.5 Downstream Targets of Germline and Neuronal GLP-1 in Regulating Transgenerational Toxicity of Pollutants

In the germline, insulin ligands INS-3, INS-39, and DAF-28 acted as downstream targets of GLP-1 to regulate the transgenerational pollutant toxicity. Insulin ligands INS-3, INS-39, and DAF-28 and Wnt ligand LIN-44 could function in the germline

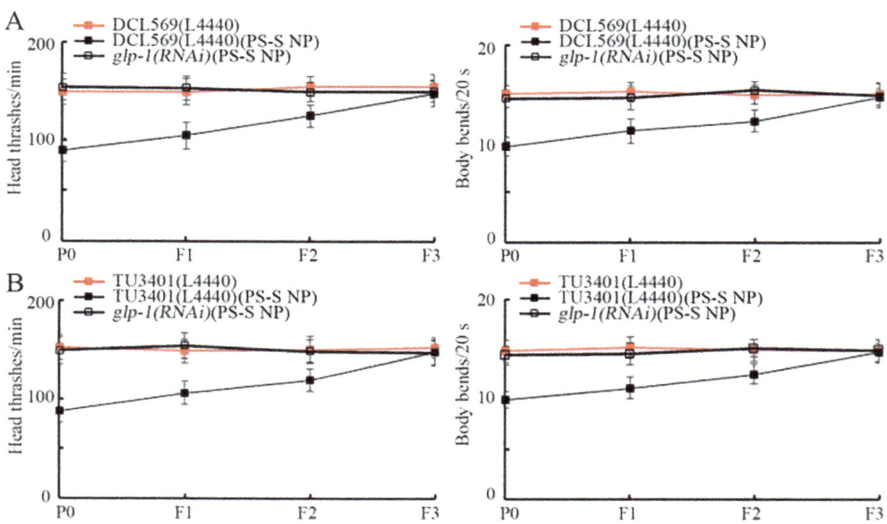

Fig. 9.3 Tissue-specific activity of GLP-1 in controlling the transgenerational PS-S NP toxicity [57]. (**a**) Germline-specific activity of GLP-1 in controlling the transgenerational toxicity of 10 μg/L PS-S NP in inhibiting locomotion. (**b**) Neuron-specific activity of GLP-1 in controlling the transgenerational toxicity of 10 μg/L PS-S NP in inhibiting locomotion. Curves of DCL569(L4440)(PS-S NP) showed a significant difference ($P < 0.01$) compared with DCL569(L4440), and curves of TU3401(L4440)(PS-S NP) showed a significant difference ($P < 0.01$) compared with TU3401(L4440). After the PS-S NP exposure, the curves of *glp-1(RNAi)* showed a significant difference ($P < 0.01$) compared with DCL569(L4440) or TU3401(L4440)

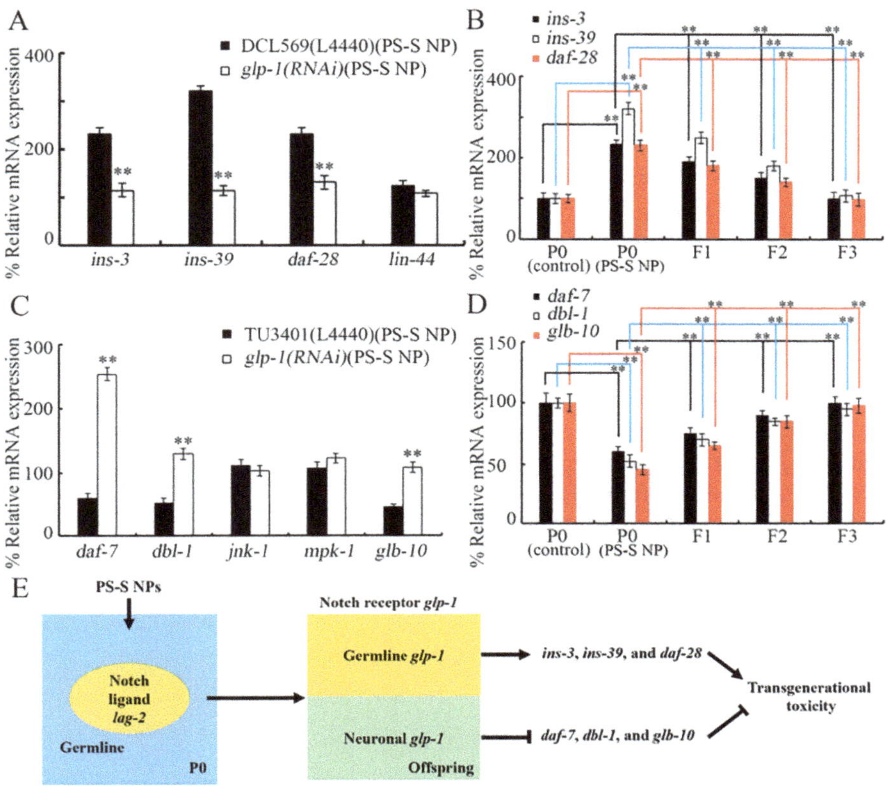

Fig. 9.4 Identification of downstream targets of the germline and the neuronal GLP-1 in controlling transgenerational PS-S NP toxicity [57]. (**a**) Effect of germline *glp-1* RNAi on expressions of *ins-3*, *ins-39*, *daf-28*, and *lin-44* in 10 μg/L PS-S NP–exposed nematodes at P0-G. $^{**}P < 0.01$ vs. DCL569(L4440). (**b**) Transgenerational expressions of *ins-3*, *ins-39*, and *daf-28* in 10 μg/L PS-S NP–exposed nematodes. $^{**}P < 0.01$. (**c**) Effect of neuronal *glp-1* RNAi on expressions of *daf-7*, *dbl-1*, *jnk-1*, *mpk-1*, and *glb-10* in 10 μg/L PS-S NP–exposed nematodes at P0-G. $^{**}P < 0.01$ vs. TU3401(L4440). (**d**) Transgenerational expressions of *daf-7*, *dbl-1*, and *glb-10* in 10 μg/L PS-S NP–exposed nematodes. $^{**}P < 0.01$. (**e**) Molecular basis for germline Notch signal in regulating transgenerational PS-S NP toxicity in nematodes

to control transgenerational PS-NP toxicity [53, 54]. Nevertheless, after PS-S NP exposure, germline RNAi of *glp-1* only decreased *ins-3*, *ins-39*, and *daf-28* expressions, but did not affect *lin-44* expression (Fig. 9.4) [57]. After PS-S NP exposure, transgenerational activation of INS-3, INS-39, and DAF-28 was induced (Fig. 9.4) [57]. That is, in the germline, activation of GLP-1 mediates transgenerational pollutant toxicity by causing the following transgenerational activation of insulin ligands.

In the neurons, DAF-7, DBL-1, and GLB-10 acted as downstream targets of GLP-1 to regulate the transgenerational pollutant toxicity. TGF-β ligands (DAF-7 and DBL-1), JNK-1/JNK MAPK, MPK-1/ERK MAPJ, and GLB-10 functioned in the neurons to regulate the pollutant toxicity [60–64]. In PS-S NP–exposed nematodes, among these genes, neuronal RNAi of *glp-1* caused activation of DAF-7, DBL-1, and GLB-10 (Fig. 9.4) [57]. After PS-S NP exposure, transgenerational inhibition of DAF-7, DBL-1, and GLB-10 was induced (Fig. 9.4) [57]. Therefore, in the neurons, activation of GLP-1 mediates transgenerational pollutant toxicity by causing the following transgenerational inhibition of DAF-7, DBL-1, and GLB-10.

9.6 Transgenerational Communication Between Notch Ligand and Its Receptor in Regulating Transgenerational Toxicity of Pollutants

To determine the potential transgenerational communication between Notch ligand and its receptor in regulating transgenerational toxicity of pollutants, germline RNAi of *lag-2* was performed at the P0-G, and the expression of *glp-1* was assessed at the F1-G [57]. After *lag-2* RNAi at the P0-G, *glp-1* expression at the F1-G could be decreased (Fig. 9.2) [57]. This implied that the activated Notch ligand LAG-2 by pollutants could potentially cause the further activation of Notch receptor GLP-1 in the offspring.

9.7 Perspectives

So far, activation of both Notch ligand LAG-2 in the germline and Notch GLP-1 was found to mediate the transgenerational pollutant toxicity. Nevertheless, considering that the selected example of pollutant was nanoplastic, it is supposed that exposure to other pollutants with more severe toxicity will cause transgenerational response of other Notch ligands (such as APX-1) and Notch receptors (such as LIN-12). Previous observation implied that LIN-12 and GLP-1, two Notch receptors, have overlapping functions [65].

During the control of transgenerational pollutant toxicity, TAF-1 and HDA-3 were identified as upregulators of germline LAG-2. This suggested that, in the germline, epigenetic regulation is an important force to drive transgenerational activation of germline Notch ligand LAG-2 to mediate the transgenerational pollutant toxicity. During the developmental control, some other molecular signals were identified as upregulators for Notch ligands, such as HOX protein [66]. That is, the other upregulators for germline Notch ligand LAG-2 in regulating transgenerational pollutant toxicity still need to be further identified.

In *C. elegans*, GLP-1 could function in both germline and neurons to regulate the transgenerational pollutant toxicity. During the control of this process, insulin ligands were identified as downstream targets of germline GLP-1, and TGF-β ligands and GLB-10 were identified as downstream targets of neuronal GLP-1.

However, in the germline, besides insulin ligands, some other molecular ligands (such as FGF ligands and Hedgehog ligands) were also involved in controlling transgenerational pollutant toxicity [45, 49, 55]. In addition, certain downstream targets were also identified for Notch receptors during the developmental control [67]. Therefore, the other downstream targets for germline GLP-1 and neuronal GLP-1 may still exist during the control of transgenerational pollutant toxicity.

Certain evidence was raised for the potential transgenerational communication between germline Notch ligand LAG-2 and its receptor GLP-1. Besides this, some other molecular mechanisms may also exist for the involvement of Notch signaling in regulating the transgenerational pollutant toxicity. The first possibility is that, in the germline, LAG-2 activates its receptor GLP-1, and germline GLP-1 regulates the transgenerational pollutant toxicity by activating or inhibiting transgenerational communication of other molecular signals. The second possibility is that maternal GLP-1 directly affects the activity of GLP-1 in the offspring of pollutant-exposed nematodes to influence the toxicity induction of pollutants in the offspring. During the developmental control, activity of Notch receptors can be regulated by maternal GLP-1 [68]. The third possibility is that certain metabolic signals may modulate the function of Notch signaling in regulating transgenerational pollutant toxicity. During the control of germline proliferation, function of GLP-1/Notch signaling could be affected by nucleotide levels [69].

References

1. Artavanis-Tsakonas S, Matsuno K, Fortini ME (1995) Notch signaling. Science 268:225–232
2. Austin J, Kimble J (1987) *glp-1* is required in the germ line for regulation of the decision between mitosis and meiosis in *C. elegans*. Cell 51:589–599
3. Austin J, Kimble J (1989) Transcript analysis of *glp-1* and *lin-12*, homologous genes required for cell interactions during development of *C. elegans*. Cell 58:565–571
4. Berry LW, Westlund B, Schedl T (1997) Germ-line tumor formation caused by activation of *glp-1*, a *Caenorhabditis elegans* member of the notch family of receptors. Development 124:925–936
5. Wilkinson HA, Fitzgerald K, Greenwald I (1994) Reciprocal changes in expression of the receptor *lin-12* and its ligand *lag-2* prior to commitment in a *C. elegans* cell fate decision. Cell 79:1187–1198
6. Park SK, Choi VN, Hwang BJ (2013) LIN-12/notch regulates *lag-1* and *LIN-12* expression during anchor cell/ventral uterine precursor cell fate specification. Mol Cells 35:249–254
7. Komatsu H, Chao MY, Larkins-Ford J, Corkins ME, Somers GA, Tucey T, Dionne HM, White JQ, Wani K, Boxem M, Hart AC (2008) OSM-11 facilitates LIN-12 notch signaling during *Caenorhabditis elegans* vulval development. PLoS Biol 6:e196
8. Hermann GJ, Leung B, Priess JR (2000) Left-right asymmetry in *C. elegans* intestine organogenesis involves a LIN-12/notch signaling pathway. Development 127:3429–3440
9. Chao MY, Larkins-Ford J, Tucey TM, Hart AC (2005) *lin-12* notch functions in the adult nervous system of *C. elegans*. BMC Neurosci 6:45
10. Qiao L, Lissemore JL, Shu P, Smardon A, Gelber MB, Maine EM (1995) Enhancers of *glp-1*, a gene required for cell-signaling in *Caenorhabditis elegans*, define a set of genes required for germline development. Genetics 141:551–569

11. Nadarajan S, Govindan JA, McGovern M, Hubbard EJ, Greenstein D (2009) MSP and GLP-1/notch signaling coordinately regulate actomyosin-dependent cytoplasmic streaming and oocyte growth in *C. elegans*. Development 136:2223–2234
12. Henderson ST, Gao D, Lambie EJ, Kimble J (1994) *Lag-2* may encode a signaling ligand for the GLP-1 and LIN-12 receptors of *C. elegans*. Development 120:2913–2924
13. Gao D, Kimble J (1995) APX-1 can substitute for its homolog LAG-2 to direct cell interactions throughout *Caenorhabditis elegans* development. Proc Natl Acad Sci USA 92:9839–9842
14. Zhang X, Greenwald I (2011) Spatial regulation of *lag-2* transcription during vulval precursor cell fate patterning in *Caenorhabditis elegans*. Genetics 188:847–858
15. Hale JJ, Amin NM, George C, Via Z, Shi H, Liu J (2014) A role of the LIN-12/notch signaling pathway in diversifying the non-striated egg-laying muscles in *C. elegans*. Dev Biol 389:137–148
16. McGovern M, Castaneda PG, Pekar O, Vallier LG, Cram EJ, Hubbard EJA (2018) The DSL ligand APX-1 is required for normal ovulation in *C. elegans*. Dev Biol 435:162–169
17. Bertrand V, Bisso P, Poole RJ, Hobert O (2011) Notch-dependent induction of left/right asymmetry in *C. elegans* interneurons and motoneurons. Curr Biol 21:1225–1231
18. Chuang CH, Aprison EZ, Ruvinsky I, Bowerman B (2025) Distinct requirements for the *C. elegans* Delta ligand APX-1 in embryonic viability and adult fertility. G3 15:jkaf229
19. Mello CC, Draper BW, Priess JR (1994) The maternal genes *apx-1* and *glp-1* and establishment of dorsal-ventral polarity in the early *C. elegans* embryo. Cell 77:95–106
20. Singh K, Chao MY, Somers GA, Komatsu H, Corkins ME, Larkins-Ford J, Tucey T, Dionne HM, Walsh MB, Beaumont EK, Hart DP, Lockery SR, Hart AC (2011) *C. elegans* notch signaling regulates adult chemosensory response and larval molting quiescence. Curr Biol 21:825–834
21. Ouellet J, Li S, Roy R (2008) Notch signalling is required for both dauer maintenance and recovery in *C. elegans*. Development 135:2583–2592
22. Liu H-L, Tian L-J, Wang D-Y (2021) Notch receptor GLP-1 regulates toxicity of simulated microgravity stress by activating germline-intestine communication of insulin signaling in *C. elegans*. Biochem Biophys Res Commun 534:248–253
23. Wang D-Y (2018) Nanotoxicology in *Caenorhabditis elegans*. Springer Nature Singapore Pte Ltd., Singapore
24. Wang D-Y (2020) Exposure toxicology in *Caenorhabditis elegans*. Springer Nature Singapore Pte Ltd, Singapore
25. Wang D-Y (2022) Toxicology at environmentally relevant concentrations in *Caenorhabditis elegans*. Springer Nature Singapore Pte Ltd, Singapore
26. Wang D-Y (2019) Molecular toxicology in *Caenorhabditis elegans*. Springer Nature Singapore Pte Ltd., Singapore
27. Wang D-Y (2019) Target organ toxicology in *Caenorhabditis elegans*. Springer Nature Singapore Pte Ltd., Singapore
28. Hua X, Wang D-Y (2022) Cellular uptake, transport, and organelle response after exposure to microplastics and nanoplastics: current knowledge and perspectives for environmental and health risks. Rev Environ Contam Toxicol 260:12
29. Qu M, Kong Y, Yuan Y-J, Wang D-Y (2019) Neuronal damage induced by nanopolystyrene particles in nematode *Caenorhabditis elegans*. Environ Sci Nano 6:2591–2601
30. Qu M, Qiu Y-X, Kong Y, Wang D-Y (2019) Amino modification enhances reproductive toxicity of nanopolystyrene on gonad development and reproductive capacity in nematode *Caenorhabditis elegans*. Environ Pollut 254:112978
31. Qu M, Nida A, Kong Y, Du H-H, Xiao G-S, Wang D-Y (2019) Nanopolystyrene at predicted environmental concentration enhances microcystin-LR toxicity by inducing intestinal damage in *Caenorhabditis elegans*. Ecotoxicol Environ Saf 183:109568
32. Qu M, Xu K-N, Li Y-H, Wong G, Wang D-Y (2018) Using *acs-22* mutant *Caenorhabditis elegans* to detect the toxicity of nanopolystyrene particles. Sci Total Environ 643:119–126

33. Qiu Y-X, Liu Y-Q, Li Y-H, Li G-J, Wang D-Y (2020) Effect of chronic exposure to nanopolystyrene on nematode *Caenorhabditis elegans*. Chemosphere 256:127172
34. Qiu Y-X, Luo L-B, Yang Y-H, Kong Y, Li Y-H, Wang D-Y (2020) Potential toxicity of nanopolystyrene on lifespan and aging process of nematode *Caenorhabditis elegans*. Sci Total Environ 705:135918
35. Li D, Deng Y-J, Wang S-T, Du H-H, Xiao G-S, Wang D-Y (2020) Assessment of nanopolystyrene toxicity under fungal infection condition in *Caenorhabditis elegans*. Ecotoxicol Environ Saf 197:110625
36. Liu H-L, Shao H-M, Guo Z-J, Wang D-Y (2020) Nanopolystyrene exposure activates a fat metabolism related signaling-mediated protective response in *Caenorhabditis elegans*. NanoImpact 17:100204
37. Dong S-S, Qu M, Rui Q, Wang D-Y (2018) Combinational effect of titanium dioxide nanoparticles and nanopolystyrene particles at environmentally relevant concentrations on nematodes *Caenorhabditis elegans*. Ecotoxicol Environ Saf 161:444–450
38. Zhao Y-Y, Li D, Rui Q, Wang D-Y (2020) Toxicity induction of nanopolystyrene under microgravity stress condition in *Caenorhabditis elegans*. Sci Total Environ 703:135623
39. Liu H-L, Wang D-Y (2021) Intestinal mitochondrial unfolded protein response induced by nanoplastic particles in *Caenorhabditis elegans*. Chemosphere 267:128917
40. Tang M, Ding G, Lu X, Huang Q, Du H, Xiao G, Wang D (2022) Exposure to nanoplastic particles enhances *Acinetobacter* survival, biofilm formation, and serum resistance. Nano 12:4222
41. Zhao L, Qu M, Wong G, Wang D-Y (2017) Transgenerational toxicity of nanopolystyrene particles in the range of μg/L in nematode *Caenorhabditis elegans*. Environ Sci Nano 4:2356–2366
42. Sun L-M, Liao K, Wang D-Y (2021) Comparison of transgenerational reproductive toxicity induced by pristine and amino modified nanoplastics in *Caenorhabditis elegans*. Sci Total Environ 768:144362
43. Liu H-L, Tian L-J, Wang S-T, Wang D-Y (2021) Size-dependent transgenerational toxicity induced by nanoplastics in nematode *Caenorhabditis elegans*. Sci Total Environ 790:148217
44. Zhao Y, Chen J, Wang R, Pu X, Wang D (2023) A review of transgenerational and multigenerational toxicology in the *in vivo* model animal *Caenorhabditis elegans*. J Appl Toxicol 43:122–145
45. Wu J-W, Shao Y-T, Hua X, Wang D-Y (2024) Activated hedgehog and insulin ligands by decreased transcriptional factor DAF-16 mediate transgenerational nanoplastic toxicity in *Caenorhabditis elegans*. J Hazard Mater 480:135909
46. Hua X, Zhang L, Wang D-Y (2024) Nanoplastic at environmentally relevant concentrations activates germline *mir-240-rab-5* signaling cascade to affect secreted ligands associated with transgenerational toxicity induction in *C. elegans*. Environ Sci Nano 11:3615
47. Liu Z-Y, Wang Y-X, Bian Q, Wang D-Y (2024) Transgenerational response of germline nuclear hormone receptor genes to nanoplastics at predicted environmental doses in *Caenorhabditis elegans*. Toxics 12:420
48. Wu J-W, Shao Y-T, Hua X, Wang Y-X, Wang D-Y (2024) Nanoplastic at environmentally relevant concentrations induces toxicity across multiple generations associated with inhibition in germline G protein-coupled receptor CED-1 in *Caenorhabditis elegans*. Chemosphere 364:143011
49. Hua X, Cao C, Zhang L, Wang D-Y (2023) Activation of FGF signal in germline mediates transgenerational toxicity of polystyrene nanoparticles at predicted environmental concentrations in *Caenorhabditis elegans*. J Hazard Mater 451:131174
50. Xu R-R, Hua X, Rui Q, Wang D-Y (2022) Polystyrene nanoparticles caused dynamic alteration in mitochondrial unfolded protein response from parents to the offspring in *C. elegans*. Chemosphere 308:136154
51. Zhang L, Wang S-T, Zhao Y, Bi Q, Wang D-Y (2022) Increase in germline methyltransferases governing methylation of histone H3K9 is associated with transgenerational nanoplastic toxicity in *Caenorhabditis elegans*. Environ Sci Nano 9:265–274

52. Hua X, Zhao Y, Yuan Y-J, Zhang L, Bian Q, Wang D-Y (2022) Nanoplastics cause transgenerational toxicity through inhibiting germline microRNA *mir-38* in *C. elegans*. J Hazard Mater 437:129302
53. Liu H-L, Zhao Y-L, Hua X, Wang D-Y (2022) Induction of transgenerational toxicity is associated with the activated germline insulin signals in nematodes exposed to nanoplastic at predicted environmental concentrations. Ecotoxicol Environ Saf 243:114022
54. Xu R-R, Hua X, Rui Q, Wang D-Y (2022) Alteration in Wnt signaling mediates induction of transgenerational toxicity of polystyrene nanoplastics in *C. elegans*. NanoImpact 28:100425
55. Zhao Y, Hua X, Bian Q, Wang D-Y (2022) Nanoplastic exposure at predicted environmental concentrations induces activation of germline Ephrin signal associated with toxicity formation in the *Caenorhabditis elegans* offspring. Toxics 10:699
56. Qu M, Wang D-Y (2020) Toxicity comparison between pristine and sulfonate modified nanopolystyrene particles in affecting locomotion behavior, sensory perception, and neuronal development in *Caenorhabditis elegans*. Sci Total Environ 703:134817
57. He W-M, Gu A-H, Wang D-Y (2023) Sulfonate-modified polystyrene nanoparticle at precited environmental concentrations induces transgenerational toxicity associated with increase in germline notch signal of *Caenorhabditis elegans*. Toxics 11:511
58. Song M-X, Ruan Q-L, Wang D-Y (2024) Comparison of transgenerational toxicity on locomotion behavior and development of D-type motor neurons between pristine and amino modified nanoplastics at environmentally relevant doses in *C. elegans*. Toxics 12:555
59. Hua X, Wang D-Y (2024) Transgenerational response of germline histone acetyltransferases and deacetylases to nanoplastics at predicted environmental doses in *Caenorhabditis elegans*. Sci Total Environ 952:175903
60. Wang S-T, Liu H-L, Zhao Y-Y, Rui Q, Wang D-Y (2020) Dysregulated *mir-354* enhanced the protective response to nanopolystyrene by affecting the activity of TGF-β signaling pathway in nematode *Caenorhabditis elegans*. NanoImpact 20:100256
61. Liu H-L, Zhang R-J, Wang D-Y (2020) Response of DBL-1/TGF-β signaling-mediated neuron-intestine communication to nanopolystyrene in nematode *Caenorhabditis elegans*. Sci Total Environ 745:1141047
62. Qu M, Li D, Zhao Y-L, Yuan Y-J, Wang D-Y (2020) Exposure to low-dose nanopolystyrene induces the response of neuronal JNK MAPK signaling pathway in nematode *Caenorhabditis elegans*. Environ Sci Euro 32:58
63. Qu M, Li D, Qiu Y-X, Wang D-Y (2020) Neuronal ERK MAPK signaling in response to low-dose nanopolystyrene exposure by suppressing insulin peptide expression in *Caenorhabditis elegans*. Sci Total Environ 724:138378
64. Liu H-L, Zhao Y-Y, Bi Q, Rui Q, Wang D-Y (2021) Dysregulated *mir-76* mediated a protective response to nanopolystyrene by modulating heme homeostasis related molecular signaling in nematode *Caenorhabditis elegans*. Ecotoxicol Environ Saf 212:112018
65. Lambie EJ, Kimble J (1991) Two homologous regulatory genes, *lin-12* and *glp-1*, have overlapping functions. Development 112:231–240
66. Takács-Vellai K, Vellai T, Chen EB, Zhang Y, Guerry F, Stern MJ, Müller F (2007) Transcriptional control of notch signaling by a HOX and a PBX/EXD protein during vulval development in *C. elegans*. Dev Biol 302:661–669
67. Kershner AM, Shin H, Hansen TJ, Kimble J (2014) Discovery of two GLP-1/notch target genes that account for the role of GLP-1/notch signaling in stem cell maintenance. Proc Natl Acad Sci USA 111:3739–3744
68. Moskowitz IP, Rothman JH (1996) *lin-12* and *glp-1* are required zygotically for early embryonic cellular interactions and are regulated by maternal GLP-1 signaling in *Caenorhabditis elegans*. Development 122:4105–4117
69. Chi C, Ronai D, Than MT, Walker CJ, Sewell AK, Han M (2016) Nucleotide levels regulate germline proliferation through modulating GLP-1/notch signaling in *C. elegans*. Genes Dev 30:307–320

Chapter 10
Regulation of Transgenerational Toxicity by Transgenerational Communication of Germline Ephrin Signals

Abstract In *Caenorhabditis elegans*, exposure to pollutants induced transgenerational activation of germline Ephrin ligand EFN-3 and Ephrin receptor VAB-1. Activation of germline EFN-3 mediated transgenerational pollutant toxicity by causing VAB-1 activation in the offspring. During control of transgenerational pollutant toxicity, KSR-1/2, CED-1, SET-6, nuclear hormone receptors, and RAB-5 were identified as upregulators for germline EFN-3. VAB-1 functions in germline and neurons to regulate transgenerational pollutant toxicity. During control of transgenerational pollutants toxicity, NDK-1 and LIN-23 were identified as targets of germline VAB-1, and DBL-1, JNK-1, MPK-1, and GLB-10 were identified as targets of neuronal VAB-1.

Keywords Transgenerational toxicology · Ephrin ligand · Ephrin receptor · *Caenorhabditis elegans*

10.1 Introduction

Ephrin signaling pathway is one of the important signaling pathways for the control of diverse processes of development. In *Caenorhabditis elegans*, the Ephrin ligands, such as EFN-3, were required for controlling developmental processes of embryonic morphogenesis, epidermal morphogenesis, and neuronal circuit assembly (such as the assembly of sensory neuronal circuit) [1–3]. In *C. elegans*, the corresponding receptor of Ephrin ligands is the Eph receptor tyrosine kinase VAB-1. The VAB-1 has been shown to be involved in controlling the developmental processes of embryonic morphogenesis, epithelial morphogenesis, oocyte meiotic maturation, neuronal regeneration, germline cell death, meiotic diapause, axon pathfinding, cell migration, cell organization, and cell movement [4–14]. Recently, it was further observed that the Ephrin receptor VAB-1 participated in control of pathogen avoidance [15] and exhibited the response to pollutants exposure [16], suggesting its role in the control of stress response.

D. Wang, *Transgenerational Toxicology in Caenorhabditis elegans*,
https://doi.org/10.1007/978-981-92-1389-4_10

C. elegans is an important animal model for evaluating toxicity at different aspects after exposure to stresses and pollutants [17, 18]. Based on the confirmed high sensitivity to environmental exposures [19], *C. elegans* is suitable for the study on molecular toxicology and target organ toxicology of stresses and pollutants at environmentally relevant doses or concentrations [20, 21]. In this chapter, we first introduced role of Ephrin ligand and its receptor in regulating transgenerational pollutant toxicity. The potential transgenerational communication between germline Ephrin ligand and its receptor and its role in modulating transgenerational pollutants toxicity were discussed. Moreover, tissue-specific activity of Ephrin receptor and downstream targets of Ephrin receptor in different tissues were introduced during the control of transgenerational pollutants toxicity.

10.2 Role of Ephrin Ligand

10.2.1 Transgenerational Response of Germline EFN-3 to Pollutants

After pollutant exposure, Ephrin ligand EFN-3 shows the transgenerational activation. Nanoplastics are a member of emergent contaminants and receive increasing attentions and wide concern for their environmental exposure risk to human health [22–26]. Nanoplastic exposure caused multiple toxicities on *C. elegans*, including intestinal oxidative stress, reproductive inhibition, neuronal damage, and metabolic dysfunction [27–39]. The transgenerational toxicity on locomotion and reproduction could be further caused by exposure to nanoplastics, such as polystyrene nanoparticle (PS-NP) [40–43]. The PS-NP was thus used as the example of pollutants. Ephrin ligand EFN-3 could be activated by 0.1–10 μg/L PS-NP (Fig. 10.1) [44]. After exposure to 1 μg/L pristine PS-NP, activated EFN-3 could be observed till to the F2-G (Fig. 10.1) [44]. After exposure to 10 μg/L pristine PS-NP, activated EFN-3 could be observed till to the F3-G [45]. Moreover, after exposure to 10 μg/L amino-modified PS-NP, activated EFN-3 could be detected till to the F4-G [45].

10.2.2 Function of Germline EFN-3 in Regulating Transgenerational Toxicity of Pollutants

Based on functional analysis, germline RNAi of *efn-3* conferred a resistance to the transgenerational pollutants toxicity. Transgenerational neurotoxicity and transgenerational reproductive toxicity of PS-NPs could be inhibited by germline RNAi of *efn-3* (Fig. 10.1) [44]. That is, transgenerational activation of Ephrin ligand EFN-3 mediates induction of transgenerational pollutant toxicity.

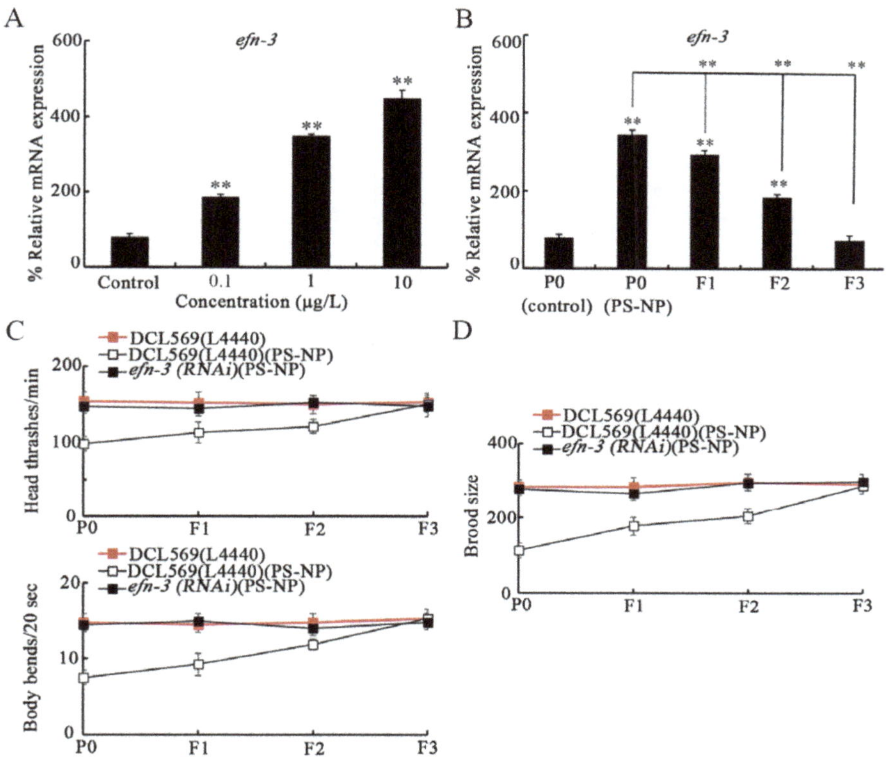

Fig. 10.1 Requirement of germline Ephrin ligand EFN-3 in controlling transgenerational PS-NP toxicity [44]. (**a**) Effect of PS-NP exposure on germline *efn-3* expression. (**b**) Transgenerational expressions of germline *efn-3* after PS-NP (1 μg/L) exposure. (**c**, **d**) Effect of germline RNAi of *efn-3* on transgenerational toxicity of PS-NP (1 μg/L) in decreasing locomotion and in inhibiting brood size. Curves of DCL569(L4440)(PS-NP) showed a significant difference ($P < 0.01$) compared to DCL569(L4440). After the PS-NP exposure, the curves of *efn-3(RNAi)* showed a significant difference ($P < 0.01$) compared to DCL569(L4440). A total of 30 intact gonads were used for the qRT-PCR assay. $^{**}P < 0.01$ vs. control or P0(control) (if not specifically indicated)

10.3 Upregulators for Germline EFN-3 in Regulating Transgenerational Toxicity of Pollutants

10.3.1 KSR-1 and KSR-2

Kinase suppressors of Ras, KSR-1 and KSR-2, could function in the germline to regulate transgenerational pollutants toxicity, and their function was inhibited by upstream NDK-1 [46]. After PS-NP exposure, EFN-3 activation was caused by RNAi of *ksr-1* and *ksr-2* [44]. Germline RNAi of *ksr-1* and *ksr-2* caused the susceptibility to transgenerational neurotoxicity and transgenerational reproductive

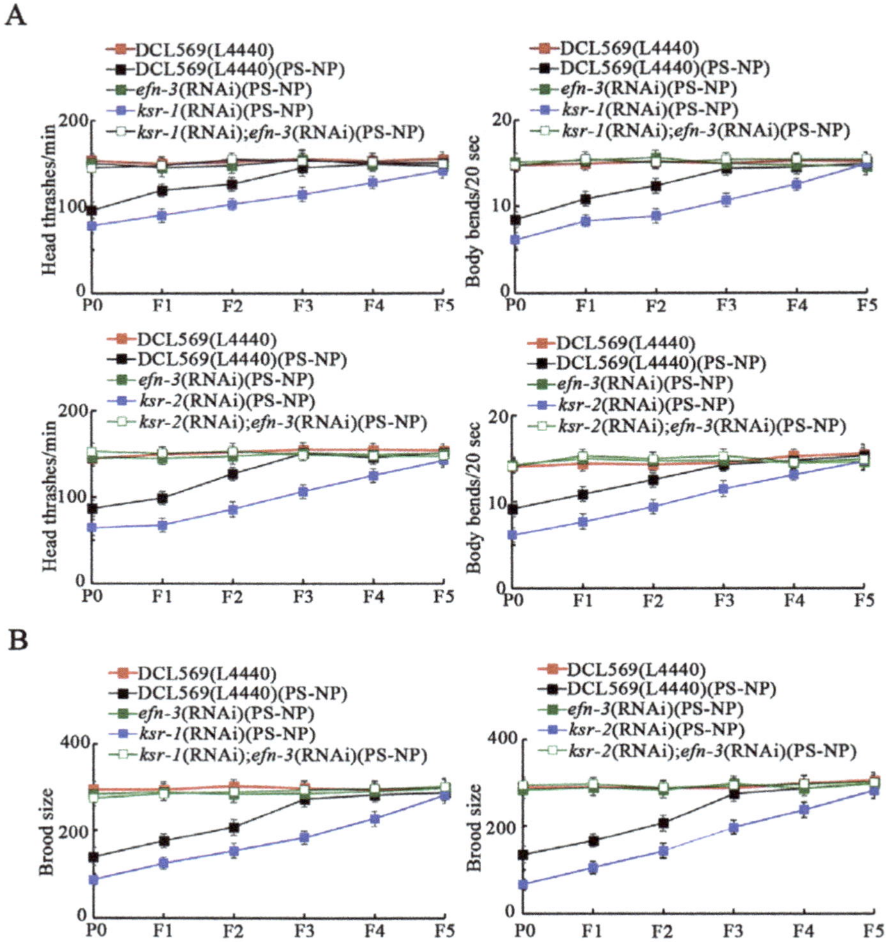

Fig. 10.2 Genetic interaction between EFN-3 and KSR-1/2 in controlling transgenerational PS-NP toxicity in decreasing locomotion (**a**) and in reducing brood size (**b**) [44]. Curves of DCL569(L4440)(PS-NP) showed a significant difference ($P < 0.01$) compared to DCL569(L4440). After the PS-NP exposure, the curves of *efn-3(RNAi)*, *ksr-1(RNAi)*, and *ksr-2(RNAi)* showed a significant difference ($P < 0.01$) compared to DCL569(L4440). After the PS-NP exposure, the curves of *ksr-1(RNAi)efn-3(RNAi)* showed a significant difference ($P < 0.01$) compared to *ksr-1(RNAi)*, and the curves of *ksr-2(RNAi);efn-3(RNAi)* showed a significant difference ($P < 0.01$) compared to *ksr-2(RNAi)*

toxicity of PS-NPs (Fig. 10.2) [44]. More importantly, the susceptibility to transgenerational neurotoxicity and transgenerational reproductive toxicity of PS-NPs induced by germline RNAi of *ksr-1* and *ksr-2* could be inhibited by germline *efn-3*, and the *ksr-1(RNAi);efn-3(RNAi)* and *ksr-2(RNAi);efn-3(RNAi)* exhibited the resistance to transgenerational neurotoxicity and transgenerational reproductive toxicity of PS-NPs (Fig. 10.2) [44]. These evidence support the role of KSR-1 and KSR-2 as upregulators for EFN-3 in the germline to regulate the transgenerational pollutants toxicity.

10.3.2 CED-1 and SET-6

CED-1 is a germline G protein-coupled receptor (GPCR). After PS-NPs exposure, transgenerational inhibition in CED-1 expression was induced, and germline RNAi of *ced-1* caused susceptibility to transgenerational neurotoxicity and transgenerational reproductive toxicity of PS-NPs [47]. After PS-NP exposure, germline RNAi of *ced-1* could activate the EFN-3 and SET-6 (a histone methylation transferase) [47]. Moreover, after PS-NP exposure, the *efn-3* expression was decreased by germline RNAi of *set-6* [47]. Therefore, a signaling cascade CED-1-SET-6-EFN-3 was formed to regulate the transgenerational pollutants toxicity. During this process, transgenerational inhibition in CED-1 mediates induction of transgenerational pollutant toxicity by resulting in transgenerational activation of SET-6, which further causes the following transgenerational activation of EFN-3.

10.3.3 Nuclear Hormone Receptors

NHR-12, DAF-12, NHR-14, and NHR-47 are nuclear hormone receptors. Exposure to PS-NPs could cause transgenerational activation of DAF-12, NHR-14, and NHR-47 and transgenerational inhibition in NHR-12 [48]. The transgenerational neurotoxicity and transgenerational reproductive toxicity of PS-NPs were strengthened by germline RNAi of *nhr-12* and inhibited by germline RNAi of *daf-12*, *nhr-14*, and *nhr-47* [48]. Moreover, after PS-NP exposure, the germline *efn-3* expression could be activated by germline RNAi of *nhr-12* and inhibited by germline RNAi of *daf-12*, *nhr-14*, and *nhr-47* [48]. Therefore, certain nuclear hormone receptors can act as upregulators of EFN-3 in the germline to accelerate or inhibit its function in regulating transgenerational pollutant toxicity.

10.3.4 RAB-5

RAB-5 is endosomal Rab-type GTPase. Exposure to PS-NPs induced the transgenerational inhibition in RAB-5 [49]. RAB-5 negatively regulated transgenerational pollutant toxicity by acting as the direct target of *mir-240* [49]. Moreover, after PS-NP exposure, EFN-3 activation was caused by germline RNAi of *rab-5* [49]. These observations suggested the role of RAB-5 as an upregulator for germline EFN-3 to regulate the transgenerational pollutant toxicity.

10.4 Role of Ephrin Receptor

10.4.1 Transgenerational Response of VAB-1 to Pollutants

After pollutant exposure, transgenerational activation of VAB-1 can be induced. With PS-NP as the example of pollutants, VAB-1 activation was induced by 0.1–10 μg/L PS-NPs (Fig. 10.3) [44]. After exposure to 1 μg/L PS-NPs, the VAB-1 activation could be detected till to the F2-G (Fig. 10.3) [44].

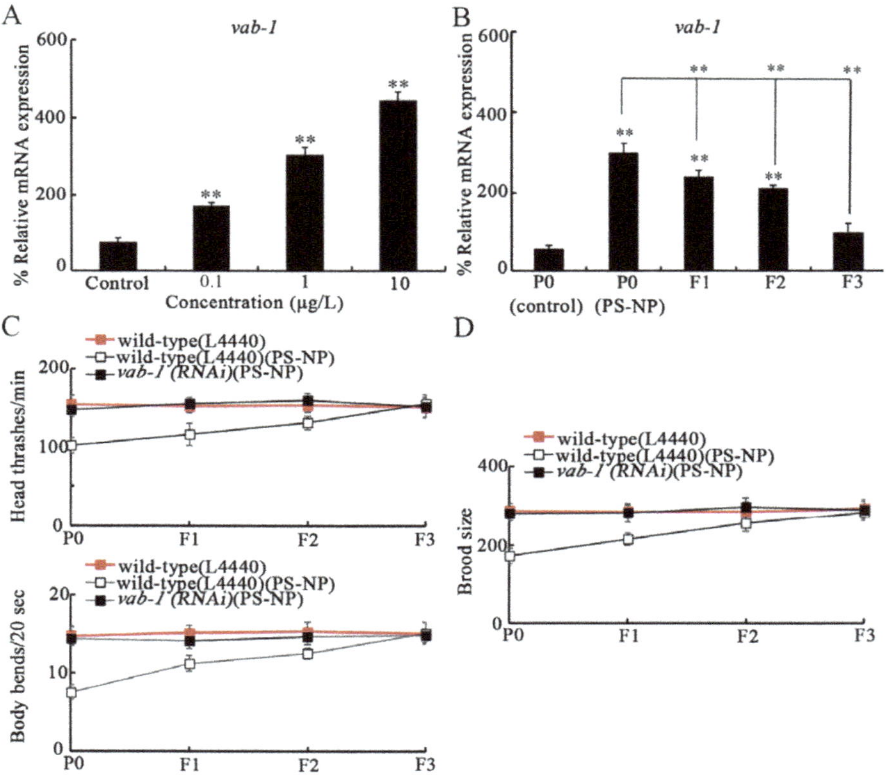

Fig. 10.3 Involvement of VAB-1 in controlling transgenerational PS-NP toxicity [44]. (**a**) Effect of PS-NP exposure on *vab-1* expression. (**b**) Transgenerational expressions of *vab-1* after PS-NP (1 μg/L) exposure. (**c**, **d**) Effect of RNAi of *vab-1* on transgenerational PS-NP toxicity in decreasing locomotion and in inhibiting brood size. Curves of wild-type (L4440)(PS-NP) showed a significant difference ($P < 0.01$) compared to wild-type (L4440). After the PS-NP exposure, the curves of *vab-1(RNAi)* showed a significant difference ($P < 0.01$) compared to wild-type (L4440). $^{**}P < 0.01$ vs control or P0(control) (if not specifically indicated)

10.4.2 Function of VAB-1 in Regulating Transgenerational Toxicity of Pollutants

Based on RNAi analysis, RNAi of *vab-1* could lead to the resistance to transgenerational pollutant toxicity. The transgenerational neurotoxicity and transgenerational reproductive toxicity of PS-NPs were suppressed by RNAi of *vab-1* (Fig. 10.3) [44]. This suggested that transgenerational activation of VAB-1 functions together with transgenerational activation of EFN-3 to mediate the induction of transgenerational pollutant toxicity.

10.5 Transgenerational Communication Between Ephrin Ligand and Its Receptor in Regulating Transgenerational Toxicity of Pollutants

Some evidence has been raised to indicate the potential transgenerational communication between Ephrin ligand EFN-3 and its receptor VAB-1 in regulating transgenerational pollutants toxicity. In *C. elegans*, germline EFN-3 overexpression could induce the susceptibility to transgenerational neurotoxicity and transgenerational reproductive toxicity of PS-NPs (Fig. 10.4) [44]. Firstly, after the germline EFN-3 overexpression in PS-NP-exposed nematodes (P0-G), VAB-1 activation could be detected at the F1-G (Fig. 10.4) [44]. Secondly, the germline EFN-3 overexpression caused susceptibility to transgenerational neurotoxicity and transgenerational reproductive toxicity of PS-NPs could be inhibited by RNAi of *vab-1* at the F1-G, and the nematodes showed the resistance to transgenerational PS-NP toxicity from the F1-G (Fig. 10.4) [44]. These observations suggested that activation of EFN-3 mediates transgenerational pollutants toxicity by causing the following transgenerational activation of VAB-1.

10.6 Tissue-Specific Activity of VAB-1 in Regulating Transgenerational Toxicity of Pollutants

The Ephrin receptor VAB-1 can function in both germline and neurons to regulate the transgenerational pollutant toxicity. The transgenerational neurotoxicity and transgenerational reproductive toxicity of PS-NPs could be inhibited by germline RNAi of *vab-1* (Fig. 10.5) [44]. Similarly, the transgenerational neurotoxicity and transgenerational reproductive toxicity of PS-NPs were also suppressed by neuronal RNAi of *vab-1* (Fig. 10.5) [44].

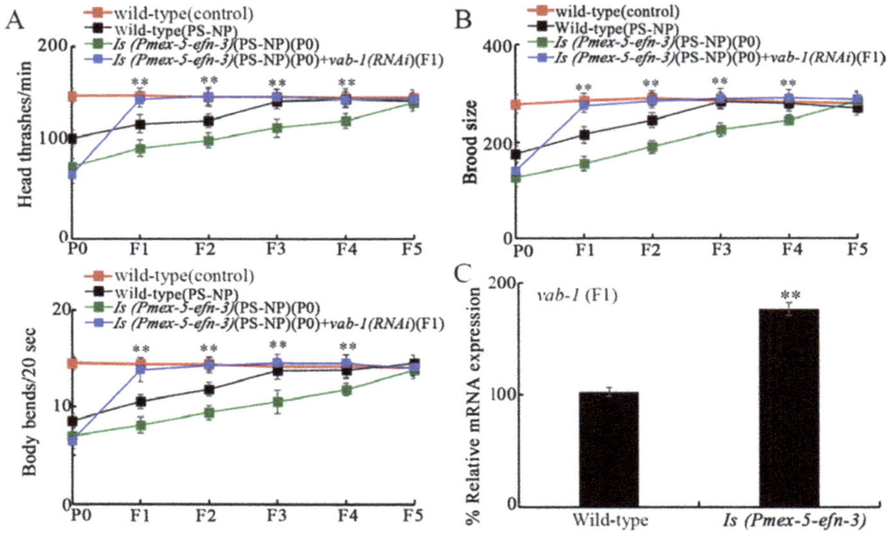

Fig. 10.4 Ephrin ligand EFN-3 controlled transgenerational PS-NP toxicity by affecting function of its receptor VAB-1 transgenerationally [44]. (**a**, **b**) Effect of *vab-1* RNAi at F1-G on locomotion and brood size in 1 μg/L PS-NP exposed *Is(Pmex-5-efn-3)* nematodes. Curves of wild-type (PS-NP) showed a significant difference ($P < 0.01$) compared to wild-type (control). Curves of *Is(Pmex-5-efn-3)*(PS-NP)(P0-G) showed a significant difference ($P < 0.01$) compared to wild-type (PS-NP). $^{**}P < 0.01$ vs. *Is(Pmex-5-efn-3)*(PSNP)(P0). (**c**) Effect of germline EFN-3 overexpression at P0-G on *vab-1* expression at F1-G after PS-NP (1 μg/L) exposure. $^{**}P < 0.01$ vs. wild type

10.7 Downstream Targets of Germline and Neuronal VAB-1 in Regulating Transgenerational Toxicity of Pollutants

In the germline, NDK-1 and LIN-23 were identified as downstream targets of VAB-1 to regulate the transgenerational pollutant toxicity. NHL-2, NDK-1, PAT-12, LIN-23, and WRT-3 could function in the germline to regulate the pollutant toxicity [50]. After PS-NP exposure, inhibition in NDK-1 and LIN-23 was caused by germline RNAi of *vab-1* (Fig. 10.6) [44]. Therefore, these observations implied that activation of germline VAB-1 mediates the transgenerational pollutant toxicity by causing transgenerational activation of NDK-1 and LIN-23. During the control of PS-NP toxicity, NDK-1 and LIN-23 also acted as the downstream targets of germline *mir-38* [50].

In the neurons, TGF-β ligand DBL-1, JNK-1/JNK MAPK, MPK-1/ERK MAPK, and GLB-10 were identified as downstream targets of VAB-1 to regulate the transgenerational pollutant toxicity. DAF-7, DBL-1, JNK-1, MPK-1, and GLB-10 could function in the neurons to regulate pollutant toxicity [51–55]. RNAi of these genes induced the susceptibility to PS-NP toxicity [51–55]. Moreover, after PS-NP exposure, activation of DBL-1, JNK-1, MPK-1, and GLB-10 was caused by neuronal

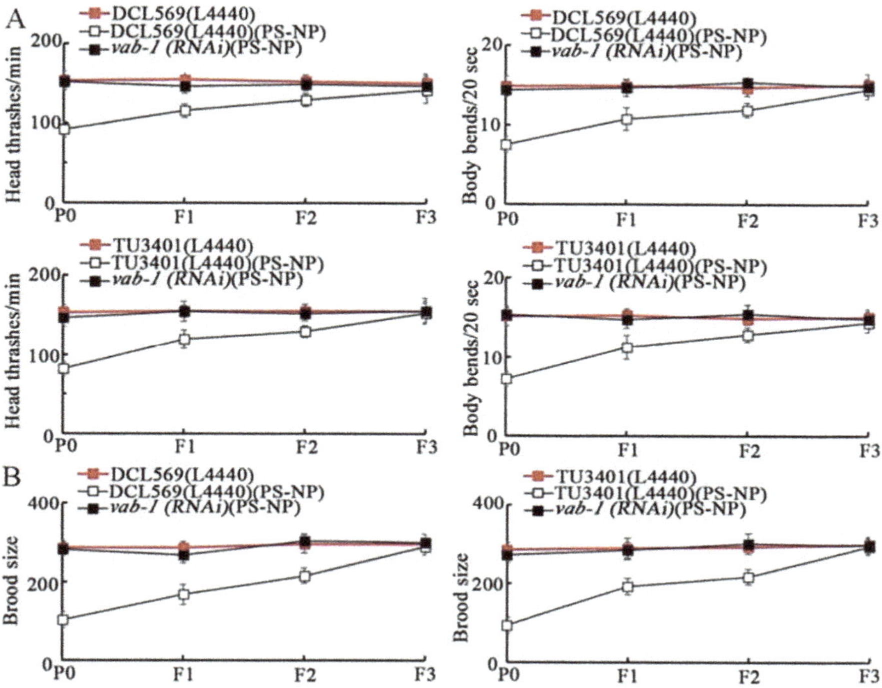

Fig. 10.5 Tissue-specific activity of VAB-1 in controlling transgenerational toxicity of PS-NP (1 μg/L) in decreasing locomotion (**a**) and in reducing brood size (**b**) [44]. Curves of DCL569(L4440)(PS-NP) showed a significant difference ($P < 0.01$) compared to DCL569(L4440). Curves of TU3401(L4440)(PS-NP) showed a significant difference ($P < 0.01$) compared to TU341(L4440). After PS-NP exposure, the curves of *vab-1(RNAi)* showed a significant difference ($P < 0.01$) compared to DCL569(L4440) or TU3401(L4440)

RNAi of *vab-1* (Fig. 10.6) [44]. Therefore, activation of neuronal VAB-1 can mediate the transgenerational pollutant toxicity by causing transgenerational inhibition in DBL-1, JNK-1, MPK-1, and GLB-10.

10.8 Perspectives

In Ephrin signaling pathway, transgenerational activation of germline Ephrin ligand EFN-3 mediated transgenerational pollutants toxicity. Nevertheless, it remains unclear whether other members of Ephrin ligands can participate in control of transgenerational pollutant toxicity. In *C. elegans*, KSR-1/2, CED-1, SET-6, nuclear hormone receptors, and RAB-5 were identified as upregulators for germline EFN-3 in regulating transgenerational pollutant toxicity. The role of SET-6 as the upregulator of germline EFN-3 suggests the role of germline EFN-3 in regulating

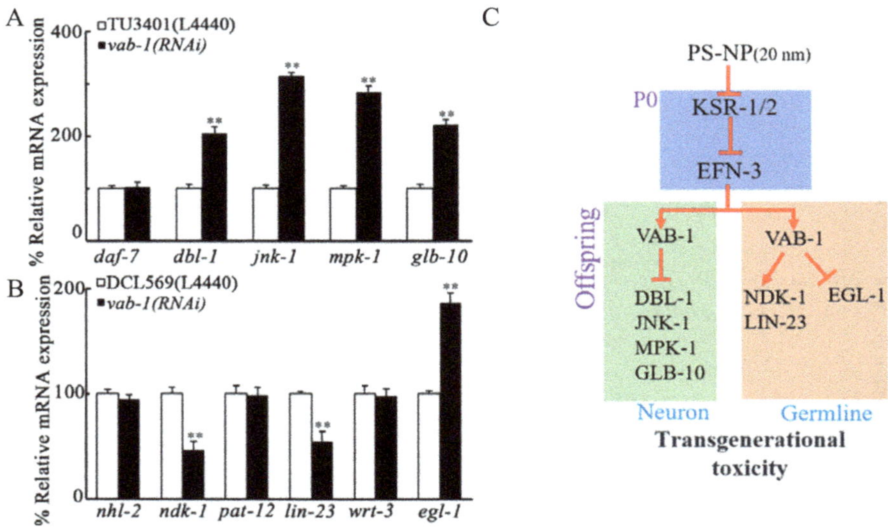

Fig. 10.6 Identification of downstream targets of neuronal and germline VAB-1 in controlling PS-NP toxicity [44]. (**a**) Effect of neuronal *vab-1* RNAi on gene expression in 1 μg/L PS-NP-exposed nematodes. (**b**) Effect of germline *vab-1* RNAi on gene expression in 1 μg/L PS-NP-exposed nematodes. $^{**}P < 0.01$ vs. TU3401(L4440) or DCL569(L4440). (**c**) A diagram showing the molecular basis for germline Ephrin signal in controlling transgenerational PS-NP toxicity in nematodes

transgenerational pollutant toxicity is under the control of certain forms of epigenetic regulation. In addition, RAB-5 was the direct target of *mir-240*, which reflects another form of upstream epigenetic regulation.

Moreover, transgenerational activation of Ephrin receptor VAB-1 mediated transgenerational pollutant toxicity, and the activation of VAB-1 in the offspring could be induced by activated germline EFN-3 in pollutant-exposed nematodes (P0-G). During the developmental control, PTP-3 and VAB-1 had partly redundant functions [56]. It needs to be determined whether some proteins also have partly redundant functions with VAB-1 in regulating transgenerational pollutant toxicity.

VAB-1 could function in germline and neurons to regulate the transgenerational pollutant toxicity. During this process, in the neurons, DBL-1, JNK-1, MPK-1, and GLB-10 were identified as downstream targets of VAB-1, and in the germline, NDK-1 and LIN-23 were identified as downstream targets of VAB-1. During the control of pollutant toxicity, NDK-1 could further acted upstream of KSR-1 and KSR-2. Thus, a feedback regulatory loop was formed between signaling cascade NDK-1-KSR-1/2-EFN-3 and signaling cascade VAB-1-NDK-1 in the germline to regulate the transgenerational pollutant toxicity. During the developmental control, some other proteins were also identified as downstream targets for VAB-1 [57, 58], and their possible roles in regulating transgenerational pollutant toxicity and relationship with VAB-1 during this process need to be further examined.

References

1. Chin-Sang ID, Moseley SL, Ding M, Harrington RJ, George SE, Chisholm AD (2002) The divergent *C. elegans* ephrin EFN-4 functions in embryonic morphogenesis in a pathway independent of the VAB-1 Eph receptor. Development 129:5499–5510
2. Chin-Sang ID, George SE, Ding M, Moseley SL, Lynch AS, Chisholm AD (1999) The ephrin VAB-2/EFN-1 functions in neuronal signaling to regulate epidermal morphogenesis in *C. elegans*. Cell 99:781–790
3. Kalichamy K, Arnold M, Hill T, Johnson A, McCardel K, Taylor WK, Upton B, Hudson ML (2025) The *vab-1*/Eph receptor and ephrins *efn-1* and *efn-4* are required for *C. elegans* chemosensory and thermosensory circuit assembly and function. BMC Neurosci 26:65
4. Wang X, Roy PJ, Holland SJ, Zhang LW, Culotti JG, Pawson T (1999) Multiple ephrins control cell organization in *C. elegans* using kinase-dependent and -independent functions of the VAB-1 Eph receptor. Mol Cell 4:903–913
5. Tucker M, Han M (2008) Muscle cell migrations of *C. elegans* are mediated by the alpha-integrin INA-1, Eph receptor VAB-1, and a novel peptidase homologue MNP-1. Dev Biol 318:215–223
6. George SE, Simokat K, Hardin J, Chisholm AD (1998) The VAB-1 Eph receptor tyrosine kinase functions in neural and epithelial morphogenesis in *C. elegans*. Cell 92:633–643
7. Bernadskaya YY, Wallace A, Nguyen J, Mohler WA, Soto MC (2012) UNC-40/DCC, SAX-3/Robo, and VAB-1/Eph polarize F-actin during embryonic morphogenesis by regulating the WAVE/SCAR actin nucleation complex. PLoS Genet 8:e1002863
8. Ikegami R, Simokat K, Zheng H, Brown L, Garriga G, Hardin J, Culotti J (2012) Semaphorin and Eph receptor signaling guide a series of cell movements for ventral enclosure in *C. elegans*. Curr Biol 22:1–11
9. Miller MA, Ruest PJ, Kosinski M, Hanks SK, Greenstein D (2003) An Eph receptor sperm-sensing control mechanism for oocyte meiotic maturation in *Caenorhabditis elegans*. Genes Dev 17:187–200
10. Wu Z, Ghosh-Roy A, Yanik MF, Zhang JZ, Jin Y, Chisholm AD (2007) *Caenorhabditis elegans* neuronal regeneration is influenced by life stage, ephrin signaling, and synaptic branching. Proc Natl Acad Sci USA 104:15132–15137
11. Li X, Johnson RW, Park D, Chin-Sang I, Chamberlin HM (2012) Somatic gonad sheath cells and Eph receptor signaling promote germ-cell death in *C. elegans*. Cell Death Differ 19:1080–1089
12. Govindan JA, Cheng H, Harris JE, Greenstein D (2006) Galphao/i and Galphas signaling function in parallel with the MSP/Eph receptor to control meiotic diapause in *C. elegans*. Curr Biol 16:1257–1268
13. Zallen JA, Kirch SA, Bargmann CI (1999) Genes required for axon pathfinding and extension in the *C. elegans* nerve ring. Development 126:3679–3692
14. Mohamed AM, Chin-Sang ID (2006) Characterization of loss-of-function and gain-of-function Eph receptor tyrosine kinase signaling in *C. elegans* axon targeting and cell migration. Dev Biol 290:164–176
15. Seto RJ, Brown R, Kaletsky R, Parsons LR, Moore RS, Balch JM, Gitai Z, Murphy CT (2025) *C. elegans* transgenerational avoidance of *P. fluorescens* is mediated by the Pfs1 sRNA and *vab-1*. Sci Adv 11:eadt3850
16. Zhang J, Lv C, Yu Z, Zhou Y (2024) Step-wise reproductive toxicities of imidazolium- and pyridinium-based ionic liquids on *Caenorhabditis elegans*. J Hazard Mater 480:136458
17. Wang D-Y (2018) Nanotoxicology in *Caenorhabditis elegans*. Springer Nature Singapore Pte Ltd., Singapore
18. Wang D-Y (2020) Exposure toxicology in *Caenorhabditis elegans*. Springer Nature Singapore Pte Ltd, Singapore
19. Wang D-Y (2022) Toxicology at environmentally relevant concentrations in *Caenorhabditis elegans*. Springer Nature Singapore Pte Ltd., Singapore

20. Wang D-Y (2019) Molecular toxicology in *Caenorhabditis elegans*. Springer Nature Singapore Pte Ltd., Singapore
21. Wang D-Y (2019) Target organ toxicology in *Caenorhabditis elegans*. Springer Nature Singapore Pte Ltd., Singapore
22. Lamoree MH, van Boxel J, Nardella F, Houthuijs KJ, Brandsma SH, Béen F, van Duursen MBM (2025) Health impacts of microplastic and nanoplastic exposure. Nat Med 31:2873–2887
23. Feng Y, Tu C, Li R, Wu D, Yang J, Xia Y, Peijnenburg WJGM, Luo Y (2023) A systematic review of the impacts of exposure to micro- and nano-plastics on human tissue accumulation and health. Eco-Environ Health 2:195–207
24. Ramsperger AFRM, Bergamaschi E, Panizzolo M, Fenoglio I, Barbero F, Peters R, Undas A, Purker S, Giese B, Lalyer CR, Tamargo A, Moreno-Arribas MV, Grossart HP, Kühnel D, Dietrich J, Paulsen F, Afanou AK, Zienolddiny-Narui S, Eriksen Hammer S, Kringlen Ervik T, Graff P, Brinchmann BC, Nordby KC, Wallin H, Nassi M, Benetti F, Zanella M, Brehm J, Kress H, Löder MGJ, Laforsch C (2023) Nano- and microplastics: a comprehensive review on their exposure routes, translocation, and fate in humans. NanoImpact 29:100441
25. Vogel A, Tentschert J, Pieters R, Bennet F, Dirven H, van den Berg A, Lenssen E, Rietdijk M, Broßell D, Haase A (2024) Towards a risk assessment framework for micro- and nanoplastic particles for human health. Part Fibre Toxicol 21:48
26. Zheng K, Wang P, Lou X, Zhou Z, Zhou L, Hu Y, Luan Y, Quan C, Fang J, Zou H, Gao X (2024) A review of airborne micro- and nano-plastics: sampling methods, analytical techniques, and exposure risks. Environ Pollut 363:125074
27. Hua X, Wang D-Y (2022) Cellular uptake, transport, and organelle response after exposure to microplastics and nanoplastics: current knowledge and perspectives for environmental and health risks. Rev Environ Contam Toxicol 260:12
28. Qu M, Kong Y, Yuan Y-J, Wang D-Y (2019) Neuronal damage induced by nanopolystyrene particles in nematode *Caenorhabditis elegans*. Environ Sci Nano 6:2591–2601
29. Qu M, Qiu Y-X, Kong Y, Wang D-Y (2019) Amino modification enhances reproductive toxicity of nanopolystyrene on gonad development and reproductive capacity in nematode *Caenorhabditis elegans*. Environ Pollut 254:112978
30. Qu M, Nida A, Kong Y, Du H-H, Xiao G-S, Wang D-Y (2019) Nanopolystyrene at predicted environmental concentration enhances microcystin-LR toxicity by inducing intestinal damage in *Caenorhabditis elegans*. Ecotoxicol Environ Saf 183:109568
31. Qu M, Xu K-N, Li Y-H, Wong G, Wang D-Y (2018) Using *acs-22* mutant *Caenorhabditis elegans* to detect the toxicity of nanopolystyrene particles. Sci Total Environ 643:119–126
32. Qiu Y-X, Liu Y-Q, Li Y-H, Li G-J, Wang D-Y (2020) Effect of chronic exposure to nanopolystyrene on nematode *Caenorhabditis elegans*. Chemosphere 256:127172
33. Qiu Y-X, Luo L-B, Yang Y-H, Kong Y, Li Y-H, Wang D-Y (2020) Potential toxicity of nanopolystyrene on lifespan and aging process of nematode *Caenorhabditis elegans*. Sci Total Environ 705:135918
34. Li D, Deng Y-J, Wang S-T, Du H-H, Xiao G-S, Wang D-Y (2020) Assessment of nanopolystyrene toxicity under fungal infection condition in *Caenorhabditis elegans*. Ecotoxicol Environ Saf 197:110625
35. Liu H-L, Shao H-M, Guo Z-J, Wang D-Y (2020) Nanopolystyrene exposure activates a fat metabolism related signaling-mediated protective response in *Caenorhabditis elegans*. NanoImpact 17:100204
36. Dong S-S, Qu M, Rui Q, Wang D-Y (2018) Combinational effect of titanium dioxide nanoparticles and nanopolystyrene particles at environmentally relevant concentrations on nematodes *Caenorhabditis elegans*. Ecotoxicol Environ Saf 161:444–450
37. Zhao Y-Y, Li D, Rui Q, Wang D-Y (2020) Toxicity induction of nanopolystyrene under microgravity stress condition in *Caenorhabditis elegans*. Sci Total Environ 703:135623
38. Liu H-L, Wang D-Y (2021) Intestinal mitochondrial unfolded protein response induced by nanoplastic particles in *Caenorhabditis elegans*. Chemosphere 267:128917
39. Tang M, Ding G, Lu X, Huang Q, Du H, Xiao G, Wang D (2022) Exposure to nanoplastic particles enhances *Acinetobacter* survival, biofilm formation, and serum resistance. Nano 12:4222

40. Zhao L, Qu M, Wong G, Wang D-Y (2017) Transgenerational toxicity of nanopolystyrene particles in the range of μg/L in nematode *Caenorhabditis elegans*. Environ Sci Nano 4:2356–2366
41. Sun L-M, Liao K, Wang D-Y (2021) Comparison of transgenerational reproductive toxicity induced by pristine and amino modified nanoplastics in *Caenorhabditis elegans*. Sci Total Environ 768:144362
42. Liu H-L, Tian L-J, Wang S-T, Wang D-Y (2021) Size-dependent transgenerational toxicity induced by nanoplastics in nematode *Caenorhabditis elegans*. Sci Total Environ 790:148217
43. Zhao Y, Chen J, Wang R, Pu X, Wang D (2023) A review of transgenerational and multigenerational toxicology in the *in vivo* model animal *Caenorhabditis elegans*. J Appl Toxicol 43:122–145
44. Zhao Y, Hua X, Bian Q, Wang D-Y (2022) Nanoplastic exposure at predicted environmental concentrations induces activation of germline Ephrin signal associated with toxicity formation in the *Caenorhabditis elegans* offspring. Toxics 10:699
45. Song M-X, Ruan Q-L, Wang D-Y (2024) Comparison of transgenerational neurotoxicity between pristine and amino-modified nanoplastics *in C. elegans*. Toxics 12:555
46. Hua X, Zhao Y, Yuan Y-J, Zhang L, Bian Q, Wang D-Y (2022) Nanoplastics cause transgenerational toxicity through inhibiting germline microRNA *mir-38* in *C. elegans*. J Hazard Mater 437:129302
47. Wu J-W, Shao Y-T, Hua X, Wang Y-X, Wang D-Y (2024) Nanoplastic at environmentally relevant concentrations induces toxicity across multiple generations associated with inhibition in germline G protein-coupled receptor CED-1 in *Caenorhabditis elegans*. Chemosphere 364:143011
48. Liu Z-Y, Wang Y-X, Bian Q, Wang D-Y (2024) Transgenerational response of germline nuclear hormone receptor genes to nanoplastics at predicted environmental doses in *Caenorhabditis elegans*. Toxics 12:420
49. Hua X, Zang L, Wang D-Y (2024) Nanoplastic at environmentally relevant concentrations activates germline *mir-240-rab-5* signaling cascade to affect secreted ligands associated with transgenerational toxicity induction in *C. elegans*. Environ Sci Nano 11:3615
50. Yang Y-H, Wu Q-L, Wang D-Y (2020) Epigenetic response to nanopolystyrene in germline of nematode *Caenorhabditis elegans*. Ecotoxicol Environ Saf 206:111404
51. Wang S-T, Liu H-L, Zhao Y-Y, Rui Q, Wang D-Y (2020) Dysregulated *mir-354* enhanced the protective response to nanopolystyrene by affecting the activity of TGF-β signaling pathway in nematode *Caenorhabditis elegans*. NanoImpact 20:100256
52. Liu H-L, Zhang R-J, Wang D-Y (2020) Response of DBL-1/TGF-β signaling-mediated neuron-intestine communication to nanopolystyrene in nematode *Caenorhabditis elegans*. Sci Total Environ 745:1141047
53. Qu M, Li D, Zhao Y-L, Yuan Y-J, Wang D-Y (2020) Exposure to low-dose nanopolystyrene induces the response of neuronal JNK MAPK signaling pathway in nematode *Caenorhabditis elegans*. Environ Sci Euro 32:58
54. Qu M, Li D, Qiu Y-X, Wang D-Y (2020) Neuronal ERK MAPK signaling in response to low-dose nanopolystyrene exposure by suppressing insulin peptide expression in *Caenorhabditis elegans*. Sci Total Environ 724:138378
55. Liu H-L, Zhao Y-Y, Bi Q, Rui Q, Wang D-Y (2021) Dysregulated *mir-76* mediated a protective response to nanopolystyrene by modulating heme homeostasis related molecular signaling in nematode *Caenorhabditis elegans*. Ecotoxicol Environ Saf 212:112018
56. Harrington RJ, Gutch MJ, Hengartner MO, Tonks NK, Chisholm AD (2002) The *C. elegans* LAR-like receptor tyrosine phosphatase PTP-3 and the VAB-1 Eph receptor tyrosine kinase have partly redundant functions in morphogenesis. Development 129:2141–2153
57. Mohamed AM, Boudreau JR, Yu FP, Liu J, Chin-Sang ID (2012) The *Caenorhabditis elegans* Eph receptor activates NCK and N-WASP, and inhibits Ena/VASP to regulate growth cone dynamics during axon guidance. PLoS Genet 8:e1002513
58. Brisbin S, Liu J, Boudreau J, Peng J, Evangelista M, Chin-Sang I (2009) A role for *C. elegans* Eph RTK signaling in PTEN regulation. Dev Cell 17:459–469

Chapter 11
Regulation of Transgenerational Toxicity by Transgenerational Communication of Germline Neurotransmitter Signals

Abstract Neurotransmitters are important chemical messengers to transduce the neuronal information. Exposure to pollutants could induce transgenerational response of certain neurotransmitters. Accompanied with this, transgenerational response of genes regulating neurotransmitter synthesis, such as germline *tbh-1* governing octopamine synthesis, was induced by pollutant exposure. Inhibition in germline TBH-1 functioned together with change of octopamine receptors of SER-6 and OCTR-1 in the offspring to regulate transgenerational pollutant toxicity by causing transgenerational inhibition in SER-6 and transgenerational activation of OCTR-1. In the germline, nuclear hormone receptor DAF-12 acted as upregulator for TBH-1 to regulate transgenerational pollutant toxicity. In the intestine, ELT-2 was identified as target of SER-6 an OCTR-1 to regulate transgenerational pollutant toxicity.

Keywords Transgenerational toxicology · Neurotransmitter signal · *Caenorhabditis elegans*

11.1 Introduction

The neurotransmitters belong to the chemical messenger, which are normally considered to be transmitted across the synapses. *Caenorhabditis elegans* neurotransmitters contain glutamate, GABA, acetylcholine, dopamine, serotonin, octopamine, tyramine, and certain neuropeptides. After the synthesis, neurotransmitters will be released and bind to their corresponding receptors to exhibit their functions.

With the help of high sensitivity to environmental exposures and established system of exposure toxicology [1, 2], *C. elegans* is not only applied to detect toxicity of stresses and pollutants at low doses and concentrations [1, 3] but also provides a powerful platform for toxicological study, such as target organ toxicology [4]. For the study of molecular toxicology, neurotransmitters represent an important signaling involved in the control of toxicity of pollutants [5–8]. In this chapter,

D. Wang, *Transgenerational Toxicology in Caenorhabditis elegans*,
https://doi.org/10.1007/978-981-92-1389-4_11

we first introduced transgenerational response of neurotransmitters to pollutant exposure. After that, with the octopamine as the example of neurotransmitters, we introduce the involvement of neurotransmitter synthesis and its receptor in regulating the transgenerational pollutant toxicity. The downstream targets of neurotransmitter receptor in regulating transgenerational pollutant toxicity were further discussed.

11.2 Transgenerational Response of Neurotransmitters

In *C. elegans*, exposure to pollutants can induce transgenerational response of certain neurotransmitters. Different forms of transgenerational response of neurotransmitters could be caused by exposure to pollutants, such as di-n-octyl phthalate and nanoplastics [9, 10]. Exposure to nanoplastics caused different aspects of toxicity on *C. elegans*, including organ toxicity, lifespan reduction, and metabolic disruption [11–23]. Nanoplastic exposure could further result in transgenerational toxicity, and photo-aging and certain chemical modifications could strengthen the transgenerational toxicity of nanoplastics [24–27]. With photoaged polystyrene nanoparticle (PS-NP) as the example of pollutants, photoaged PS-NPs caused transgenerational response of dopamine from parental generation (P0-G) to F2-G, transgenerational response of serotonin from P0-G to F2-G, and transgenerational response of glutamate from P0-G to F1-G [10].

11.3 Role of Neurotransmitters

11.3.1 Transgenerational Response of Germline Genes Governing Neurotransmitter Synthesis to Pollutants

In *C. elegans*, pollutant exposure can induce transgenerational response of germline genes governing neurotransmitter synthesis. Multi-walled carbon nanotube (MWCNT) is a member of carbon-based nanomaterials, and its exposure also caused multiple aspects of toxicity on *C. elegans* [28–38]. MWCNTs was selected as the example of pollutants to determine effect of pollutant exposure on transgenerational expression of germline *tbh-1* governing octopamine synthesis [39]. TBH-1 is a tyramine beta-hydroxylase [40, 41]. The germline *tbh-1* was reduced by 0.1–10 μg/L MWCNTs, and transgenerational inhibition in germline TBH-1 was induced by MWCNT exposure (Fig. 11.1) [39]. After exposure to MWCNTs at the concentration of 0.1 μg/L, the induced transgenerational inhibition in germline TBH-1 could be observed till to the F2-G (Fig. 11.1) [39].

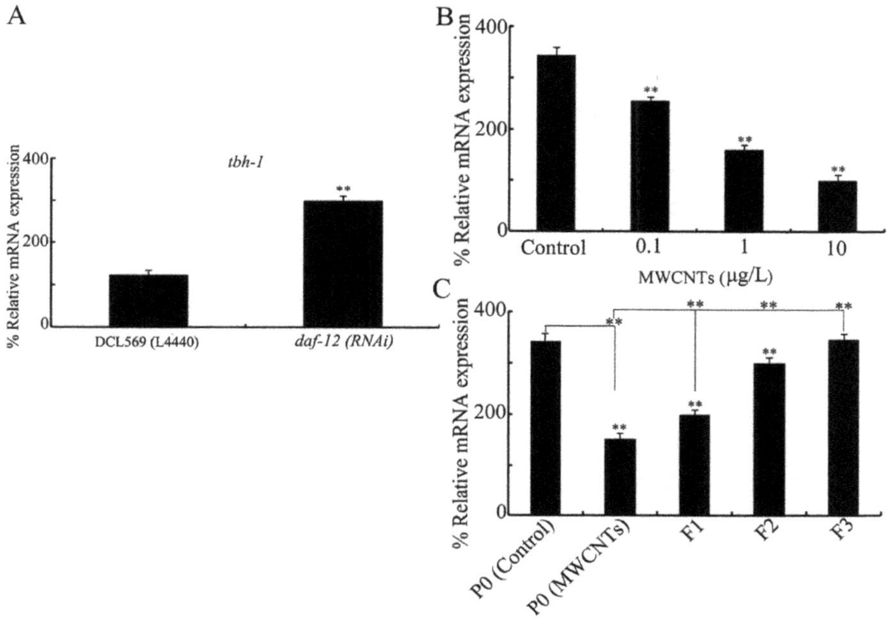

Fig. 11.1 Identification of TBH-1 as potential downstream target of germline DAF-12 to control transgenerational MWCNT toxicity [39]. (**a**) Effect of germline RNAi of *daf-12* on expression of germline *tbh-1* in MWCNT-exposed nematodes. Exposure concentration was 0.1 μg/L. (**b**) Effect of 0.1–10 μg/L MWCNT exposure on germline *tbh-1* expression. (**c**) Transgenerational expression of germline *tbh-1*. Exposure concentration was 0.1 μg/L. A total of 30 intact gonads were used for qRT-PCR. $^{**}P < 0.01$ vs DCL569 or Control (if not specially indicated)

11.3.2 Function of Germline TBH-1 in Regulating Transgenerational Toxicity of Pollutants

Based on functional analysis, germline RNAi of *tbh-1* conferred the susceptibility to transgenerational pollutant toxicity. Germline RNAi of *tbh-1* caused the susceptibility to transgenerational neurotoxicity and transgenerational reproductive toxicity of MWCNTs (Fig. 11.2) [39]. That is, transgenerational inhibition in germline TBH-1 mediates induction of transgenerational pollutant toxicity.

11.4 Upregulators for Germline TBH-1 in Regulating Transgenerational Toxicity of Pollutants

Nuclear hormone receptor DAF-12 was identified as the upregulator for germline TBH-1 to regulate transgenerational pollutant toxicity. Some experimental evidence supported this. Firstly, after MWCNT exposure, activation of germline TBH-1 was induced by germline RNAi of *daf-12* (Fig. 11.1) [39]. Secondly, genetic interaction

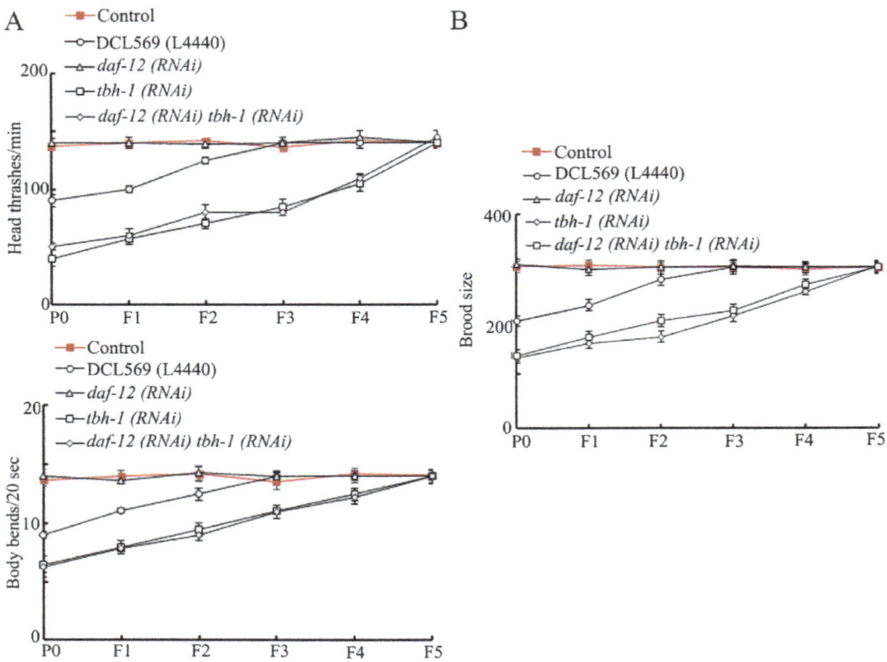

Fig. 11.2 Genetic interaction between *daf-12* and *tbh-1* in the germline to control transgenerational MWCNT toxicity in decreasing locomotion (**a**) and in inhibiting brood size (**b**) [39]. Exposure concentration was 0.1 μg/L. Curve of DCL569(L4440) showed a significant difference ($P < 0.01$) compared to control. Curves of *daf-12(RNAi)* and *tbh-1(RNAi)* showed a significant difference ($P < 0.01$) compared to DCL569(L4440). Curve of *daf-12(RNAi)tbh-1(RNAi)* showed a significant difference ($P < 0.01$) compared to *daf-12(RNAi)*

between germline DAF-12 and germline TBH-1 supported this. Germline RNAi of *daf-12* caused resistance to transgenerational MWCNT toxicity (Fig. 11.2) [39]. Moreover, resistance to transgenerational neurotoxicity and transgenerational reproductive toxicity of MWCNTs induced by germline RNAi of *daf-12* could be inhibited by germline RNAi of *tbh-1*, and the nematodes exhibited the susceptibility to transgenerational MWCNT toxicity (Fig. 11.2) [39].

11.5 Role of Neurotransmitter Receptors

11.5.1 Transgenerational Response of SER-6 and OCTR-1 to Pollutants

In *C. elegans*, octopamine receptors exhibited opposite transgenerational responses after pollutant exposure. SER-6 and OCTR-1 are octopamine receptors [42–45]. These two receptors are involved in control of stress response at the

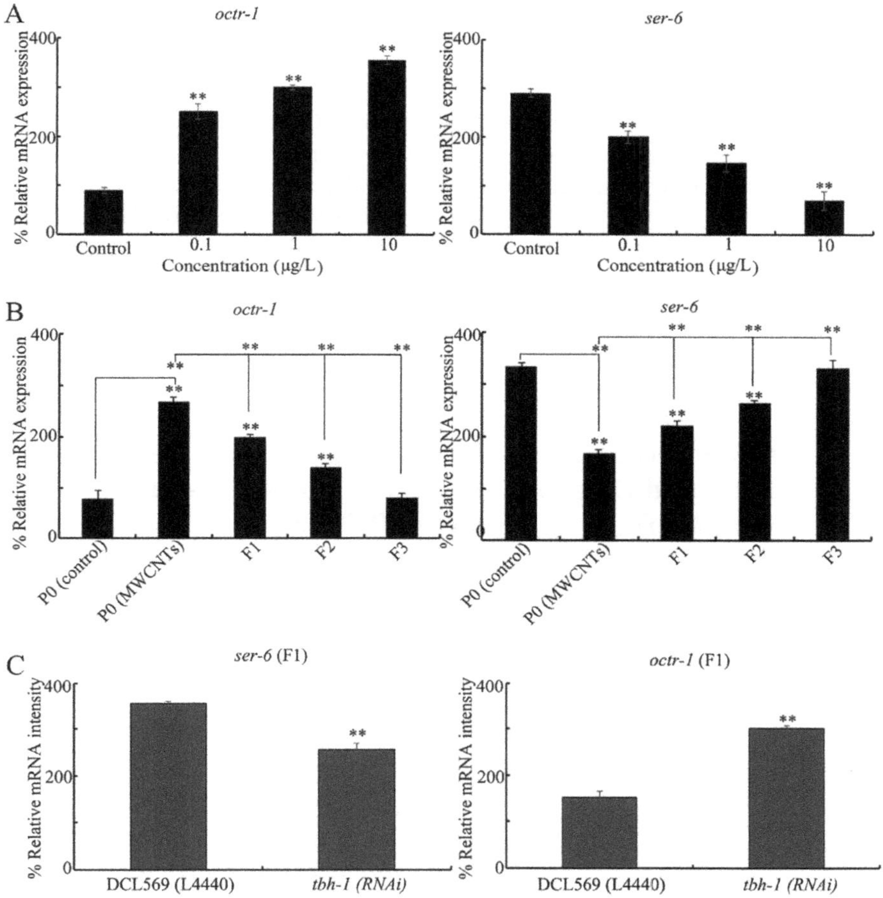

Fig. 11.3 Germline TBH-1 transgenerationally affected expressions of octopamine receptors of OCTR-1 and SER-6 [39]. (**a**) Effect of MWCNT exposure on expressions of *octr-1* and *ser-6*. (**b**) Transgenerational expressions of *octr-1* and *ser-6* after MWCNT exposure. Exposure concentration was 0.1 μg/L. (**c**) Effect of germline RNAi of *tbh-1* in MWCNT-exposed nematodes (P0) on expressions of *octr-1* and *ser-6* in F1-G. Exposure concentration was 0.1 μg/L. $^{**}P < 0.01$ vs Control or DCL569 (if not specially indicated)

P0-G, such as control of oxidative stress and pathogen infection [46–48]. With MWCNTs as examples of pollutants, activation of OCTR-1 and inhibition in SER-6 were induced by 0.1–10 μg/L MWCNTs (Fig. 11.3) [39]. After MWCNTs exposure at the concentration of 0.1 μg/L, transgenerational activation of OCTR-1 and transgenerational inhibition in SER-6 could be detected till to the F2-G (Fig. 11.3) [39].

11.5.2 Function of SER-6 and OCTR-1 in Regulating Transgenerational Toxicity of Pollutants

Functional analysis demonstrated that SER-6 and OCTR-1 played opposite functions in modulating transgenerational pollutant toxicity. RNAi of *ser-6* caused susceptibility to transgenerational neurotoxicity and transgenerational reproductive toxicity of MWCNTs (Fig. 11.4) [39]. In contrast, RNAi of *octr-1* resulted in resistance to transgenerational neurotoxicity and transgenerational reproductive toxicity of MWCNTs (Fig. 11.4) [39]. Therefore, transgenerational inhibition in SER-6 and transgenerational activation of OCTR-1 mediate the transgenerational pollutants toxicity.

11.6 Transgenerational Communication Between Germline TBH-1 and Octopamine Receptors in Regulating Transgenerational Toxicity of Pollutants

Some experimental results have indicated the potential transgenerational communication between germline TBH-1 and octopamine receptors in regulating transgenerational pollutant toxicity. Firstly, after MWCNTs exposure, germline RNAi of

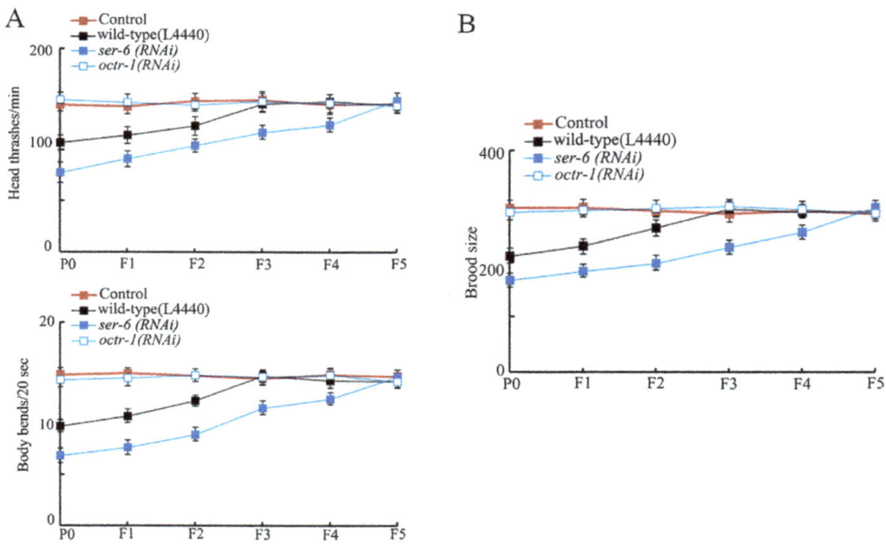

Fig. 11.4 Effect of RNAi of *ser-6* and *octr-1* on transgenerational MWCNT toxicity in decreasing locomotion (**a**) and in reducing brood size (**b**) [39]. Exposure concentration was 0.1 μg/L. Curve of wild-type (L4440) showed a significant difference ($P < 0.01$) compared to control. Curves of *ser-6(RNAi)* and *octr-1(RNAi)* showed a significant difference ($P < 0.01$) compared to wild type

tbh-1 at the P0-G caused the inhibition in SER-6 and activation of OCTR-1 at the F1-G (Fig. 11.3) [39]. In *C. elegans*, germline TBH-1 overexpression caused the opposite phenotype to that of germline RNAi of *tbh-1*, and nematodes overexpressing germline TBH-1 showed the resistance to transgenerational neurotoxicity and transgenerational reproductive toxicity of MWCNTs (Fig. 11.5) [39]. Secondly, the resistance to transgenerational neurotoxicity and transgenerational reproductive toxicity of MWCNTs induced by overexpressing germline TBH-1 could be inhibited by RNAi of *ser-6* at the F1-G (Fig. 11.5) [39]. Thirdly, the susceptibility to transgenerational neurotoxicity and transgenerational reproductive toxicity of MWCNTs induced by germline RNAi of *tbh-1* at the P0-G could be inhibited by RNAi of *octr-1* at the F1-G (Fig. 11.5) [39]. Therefore, inhibition in germline TBH-1 mediates the transgenerational pollutant toxicity by causing the following inhibition in SER-6 and activation of OCTR-1 in the offspring.

11.7 Downstream Targets of SER-6 and OCTR-1 in Regulating Transgenerational Toxicity of Pollutants

ELT-2 was identified as the downstream target of intestinal SER-6 and OCTR-1 to regulate the transgenerational pollutant toxicity. During the control of nanoplastic toxicity, SER-6 and OCTR-1 were found to play their function in intestinal cells [49]. DAF-16, PMK-1, BAR-1, and ELT-2 could function in intestinal cells to regulate the pollutant toxicity [2, 4, 50–58]. After MWCNT exposure, intestinal RNAi of *ser-6* caused inhibition in ELT-2, and intestinal RNAi of *octr-1* induced activation of ELT-2 (Fig. 11.6) [39]. In *C. elegans*, intestinal overexpression of SER-6 caused the resistance to transgenerational neurotoxicity and transgenerational reproductive toxicity of MWCNTs (Fig. 11.7) [39]. Moreover, the resistance to transgenerational neurotoxicity and transgenerational reproductive toxicity of MWCNTs induced by intestinal SER-6 overexpression could be inhibited by RNAi of *elt-2* (Fig. 11.7) [39]. Additionally, the resistance to transgenerational neurotoxicity and transgenerational reproductive toxicity of MWCNTs induced by intestinal RNAi of *octr-1* could also be inhibited by RNAi of *elt-2* (Fig. 11.7) [39]. Thus, transgenerational inhibition in SER-6 and transgenerational activation of OCTR-1 mediate transgenerational pollutant toxicity by resulting in the inhibition in their target ELT-2.

11.8 Perspectives

In *C. elegans*, after pollutant exposure, transgenerational response of certain neurotransmitters could be observed. Accompanied with this, transgenerational inhibition in TBH-1 governing octopamine synthesis was caused by pollutant exposure.

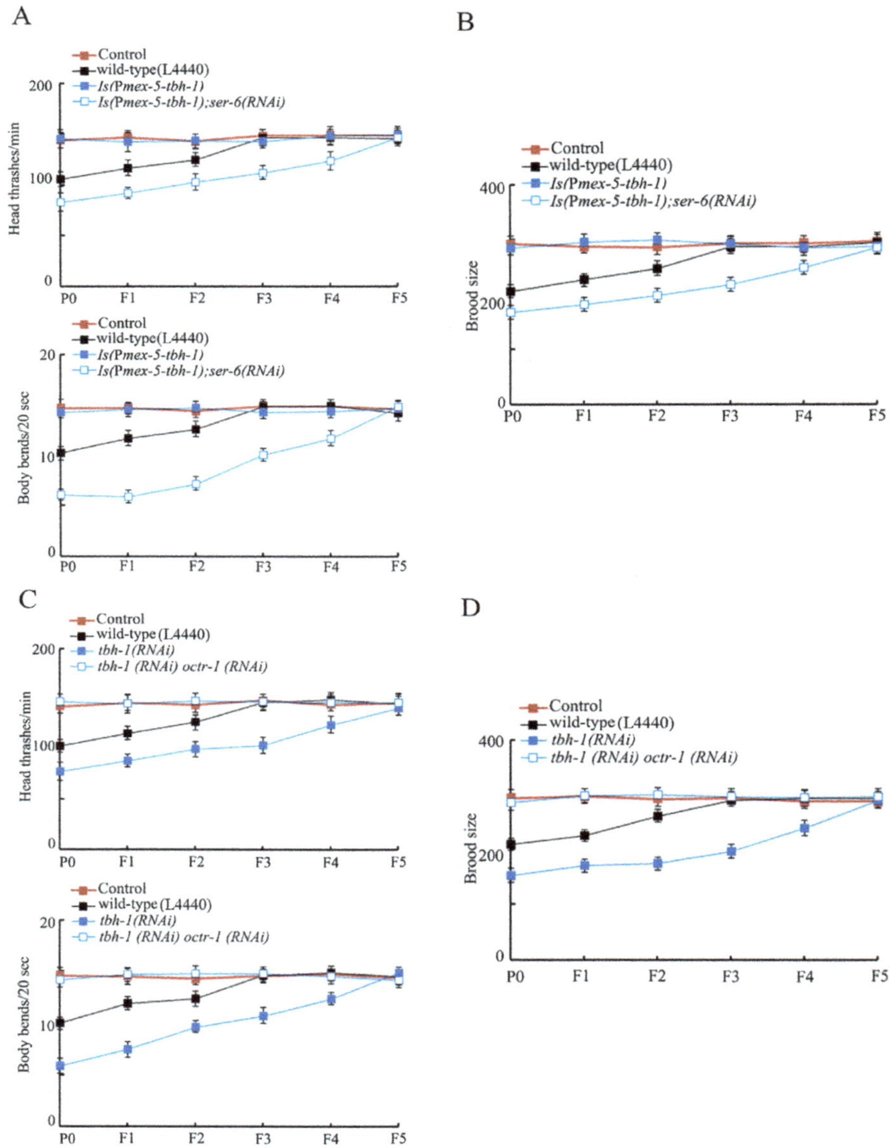

Fig. 11.5 Germline TBH-1 transgenerationally affected the function of octopamine receptors of OCTR-1 and SER-6 [39]. (**a**, **b**) Genetic interaction between TBH-1 and SER-6 in controlling transgenerational MWCNT toxicity in decreasing locomotion and in inhibiting brood size. After the MWCNT exposure in *Is(Pmex-5-tbh-1)* nematodes (P0), RNAi of *ser-6* was performed in F1-G. Curves of wild-type (L4440) showed a significant difference ($P < 0.01$) compared to control. Curves of *Is(Pmex-5-tbh-1)* showed a significant difference ($P < 0.01$) compared to wild-type (L4440). Curves of *Is(Pmex-5-tbh-1);ser-6(RNAi)* showed a significant difference ($P < 0.01$) compared to *Is(Pmex-5-tbh-1)*. (**c**, **d**) Genetic interaction between TBH-1 and OCTR-1 in controlling

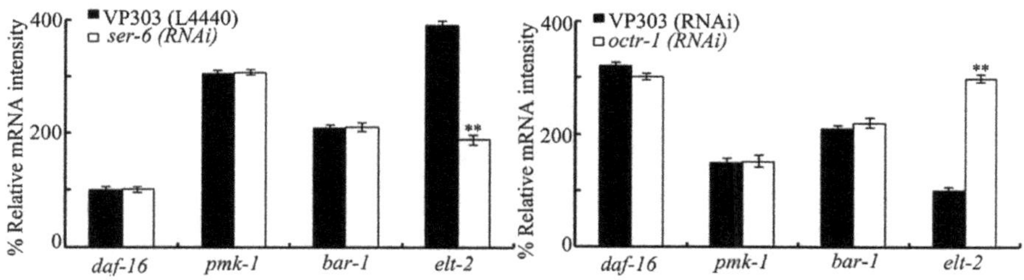

Fig. 11.6 Effect of intestinal RNAi of *ser-6* or *octr-1* on expressions of *daf-16*, *pmk-1*, *bar-1*, and *elt-2* in MWCNT-exposed nematodes [39]. Exposure concentration was 0.1 μg/L $^{**}P < 0.01$ vs VP303

Exposure to pollutants, such as lindane, also caused transgenerational response of genes regulating transport of dopamine and serotonin [59]. Therefore, the molecular basis for other neurotransmitters in possibly modulating transgenerational pollutant toxicity needs to be further examined.

Several experimental evidence demonstrated the potential transgenerational communication between germline TBH-1 and receptors SER-6 and OCTR-1 in the offspring. This implied, besides the communication between synapses formed among different neurons, transgenerational communication between octopamine and its receptors may be formed to mediate the transgenerational pollutant toxicity. During the control of transgenerational pollutant toxicity, only DAF-12 was identified as the upregulator for germline TBH-1, and only ELT-2 was identified as the downstream target of intestinal SER-6 and intestinal OCTR-1. On the one hand, more upregulators for germline TBH-1 and downstream targets for SER-6 and OCTR-1 are needed to be further identified during the control of transgenerational pollutant toxicity. On the other hand, the possible downstream targets for SET-6 and OCTR-1 in other tissues to regulate the transgenerational pollutant toxicity still remain unclear.

The obtained data suggested the role of germline TBH-1 and octopamine receptors SER-6 and OCTR-1 in regulating transgenerational reproductive toxicity and transgenerational neurotoxicity in inhibiting the locomotion induced by pollutants. In *C. elegans*, the octopamine receptors SER-6 and OCTR-1 were also required for control of sensory perception, learning, and memory [60–62]. Therefore, germline TBH-1 and octopamine receptors SER-6 and OCTR-1 may also participate in control of pollutant-induced transgenerational neurotoxicity at other aspects.

Fig. 11.5 (continued) transgenerational MWCNT toxicity in decreasing locomotion and in inhibiting brood size. After the MWCNT exposure in *tbh-1(RNAi)* nematodes (P0), RNAi of *octr-1* was performed in F1-G. Curves of wild-type (L4440) showed a significant difference ($P < 0.01$) compared to control. Curves of *tbh-1(RNAi)* showed a significant difference ($P < 0.01$) compared to wild type. Curves of *tbh-1(RNAi) octr-1(RNAi)* showed a significant difference ($P < 0.01$) compared to *tbh-1(RNAi)*. Exposure concentration was 0.1 μg/L

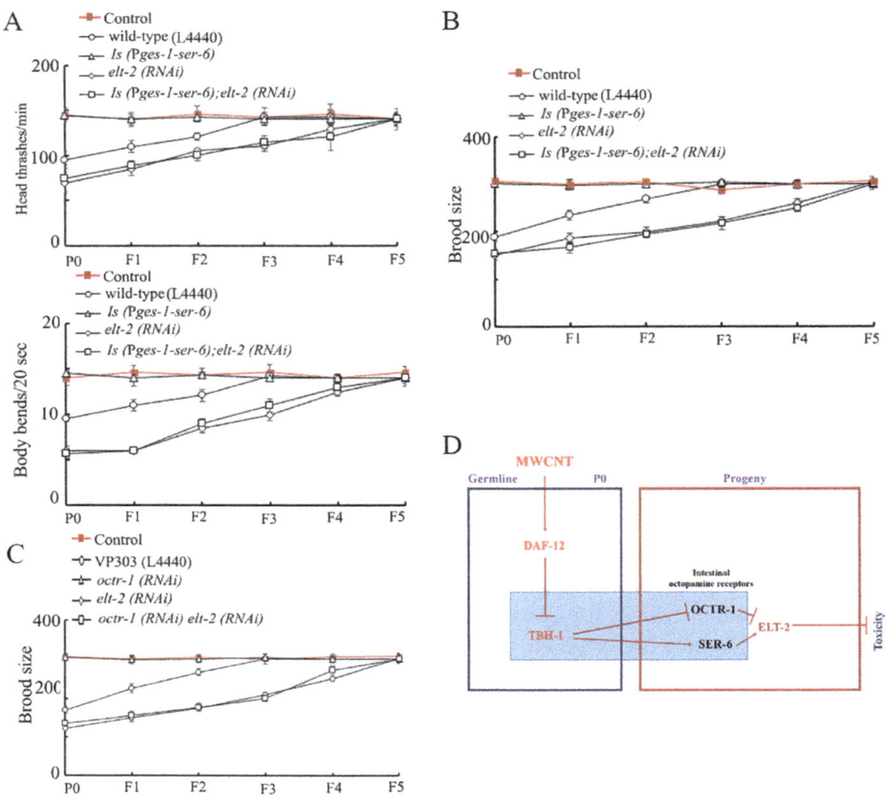

Fig. 11.7 Genetic interaction between ELT-2 and SER-6 or OCTR-1 in the intestine to control transgenerational MWCNT toxicity [39]. (**a, b**) Genetic interaction between ELT-2 and intestinal SER-6 to control transgenerational MWCNT toxicity in decreasing locomotion and in reducing brood size. Curves of wild-type (L4440) showed a significant difference ($P < 0.01$) compared to control. Curves of *Is(Pges-1-ser-6)* and *elt-2(RNAi)* showed a significant difference ($P < 0.01$) compared to wild-type (L4440). Curves of *Is(Pges-1-ser-6);elt-2(RNAi)* showed a significant difference ($P < 0.01$) compared to *Is(Pges-1-ser-6)*. (**c**) Genetic interaction between ELT-2 and OCTR-1 in the intestine to control transgenerational MWCNT toxicity in reducing brood size. Curve of VP303(L4440) showed a significant difference ($P < 0.01$) compared to control. Curves of *octr-1(RNAi)* and *elt-2(RNAi)* showed a significant difference ($P < 0.01$) compared to VP303(L4440). Curve of *octr-1(RNAi)elt-2 (RNAi)* showed a significant difference ($P < 0.01$) compared to *octr-1(RNAi)*. Exposure concentration was 0.1 μg/L. (**d**) A diagram showing molecular mechanism for octopamine-related signals in regulating transgenerational MWCNT toxicity

References

1. Wang D-Y (2022) Toxicology at environmentally relevant concentrations in *Caenorhabditis elegans*. Springer Nature Singapore Pte Ltd., Singapore
2. Wang D-Y (2020) Exposure toxicology in *Caenorhabditis elegans*. Springer Nature Singapore Pte Ltd., Singapore

3. Wang D-Y (2018) Nanotoxicology in *Caenorhabditis elegans*. Springer Nature Singapore Pte Ltd., Singapore
4. Wang D-Y (2019) Target organ toxicology in *Caenorhabditis elegans*. Springer Nature Singapore Pte Ltd, Singapore
5. Wang D-Y (2019) Molecular toxicology in *Caenorhabditis elegans*. Springer Nature Singapore Pte Ltd, Singapore
6. Wang W, Li Y-H, Wang D-Y (2025) Long-term exposure to 6-PPD quinone inhibits glutamate synthesis and glutamate receptor function associated with its toxicity induction in *Caenorhabditis elegans*. Toxics 13:434
7. Hua X, Wang D-Y (2023) Disruption of dopamine metabolism by exposure to 6-PPD quinone in *Caenorhabditis elegans*. Environ Pollut 337:122649
8. Wang S-T, Liu H-L, Qu M, Wang D-Y (2021) Response of tyramine and glutamate related signals to nanoplastic exposure in *Caenorhabditis elegans*. Ecotoxicol Environ Saf 217:112239
9. Zhang J, Wang L, Lu M, Yu Z (2023) Multi- and trans-generational effects of di-n-octyl phthalate on behavior, lifespan and reproduction of *Caenorhabditis elegans* through neural regulation and lipid metabolism. Sci Tota Environ 898:165628
10. Chen H, Gu Y, Tan S, Chen X, Jiang Y, Guo H, Chen J, Wang C, Chen C, Li H, Yu Y (2024) Photoaged nanopolystyrene affects neurotransmission to induce transgenerational neurotoxicity in *Caenorhabditis elegans*. Environ Sci Technol 58:8665–8674
11. Hua X, Wang D-Y (2022) Cellular uptake, transport, and organelle response after exposure to microplastics and nanoplastics: current knowledge and perspectives for environmental and health risks. Rev Environ Contam Toxicol 260:12
12. Qu M, Kong Y, Yuan Y-J, Wang D-Y (2019) Neuronal damage induced by nanopolystyrene particles in nematode *Caenorhabditis elegans*. Environ Sci Nano 6:2591–2601
13. Qu M, Qiu Y-X, Kong Y, Wang D-Y (2019) Amino modification enhances reproductive toxicity of nanopolystyrene on gonad development and reproductive capacity in nematode *Caenorhabditis elegans*. Environ Pollut 254:112978
14. Qu M, Nida A, Kong Y, Du H-H, Xiao G-S, Wang D-Y (2019) Nanopolystyrene at predicted environmental concentration enhances microcystin-LR toxicity by inducing intestinal damage in *Caenorhabditis elegans*. Ecotoxicol Environ Saf 183:109568
15. Qu M, Xu K-N, Li Y-H, Wong G, Wang D-Y (2018) Using *acs-22* mutant *Caenorhabditis elegans* to detect the toxicity of nanopolystyrene particles. Sci Total Environ 643:119–126
16. Qiu Y-X, Liu Y-Q, Li Y-H, Li G-J, Wang D-Y (2020) Effect of chronic exposure to nanopolystyrene on nematode *Caenorhabditis elegans*. Chemosphere 256:127172
17. Qiu Y-X, Luo L-B, Yang Y-H, Kong Y, Li Y-H, Wang D-Y (2020) Potential toxicity of nanopolystyrene on lifespan and aging process of nematode *Caenorhabditis elegans*. Sci Total Environ 705:135918
18. Li D, Deng Y-J, Wang S-T, Du H-H, Xiao G-S, Wang D-Y (2020) Assessment of nanopolystyrene toxicity under fungal infection condition in *Caenorhabditis elegans*. Ecotoxicol Environ Saf 197:110625
19. Liu H-L, Shao H-M, Guo Z-J, Wang D-Y (2020) Nanopolystyrene exposure activates a fat metabolism related signaling-mediated protective response in *Caenorhabditis elegans*. NanoImpact 17:100204
20. Dong S-S, Qu M, Rui Q, Wang D-Y (2018) Combinational effect of titanium dioxide nanoparticles and nanopolystyrene particles at environmentally relevant concentrations on nematodes *Caenorhabditis elegans*. Ecotoxicol Environ Saf 161:444–450
21. Zhao Y-Y, Li D, Rui Q, Wang D-Y (2020) Toxicity induction of nanopolystyrene under microgravity stress condition in *Caenorhabditis elegans*. Sci Total Environ 703:135623
22. Liu H-L, Wang D-Y (2021) Intestinal mitochondrial unfolded protein response induced by nanoplastic particles in *Caenorhabditis elegans*. Chemosphere 267:128917
23. Tang M, Ding G, Lu X, Huang Q, Du H, Xiao G, Wang D (2022) Exposure to nanoplastic particles enhances *Acinetobacter* survival, biofilm formation, and serum resistance. Nanomaterials 12:4222

24. Zhao L, Qu M, Wong G, Wang D-Y (2017) Transgenerational toxicity of nanopolystyrene particles in the range of μg/L in nematode *Caenorhabditis elegans*. Environ Sci Nano 4:2356–2366
25. Sun L-M, Liao K, Wang D-Y (2021) Comparison of transgenerational reproductive toxicity induced by pristine and amino modified nanoplastics in *Caenorhabditis elegans*. Sci Total Environ 768:144362
26. Liu H-L, Tian L-J, Wang S-T, Wang D-Y (2021) Size-dependent transgenerational toxicity induced by nanoplastics in nematode *Caenorhabditis elegans*. Sci Total Environ 790:148217
27. Zhao Y, Chen J, Wang R, Pu X, Wang D (2023) A review of transgenerational and multigenerational toxicology in the *in vivo* model animal *Caenorhabditis elegans*. J Appl Toxicol 43:122–145
28. Wu Q-L, Li Y-X, Li Y-P, Zhao Y-L, Ge L, Wang H-F, Wang D-Y (2013) Crucial role of biological barrier at the primary targeted organs in controlling translocation and toxicity of multi-walled carbon nanotubes in nematode *Caenorhabditis elegans*. Nanoscale 5:11166–11178
29. Nouara A, Wu Q-L, Li Y-X, Tang M, Wang H-F, Zhao Y-L, Wang D-Y (2013) Carboxylic acid functionalization prevents the translocation of multi-walled carbon nanotubes at predicted environmental relevant concentrations into targeted organs of nematode *Caenorhabditis elegans*. Nanoscale 5:6088–6096
30. Zhao Y-L, Wu Q-L, Li Y-P, Wang D-Y (2013) Translocation, transfer, and in vivo safety evaluation of engineered nanomaterials in the non-mammalian alternative toxicity assay model of nematode *Caenorhabditis elegans*. RSC Adv 3:5741–5757
31. Zhao Y-L, Wu Q-L, Li Y-P, Nouara A, Jia R-H, Wang D-Y (2014) *In vivo* translocation and toxicity of multi-walled carbon nanotubes are regulated by microRNAs. Nanoscale 6:4275–4284
32. Shakoor S, Sun L-M, Wang D-Y (2016) Multi-walled carbon nanotubes enhanced fungal colonization and suppressed innate immune response to fungal infection in nematodes. Toxicol Res 5:492–499
33. Zhi L-T, Fu W, Wang X, Wang D-Y (2016) ACS-22, a protein homologous to mammalian fatty acid transport protein 4, is essential for the control of toxicity and translocation of multi-walled carbon nanotubes in *Caenorhabditis elegans*. RSC Adv 6:4151–4159
34. Zhuang Z-H, Li M, Liu H, Luo L-B, Gu W-D, Wu Q-L, Wang D-Y (2016) Function of RSKS-1-AAK-2-DAF-16 signaling cascade in enhancing toxicity of multi-walled carbon nanotubes can be suppressed by *mir-259* activation in *Caenorhabditis elegans*. Sci Rep 6:32409
35. Zhao Y-L, Yang J-N, Wang D-Y (2016) A microRNA-mediated insulin signaling pathway regulates the toxicity of multi-walled carbon nanotubes in nematode *Caenorhabditis elegans*. Sci Rep 6:23234
36. Zhao L, Wan H-X, Liu Q-Z, Wang D-Y (2017) Multi-walled carbon nanotubes-induced alterations in microRNA *let-7* and its targets activate a protection mechanism by conferring a developmental timing control. Part Fibre Toxicol 14:27
37. Zhao Y-L, Jin L, Wang Y, Kong Y, Wang D-Y (2019) Prolonged exposure to multi-walled carbon nanotubes dysregulates intestinal *mir-35* and its direct target MAB-3 in nematode *Caenorhabditis elegans*. Sci Rep 9:12144
38. Zhao Y-Y, Xu R-R, Hua X, Rui Q, Wang D-Y (2022) Multi-walled carbon nanotubes induce transgenerational toxicity associated with activation of germline long non-coding RNA *linc-7* in *C. elegans*. Chemosphere 301:134687
39. Zhao Y-Y, Hua X, Rui Q, Wang D-Y (2023) Exposure to multi-walled carbon nanotubes causes suppression in octopamine signal associated with transgenerational toxicity induction in *C. elegans*. Chemosphere 318:137986
40. Alkema MJ, Hunter-Ensor M, Ringstad N, Horvitz HR (2005) Tyramine functions independently of octopamine in the *Caenorhabditis elegans* nervous system. Neuron 46:247–260
41. Tao J, Ma YC, Yang ZS, Zou CG, Zhang KQ (2016) Octopamine connects nutrient cues to lipid metabolism upon nutrient deprivation. Sci Adv 2:e1501372
42. Guo M, Ge M, Berberoglu MA, Zhou J, Ma L, Yang J, Dong Q, Feng Y, Wu Z, Dong Z (2018) Dissecting molecular and circuit mechanisms for inhibition and delayed response of ASI neurons during nociceptive stimulus. Cell Rep 25:1885–1897

43. Mills H, Wragg R, Hapiak V, Castelletto M, Zahratka J, Harris G, Summers P, Korchnak A, Law W, Bamber B, Komuniecki R (2012) Monoamines and neuropeptides interact to inhibit aversive behaviour in *Caenorhabditis elegans*. EMBO J 31:667–678
44. Sellegounder D, Yuan CH, Wibisono P, Liu Y, Sun J (2018) Octopaminergic signaling mediates neural regulation of innate immunity in *Caenorhabditis elegans*. MBio 9:e01645–18
45. Oakes MD, Law WJ, Clark T, Bamber BA, Komuniecki R (2017) Cannabinoids activate monoaminergic signaling to modulate key *C. elegans* behaviors. J Neurosci 37:2859–2869
46. Hoshikawa H, Uno M, Honjoh S, Nishida E (2017) Octopamine enhances oxidative stress resistance through the fasting-responsive transcription factor DAF-16/FOXO in *C. elegans*. Genes Cells 22:210–219
47. Liao CP, Chiang YC, Tam WH, Chen YJ, Chou SH, Pan CL (2022) Neurophysiological basis of stress-induced aversive memory in the nematode *Caenorhabditis elegans*. Curr Biol 32:5309–5322
48. Sun J, Singh V, Kajino-Sakamoto R, Aballay A (2011) Neuronal GPCR controls innate immunity by regulating noncanonical unfolded protein response genes. Science 332:729–732
49. Qu M, Li D, Zhao Y-L, Yuan Y-J, Wang D-Y (2020) Exposure to low-dose nanopolystyrene induces the response of neuronal JNK MAPK signaling pathway in nematode *Caenorhabditis elegans*. Environ Sci Eur 32:58
50. Zhao Y-L, Yang R-L, Rui Q, Wang D-Y (2016) Intestinal insulin signaling encodes two different molecular mechanisms for the shortened longevity induced by graphene oxide in *Caenorhabditis elegans*. Sci Rep 6:24024
51. Zhao Y-L, Zhi L-T, Wu Q-L, Yu Y-L, Sun Q-Q, Wang D-Y (2016) p38 MAPK-SKN-1/Nrf signaling cascade is required for intestinal barrier against graphene oxide toxicity in *Caenorhabditis elegans*. Nanotoxicology 10:1469–1479
52. Zhi L-T, Ren M-X, Qu M, Zhang H-Y, Wang D-Y (2016) Wnt ligands differentially regulate toxicity and translocation of graphene oxide through different mechanisms in *Caenorhabditis elegans*. Sci Rep 6:39261
53. Zhi L-T, Qu M, Ren M-X, Zhao L, Li Y-H, Wang D-Y (2017) Graphene oxide induces canonical Wnt/β-catenin signaling-dependent toxicity in *Caenorhabditis elegans*. Carbon 113:122–131
54. Shao H-M, Han Z-Y, Krasteva N, Wang D-Y (2019) Identification of signaling cascade in the insulin signaling pathway in response to nanopolystyrene particles. Nanotoxicology 13:174–188
55. Qu M, Liu Y-Q, Xu K-N, Wang D-Y (2019) Activation of p38 MAPK signaling-mediated endoplasmic reticulum unfolded protein response by nanopolystyrene particles. Adv Biosyst 3:1800325
56. Shao H-M, Wang D-Y (2020) Long-term and low-dose exposure to nanopolystyrene induces a protective strategy to maintain functional state of intestine barrier in nematode *Caenorhabditis elegans*. Environ Pollut 258:113649
57. Shao H-M, Kong Y, Wang D-Y (2020) Response of intestinal signaling communication between nucleus and peroxisome to nanopolystyrene at predicted environmental concentration. Environ Sci Nano 7:250–261
58. Hua X, Wang D-Y (2023) Exposure to 6-PPD quinone at environmentally relevant concentrations inhibits both lifespan and healthspan in *C. elegans*. Environ Sci Technol 57:19295–19303
59. Yu Y, Hua X, Chen H, Wang Z, Han Y, Chen X, Yang Y, Xiang M (2022) Glutamatergic transmission associated with locomotion-related neurotoxicity to lindane over generations in *Caenorhabditis elegans*. Chemosphere 290:133360
60. Zhao H, Shi G, Qin R, Sun Y, Guo W, Shi R, Peng M, Yang J, Zhao J, Liu Q, Xiao J, Zhang K, Liu Q, Yang W, Liu H (2025) Hungry for knowledge: octopamine signaling regulates hunger-enhanced olfactory learning. Adv Sci 15:e13842
61. O'Donnell MP, Fox BW, Chao PH, Schroeder FC, Sengupta P (2020) A neurotransmitter produced by gut bacteria modulates host sensory behaviour. Nature 583:415–420
62. Wang PZ, Ge MH, Su P, Wu PP, Wang L, Zhu W, Li R, Liu H, Wu JJ, Xu Y, Zhao JL, Li SJ, Wang Y, Chen LM, Wu TH, Wu ZX (2025) Sensory plasticity caused by up-down regulation encodes the information of short-term learning and memory. iScience 28:112215

Chapter 12
Regulation of Transgenerational Toxicity by Metabolic Signals

Abstract Metabolic signals play an important function in regulating biological processes. Exposure to pollutants could induce certain transgenerational response of metabolic molecules, such as glucose and lipid. Additionally, transgenerational response of glucose, citric acid cycle (CAC), lipid, amino acid, and retinoic acid metabolism–related signals could be induced by pollutants. These metabolism-related signals participated in control of pollutant toxicity, suggesting the role of metabolism-related signals in modulating transgenerational pollutant toxicity. Toxicity of pollutants could be inhibited by treatment with some metabolic molecules, such as pyruvate, leucine, and retinoic acid, which further suggested the important function of metabolic molecules in regulating transgenerational pollutant toxicity.

Keywords Transgenerational toxicology · Metabolic signal · Metabolic molecule · *Caenorhabditis elegans*

12.1 Introduction

Caenorhabditis elegans is helpful to detect toxic effects of stresses and pollutants at low doses, including the effect on biochemical metabolisms [1–3]. As a classic model animal, *C. elegans* is further useful to determine underlying toxicological mechanisms for the observed toxicity of stresses and pollutants [4, 5]. Especially, considering the conserved property of metabolic processes, *C. elegans* can be used to assess toxic effects of stresses and pollutants on metabolisms and to further elucidate underlying toxicological mechanisms.

In Chaps. 5, 6, 7, 8, 9, 10, and 11, we introduced and discussed role of transgenerational communication between germline secreted ligands and their receptors in regulating transgenerational toxicity of pollutants. In this chapter, we first introduced transgenerational change of biochemical metabolites, such as glucose and lipid, by pollutants. Moreover, we introduced and discussed the role of

D. Wang, *Transgenerational Toxicology in Caenorhabditis elegans*,
https://doi.org/10.1007/978-981-92-1389-4_12

metabolism-related signals and metabolic molecules in regulating transgenerational toxicity of pollutants.

12.2 Role of Glucose Metabolism–Related Signals

12.2.1 Transgenerational Response of Glucose to Pollutants

In biochemical metabolisms, glucose is the central molecule to provide immediate energy source, to act as key metabolic precursor for fatty acid synthesis and amino acid synthesis, and to link to other metabolic pathways. Exposure to pollutants can induce certain forms of transgenerational response of glucose. The 6-PPD quinone is an emergent contaminant and caused multiple toxicities on nematodes, such as lifespan reduction and organ toxicity [6–18]. The 6-PPDQ was selected as example of pollutants. At the parental generation (P0-G), 6-PPDQ exposure could induce increase in both glucose and glycogen contents [19, 20]. Moreover, 6-PPDQ exposure further caused transgenerational elevation of glucose content (Fig. 12.1) [21]. After exposure to 10 μg/L 6-PPDQ, transgenerational elevation of glucose content was observed till to the F3-G (Fig. 12.1) [21].

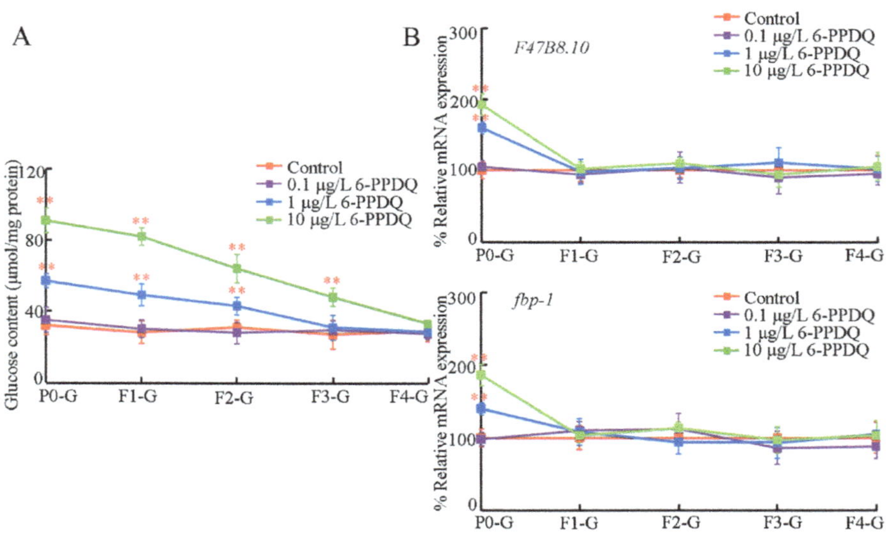

Fig. 12.1 Transgenerational changes of glucose content (**a**) and expressions of enzyme genes governing gluconeogenesis (**b**) after 6-PPDQ exposure at P0-G [21]. $^{**}P < 0.01$ vs control

12.2.2 Transgenerational Response of Glucose Metabolism–Related Signals to Pollutants

Accompanied with transgenerational glucose response induced by pollutants, usually the transgenerational response of certain glucose metabolism related signals can be formed. At the P0-G, elevation of glucose content caused by 6-PPDQ was due to both activation of enzymes governing gluconeogenesis and inhibition in enzymes governing glycolysis [19]. However, 6-PPDQ exposure did not cause the further transgenerational activations of enzymes governing gluconeogenesis (F47B8.10 and FBP-1) (Fig. 12.1) [21]. In contrast, transgenerational inhibition in enzymes governing glycolysis was induced by 6-PPDQ [21]. After exposure to 10 μg/L 6-PPDQ, transgenerational inhibition in enzymes governing glycolysis (HXK-1, HXK-3, PYK-1, and PYK-2) could be detected till to the F3-G [21]. Therefore, the observed transgenerational elevation of glucose content by 6-PPDQ was due to transgenerational inhibition in enzymes governing glycolysis.

12.2.3 Upregulators Regulating Transgenerational Response of Glucose Metabolism–Related Signals to Pollutants

Transcriptional factors DAF-16 and AAK-2/AMPK act as upregulators for the control of transgenerational response of glucose metabolism–related signals to pollutants. At the P0-G, DAF-16 and AAK-2 regulated 6-PPDQ-induced glucose elevation by affecting expression of enzyme genes governing gluconeogenesis and glycolysis. Inhibition in DAF-16 and AAK-2 was observed not only at P0-G but also in the offspring of 6-PPDQ-exposed nematodes. Exposure to 10 μg/L 6-PPDQ resulted in transgenerational inhibition in DAF-16 and AAK-2 till to the F4-G (Fig. 12.2) [21]. Moreover, RNAi of *daf-16* and *aak-2* exacerbated 6-PPDQ-induced transgenerational elevation of glucose content (Fig. 12.3) [21]. After 6-PPDQ exposure, compared to transgenerational elevation of glucose content till to the F3-G in wild type,

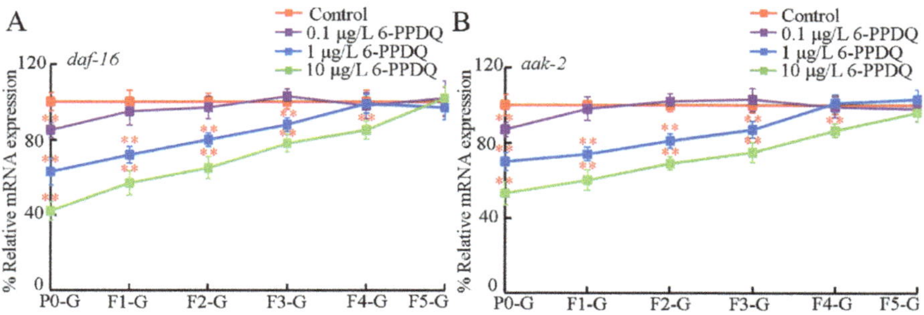

Fig. 12.2 Transgenerational alterations in expressions of *daf-16* (**a**) and *aak-2* (**b**) after 6-PPDQ exposure at P0-G [21]. $^{**}P < 0.01$ vs control

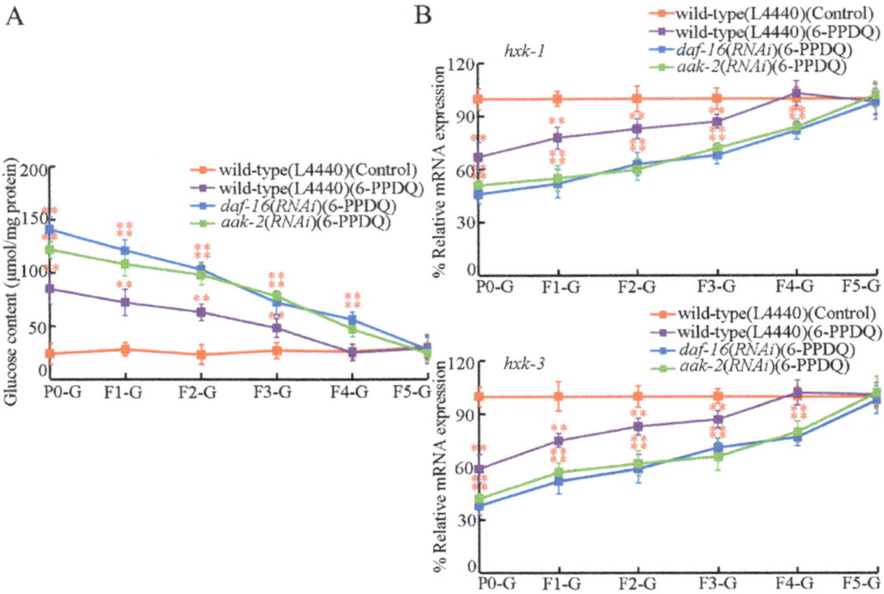

Fig. 12.3 Effect of RNAi of *daf-16* and *aak-2* on transgenerational change of glucose content (**a**) and transgenerational expressions of *hxk-1* and *hxk-3* (**b**) after 6-PPDQ exposure at P0-G [21]. Exposure concentration of 6-PPDQ was 10 μg/L. $^{**}P < 0.01$ vs wild-type (L4440)

transgenerational elevation of glucose content was formed till to the F4-G in *daf-16(RNAi)* and *aak-2(RNAi)* (Fig. 12.3) [21]. Therefore, transgenerational inhibition in DAF-16 and AAK-2 mediated transgenerational glucose elevation induced by pollutants.

12.2.4 Involvement of Glucose Metabolism–Related Signals in Regulating Transgenerational Toxicity of Pollutants

Glucose metabolism–related signals participated in control of transgenerational pollutant toxicity. Exposure to 6-PPDQ caused transgenerational neurotoxicity and transgenerational reproductive toxicity (Fig. 12.4) [21]. After 6-PPDQ exposure, transgenerational neurotoxicity and transgenerational reproductive toxicity were observed till to the F4-G, which could be strengthened by RNAi of *hxk-1*, *hxk-3*, *pyk-1*, *pyk-2*, *daf-16*, and *aak-2* (Fig. 12.4) [21]. RNAi of *hxk-1*, *hxk-3*, *pyk-1*, *pyk-2*, *daf-16*, and *aak-2* could cause the 6-PPDQ-induced transgenerational neurotoxicity and transgenerational reproductive toxicity till to the F5-G (Fig. 12.4) [21]. That is, both enzymes governing glycolysis and their upregulators (DAF-16 and AAK-2) regulated the induction of transgenerational pollutant toxicity.

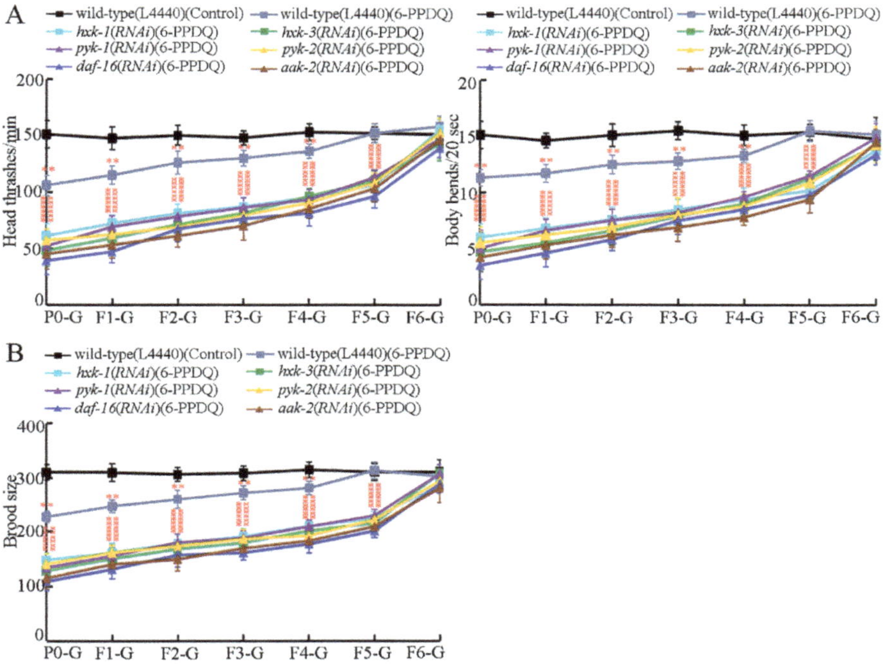

Fig. 12.4 Effect of RNAi of *daf-16*, *aak-2*, *hxk-1*, *hxk-3*, *pyk-1*, and *pyk-2* on transgenerational 6-PPDQ toxicity in decreasing locomotion (**a**) and in reducing brood size (**b**) [21]. Exposure concentration of 6-PPDQ was 10 μg/L. $^{**}P < 0.01$ vs wild-type (L4440)

12.3 Role of Citric Acid Cycle (CAC)–Related Signals

12.3.1 Transgenerational Response of CAC-Related Signals to Pollutants

The CAC located in the mitochondrial matrix acts as a central metabolic hub by harvesting high-energy electrons from acetyl-CoA and transferring them to electron carriers for ATP generation in electron transfer chain, integrating several major metabolic pathways, and proving critical source of biosynthetic precursors. Exposure to pollutants can induce transgenerational response of certain CAC-related signals. The 6-PPDQ was selected as an example of a pollutant. At the P0-G, 6-PPDQ exposure reduced contents of both intermediate metabolites in CAC and NADH and caused inhibition in CTS-1 (enzyme governing the first key reaction), inhibition in IDH-2 (enzyme governing the second key reaction), and inhibition in DLST-1 and DLD-1 (enzymes governing the key reaction) [22]. Moreover, after exposure to 10 μg/L 6-PPDQ, transgenerational inhibition in CTS-1, IDH-2, DLST-1, and DLD-1 was detected till to the F4-G (Fig. 12.5) (Y.-X. Wang, and D.-Y., persona communication).

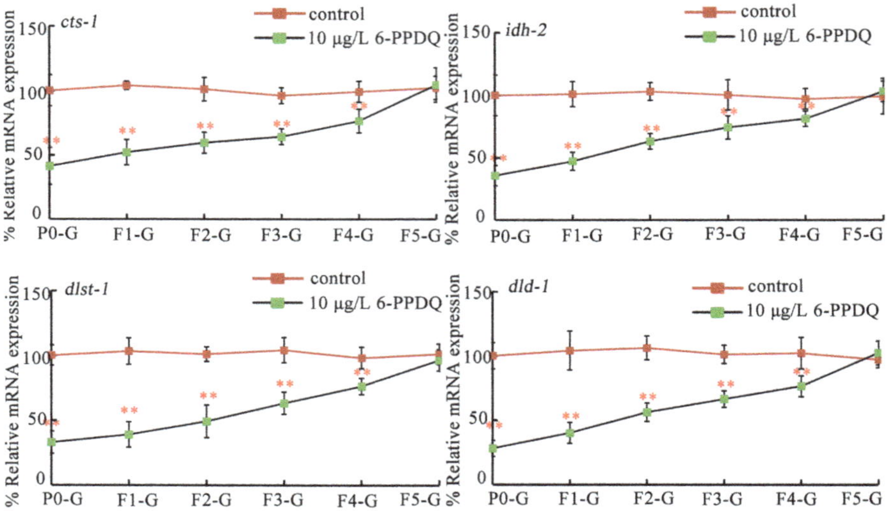

Fig. 12.5 Transgenerational response of citric acid cycle (CAC)–related signals to 6-PPDQ. $^{**}P<0.01$ vs control

12.3.2 Function of CAC-Related Signals in Regulating Toxicity of Pollutants

CAC-related signals were involved in controlling pollutant-induced mitochondrial dysfunction. Firstly, 6-PPDQ-induced reduction in NADH content was increased by RNAi of *idh-2*, *dlst-1*, and *dld-1* [22]. Secondly, 6-PPDQ-induced inhibition in GAS-1 and MEV-1, complex I/II components, was strengthened by RNAi of *cts-1*, *idh-2*, *dlst-1*, and *dld-1* [22]. Thirdly, mitochondrial dysfunction caused by 6-PPDQ (such as elevated oxygen consumption and reduced ATP content) could be exacerbated by RNAi of *cts-1*, *idh-2*, *dlst-1*, and *dld-1* [22]. These observations implied that transgenerational inhibition in CTS-1, IDH-2, DLST-1, and DLD-1 induced by pollutants possibly mediated transgenerational mitochondrial dysfunction.

CAC-related signals also participated in controlling other aspects of pollutant toxicity. The 6-PPDQ was selected as example of pollutants. The 6-PPDQ-induced intestinal oxidative stress and locomotion inhibition were exacerbated by RNAi of *cts-1*, *idh-2*, *dlst-1*, and *dld-1* [22]. This implied that transgenerational inhibition in CTS-1. IDH-2, DLST-1, and DLD-1 induced by pollutants also possibly mediated transgenerational toxicity induction, such as transgenerational neurotoxicity.

12.3.3 Pharmacological Effect of Pyruvate Treatment on CAC-Related Signals in Pollutant-Exposed Nematodes

The 6-PPDQ was selected as example of pollutants. The 6-PPDQ-induced inhibition in CTS-1, IDH-2, DLST-1, and DLD-1 could be reversed by pyruvate treatment [22]. Accompanied with this, 6-PPDQ-induced mitochondrial dysfunction, intestinal oxidative stress, and locomotion inhibition were also reversed by pyruvate treatment [22]. These observations indirectly suggested the association between transgenerational inhibition in CAC-related signals and pollutant-induced transgenerational toxicity, such as transgenerational mitochondrial dysfunction.

12.4 Role of Lipid Metabolism–Related Signals

12.4.1 Transgenerational Lipid Response to Pollutants

Exposure to pollutants could induce certain transgenerational lipid response [23, 24]. The 6-PPDQ was selected as example of pollutants. At the P0-G, 6-PPDQ exposure induced increased lipid accumulation [25]. Moreover, 6-PPDQ induced transgenerational increase in both lipid accumulation and triglyceride content [24]. After exposure to 10 μg/L 6-PPDQ, transgenerational increase in both lipid accumulation and triglyceride content could be observed till to the F4-G [24].

12.4.2 Transgenerational Response of Lipid Metabolism–Related Signals to Pollutants

1. Fatty Acid Metabolism–Related Genes

With 6-PPDQ as example of pollutants, 6-PPDQ induced transgenerational activation of enzymes (FASN-1 and POD-2) governing fatty acid synthesis and inhibition in enzymes (ACS-2 and ECH-2) governing mitochondrial fatty acid β-oxidation and enzymes (ACS-1 and ECH-3) governing peroxisome fatty acid β-oxidation (Fig. 12.6) [24]. After exposure to 10 μg/L 6-PPDQ, transgenerational activation of FASN-1 and POD-2 and transgenerational inhibition in ACS-1, ECH-2, ACS-1, and ECH-3 were observed, and some of the transgenerational response (such as the response of POD-2, ACS-2, ad ECH-2) could be detected till to the F4-G (Fig. 12.6) [24].

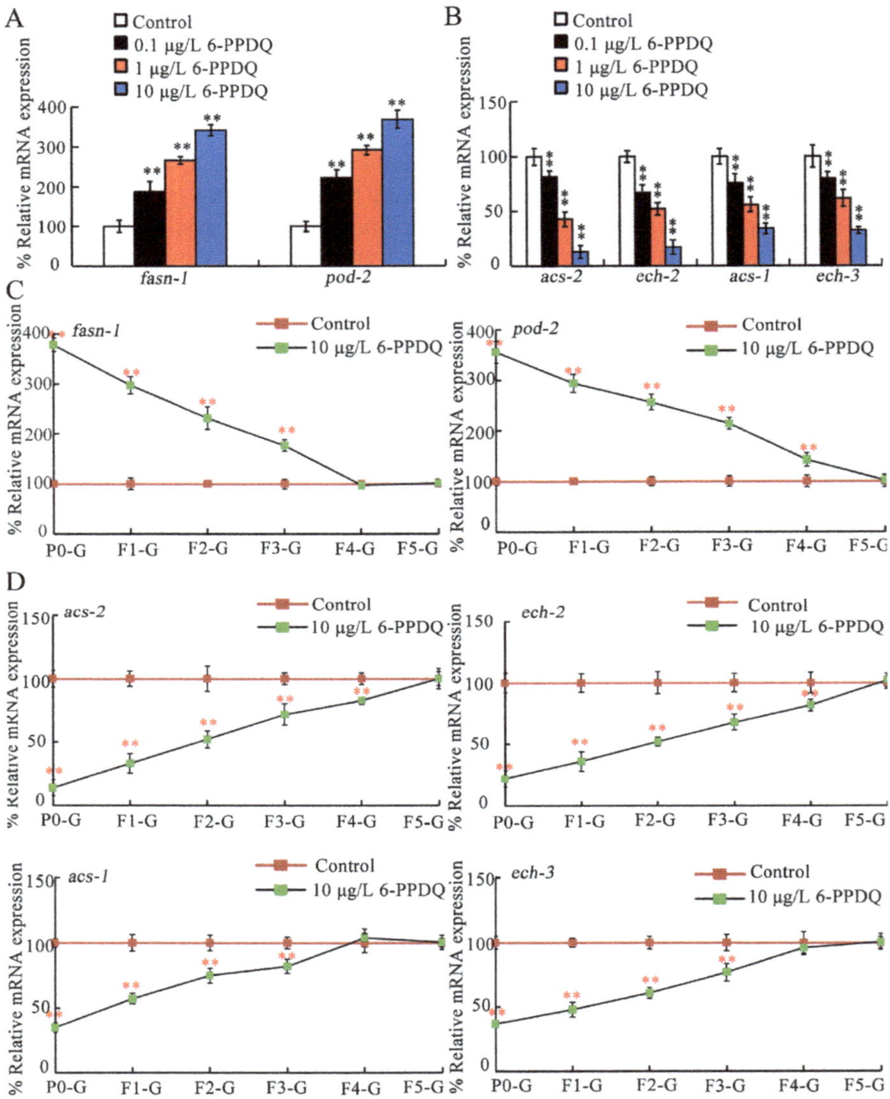

Fig. 12.6 Transgenerational expressions of genes governing fatty acid synthesis and β-oxidation after exposure to 6-PPDQ at P0-G [24]. (**a, b**) Effect of 6-PPDQ exposure on expressions of genes governing fatty acid synthesis and β-oxidation at P0-G. (**c, d**) Transgenerational expressions of *fasn-1*, *pod-2*, *acs-2*, *ech-2*, *acs-1*, and *ech-3* after exposure to 6-PPDQ at P0-G. $^{**}P < 0.01$ vs control

2. Monounsaturated Fatty Acyl-CoA Synthesis–Related Genes

With 6-PPDQ as example of pollutants, 6-PPDQ induced transgenerational activation of delta-9 desaturases (FAT-5, FAT-6, and FAT-7) governing monounsaturated fatty acyl-CoA synthesis. After exposure to 10 μg/L 6-PPDQ, transgenerational activation of FAT-5, FAT-6, and FAT-7 was observed, and transgenerational activation of FAT-6 and FAT-7 was detected till to the F4-G (Fig. 12.7) [24].

3. SBP-1 and MDT-15

SBP-1 and MDT-15 are metabolic sensors, which could be activated by 6-PPDQ at the P0-G [25]. Moreover, transgenerational activation of SBP-1 and MDT-15 could be induced by 6-PPDQ (Fig. 12.8) [24]. After exposure to 10 μg/L 6-PPDQ, transgenerational activation of SBP-1 and MDT-15 was observed till to the F4-G (Fig. 12.8) [24]. Considering the role of SBP-1 and MDT-15 in regulating lipid accumulation and expression of fatty acid metabolism–related genes [25], transgenerational activation of SBP-1 and MDT-15 mediated transgenerational increase in lipid accumulation induced by pollutants.

4. Epigenetic Regulation–Related Signals

With 6-PPDQ as the example of pollutants, 6-PPDQ could cause transgenerational response of some epigenetic regulation–related signals. SET-2 and RBR-2 are components of COMPASS complex with the function in regulating lipid accumulation [26]. SET-2 is a histone H3K4 methyltransferase, and RBR-2 is a H3K4me3/2 histone demethylase. At the P0-G, 6-PPDQ exposure activated SET-2 and inhibited RBR-2 (Fig. 12.9) [24]. After exposure to 10 μg/L 6-PPDQ, transgenerational activation of SET-2 and transgenerational inhibition in RBR-2 were induced and could be detected till to the F4-G (Fig. 12.9) [24]. The 6-PPDQ-induced transgenerational increase in lipid accumulation and triglyceride content could be strengthened by RNAi of *set-2* and suppressed by RNAi of *rbr-2* (Fig. 12.9) [24], suggesting the opposite function between SET-2 and RBR-2 in regulating 6-PPDQ-induced transgenerational lipid accumulation. Moreover, 6-PPDQ-induced activation of SBP-1 and MDT-15 could be further increased by RNAi of *set-2* and inhibited by RNAi of *rbr-2* (Fig. 12.9) [24], suggesting that SET-2 and RBR-2 oppositely regulated 6-PPDQ-induced transgenerational lipid accumulation by differentially affecting SBP-1 and MDT-15 expressions.

12.4.3 Function of Lipid Metabolism–Related Signals in Regulating Toxicity of Pollutants

Lipid metabolism–related signals were further involved in the control of pollutant toxicity. With 6-PPDQ as example of pollutants, 6-PPDQ-induced neurotoxicity, such as locomotion inhibition and damage on development of D-type neurons, was exacerbated by RNAi of *sbp-1*, *mdt-15*, and *rbr-2* (Figs. 12.10 and 12.11) [24]. Different from this, 6-PPDQ-induced neurotoxicity was suppressed by RNAi of *set-2* (Fig. 12.11) [24]. These observations implied that the lipid metabolism–related signals also regulate transgenerational toxicity induced by pollutants.

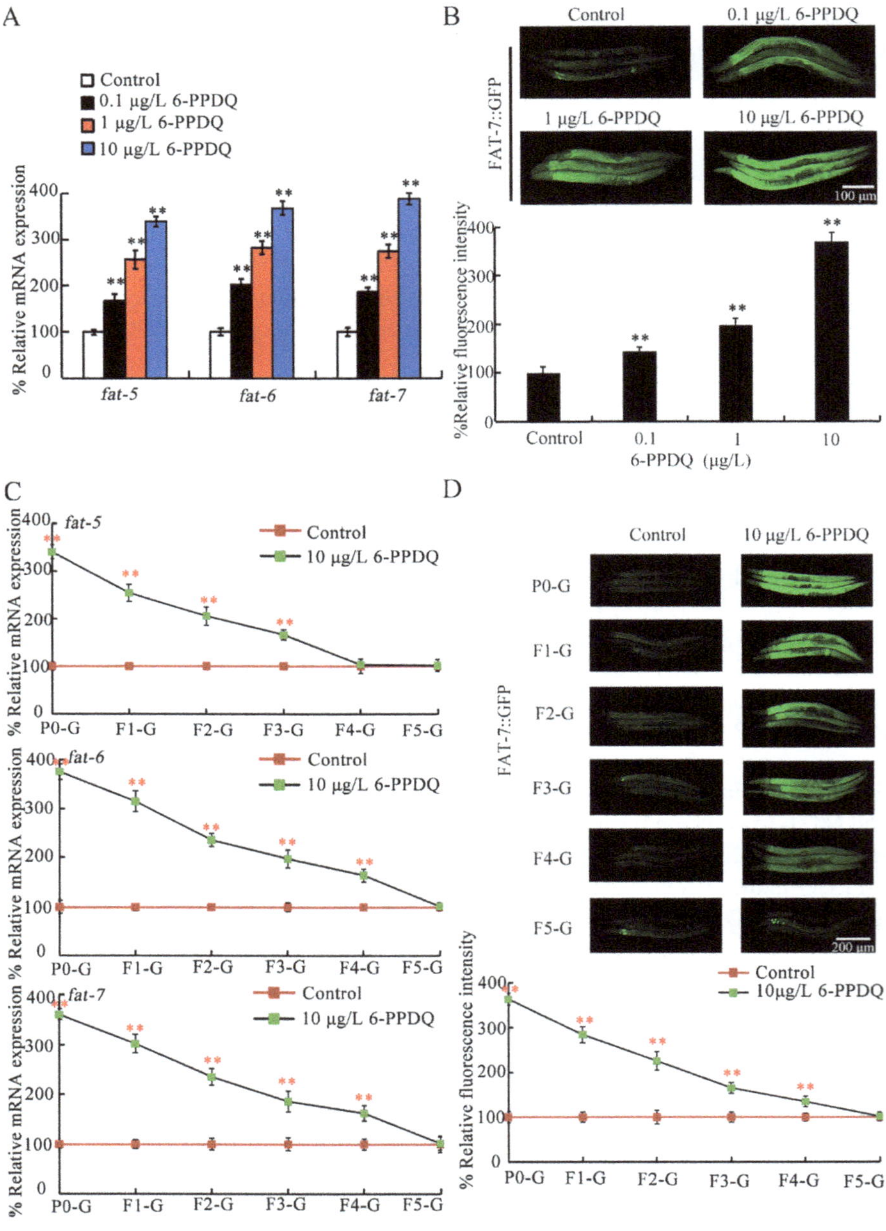

Fig. 12.7 Transgenerational expressions of genes governing monounsaturated fatty acyl-CoA synthesis after exposure to 6-PPDQ at P0-G [24]. (**a**) Effect of 6-PPDQ exposure on expressions of *fat-5*, *fat-6*, and *fat-7* at P0-G. (**b**) Effect of 6-PPDQ exposure on expression of FAT-7GFP at P0-G. (**c**) Transgenerational expressions of *fat-5*, *fat-6*, and *fat-7* after exposure to 6-PPDQ at P0-G. (**d**) Transgenerational expression of FAT-7::GFP after exposure to 6-PPDQ at P0-G. $^{**}P < 0.01$ vs control

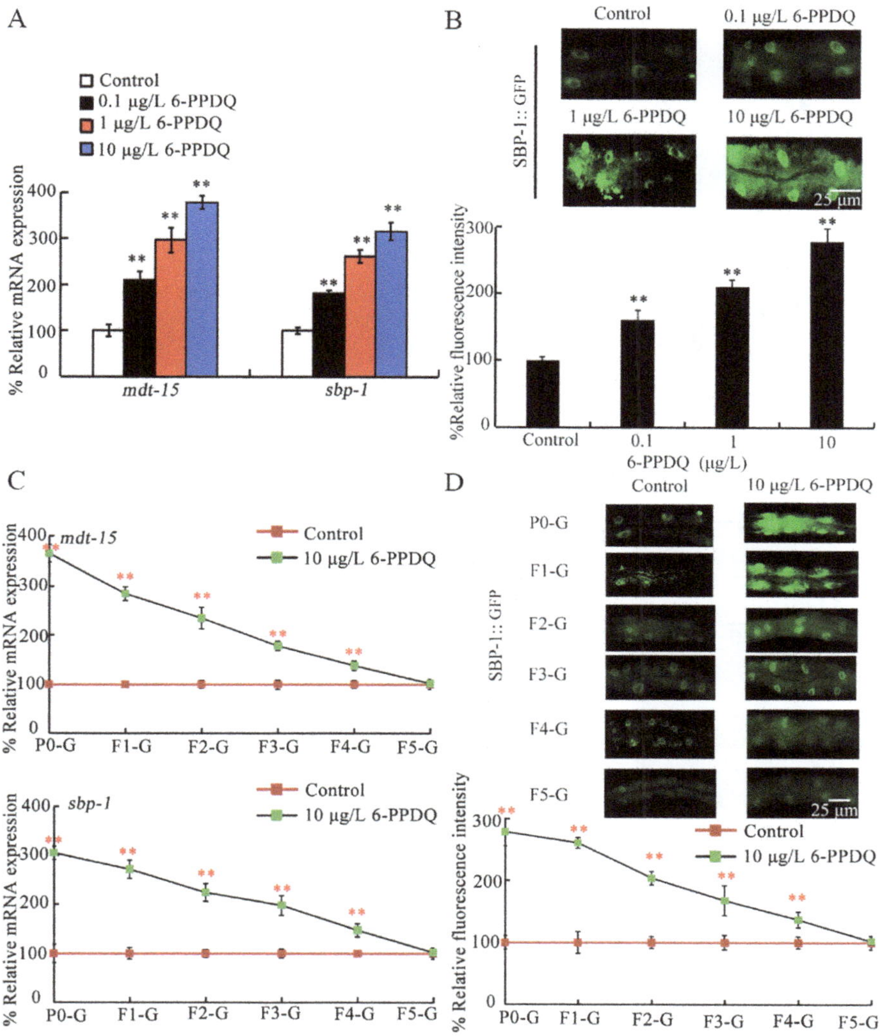

Fig. 12.8 Transgenerational expressions of metabolic sensor genes after exposure to 6-PPDQ at P0-G [24]. (**a**) Effect of 6-PPDQ exposure on expressions of *sbp-1* and *mdt-15* at P0-G. (**b**) Effect of 6-PPDQ exposure on SBP-1::GFP expression at P0-G. (**c**) Transgenerational expressions of *sbp-1* and *mdt-15* after exposure to 6-PPDQ at P0-G. (**d**) Transgenerational SBP-1::GFP expression after exposure to 6-PPDQ at P0-G. $^{**}P < 0.01$ vs control

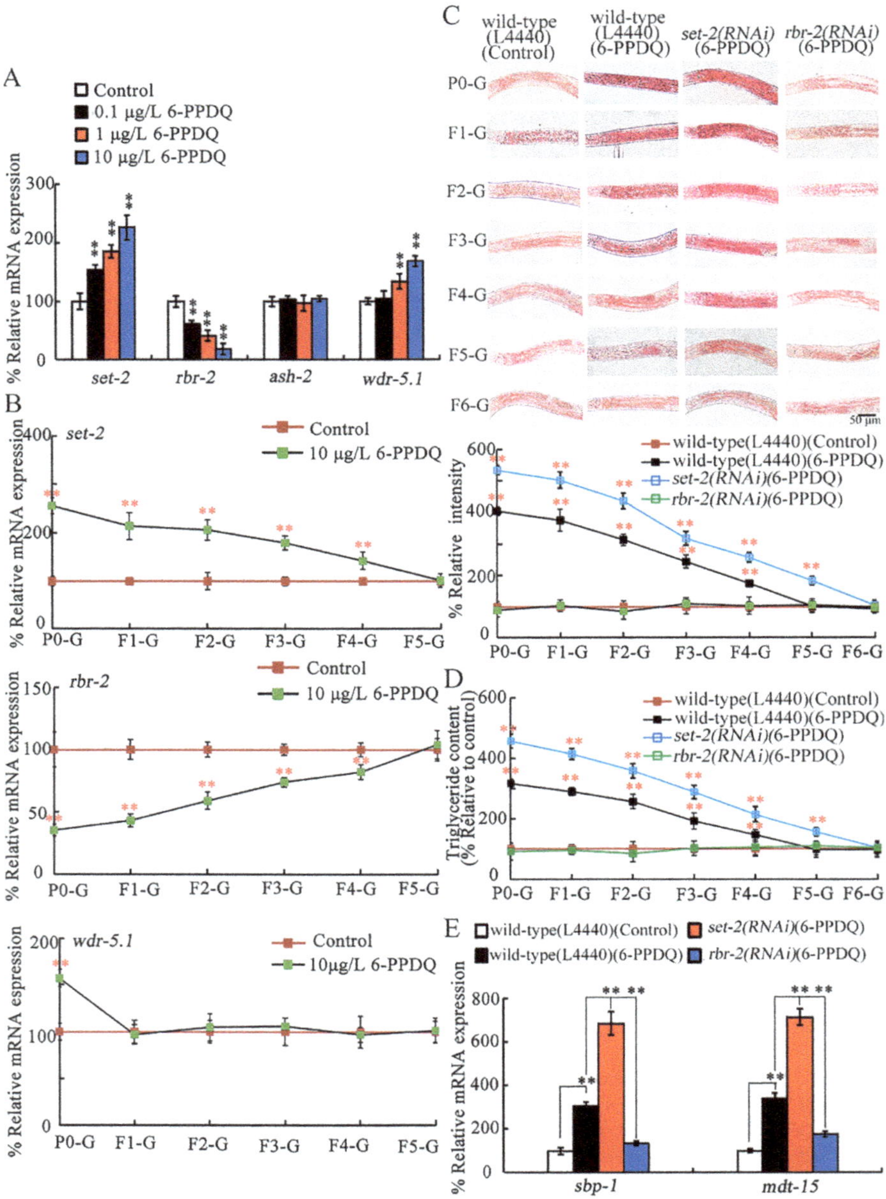

Fig. 12.9 SET-2 and RBR-2 regulated transgenerational lipid accumulation in 6-PPDQ-exposed nematodes [24]. (**a**) Effect of 6-PPDQ exposure on expressions of *set-2*, *rbr-2*, *ash-2*, and *wdr-5.1* at P0-G. $^{**}P < 0.01$ vs control. (**b**) Transgenerational expression of *set-2*, *rbr-2*, and *wdr-5.1* after 6-PPDQ exposure at P0-G. $^{**}P < 0.01$ vs control. (**c**) Effect of RNAi of *set-2* and *rbr-2* on transgenerational lipid accumulation induced by 6-PPDQ (10 μg/L). $^{**}P < 0.01$ vs wild-type (L4440) (Control). (**d**) Effect of RNAi of *set-2* and *rbr-2* on transgenerational change of triglyceride content after 6-PPDQ (10 μg/L) exposure at P0-G. $^{**}P < 0.01$ vs wild-type (L4440)(control). (**e**) Effect of RNAi of *set-2* and *rbr-2* on expressions of *sbp-1* and *mdt-15* in 6-PPDQ (10 μg/L)-exposed nematodes. $^{**}P < 0.01$

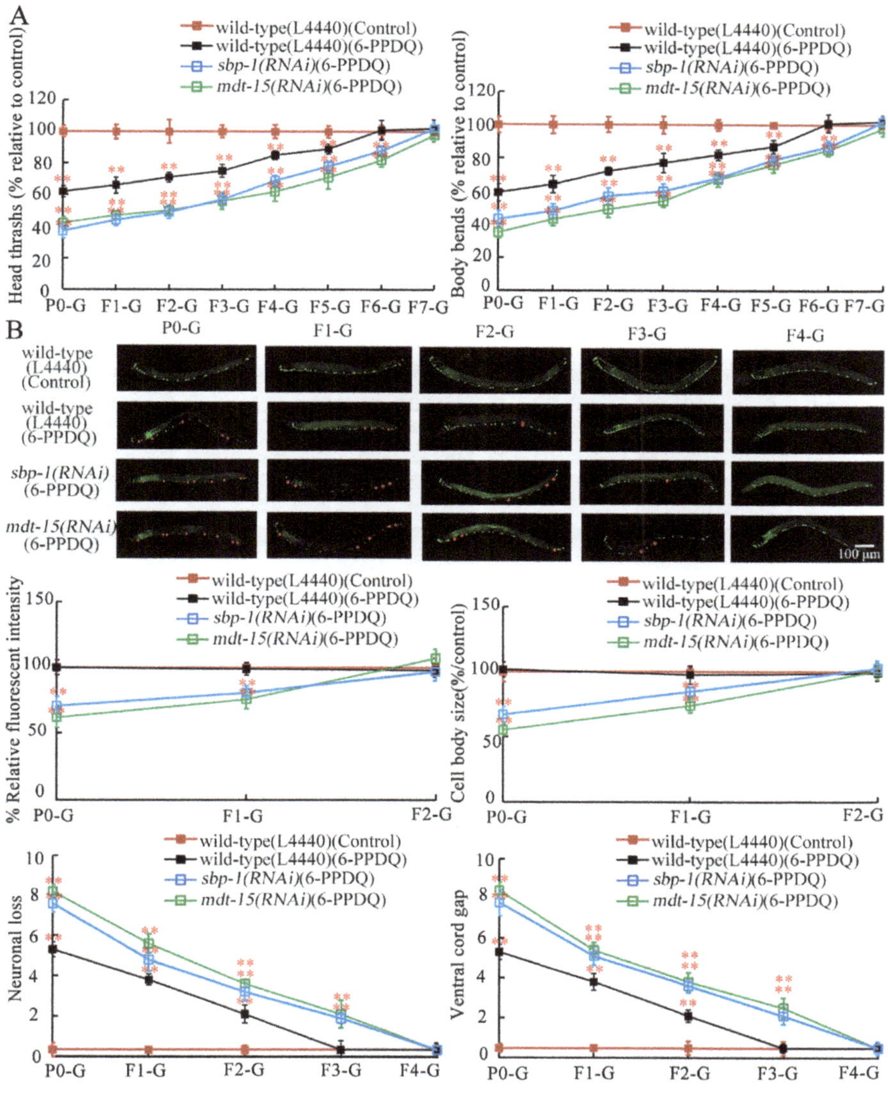

Fig. 12.10 Effect of RNAi of *sbp-1* and *mdt-15* on 6-PPDQ neurotoxicity in decreasing locomotion behavior (**a**) and in inducing neurodegeneration (**b**) [24]. In images, asterisks indicate the position with neuronal loss. Exposure concentration of 6-PPDQ was 10 μg/L. $^{**}P < 0.01$ vs wild-type (L4440)(Control). The curves of wild-type (L4440)(6-PPDQ) were significantly ($P < 0.01$) different from those of wild-type (L4440)(Control). The curves of *sbp-1(RNAi)*(6-PPDQ) and *mdt-15(RNAi)*(6-PPDQ) were significantly ($P < 0.01$) different from those of wild-type (L4440) (6-PPDQ)

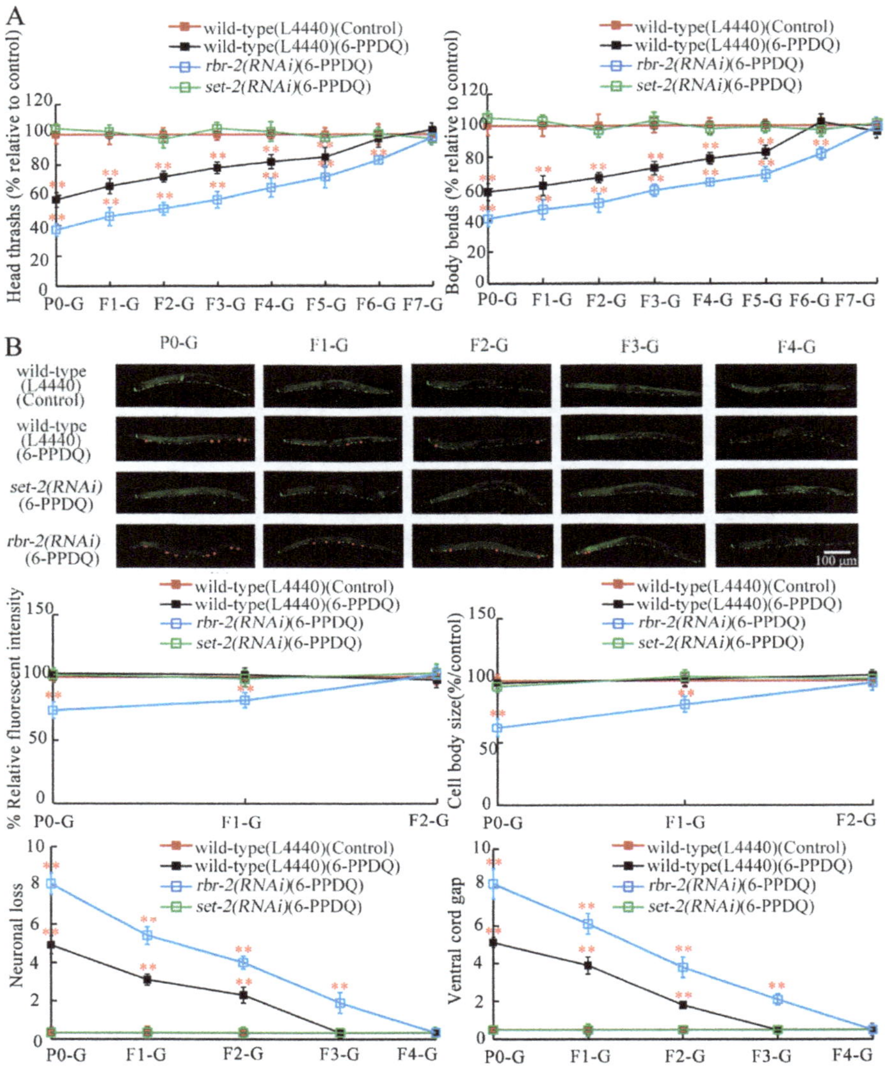

Fig. 12.11 Effect of RNAi of *set-2* and *rbr-2* on 6-PPDQ neurotoxicity in decreasing locomotion behavior (**a**) and in inducing neurodegeneration (**b**) [24]. In images, asterisks indicate the position with neuronal loss. Exposure concentration of 6-PPDQ was 10 μg/L. $^{**}P < 0.01$ vs wild-type (L4440)(Control). The curves of wild-type (L4440)(6-PPDQ) were significantly ($P < 0.01$) different from those of wild-type (L4440)(Control). The curves of *set-2(RNAi)*(6-PPDQ) and *rbr-2(RNAi)*(6-PPDQ) were significantly ($P < 0.01$) different from those of wild-type (L4440)(6-PPDQ)

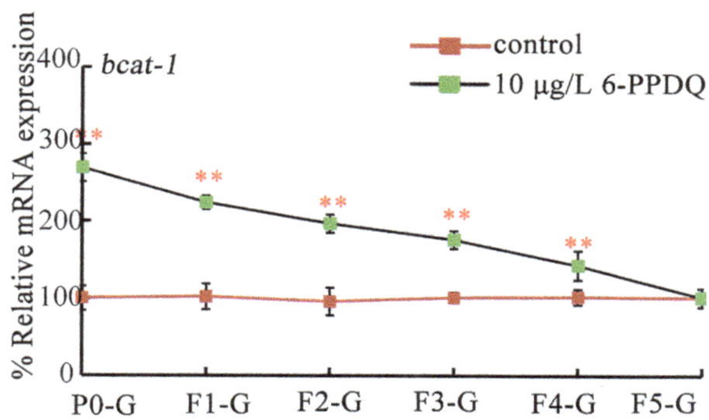

Fig. 12.12 Transgenerational response of leucine metabolism–related signal to 6-PPDQ. **$P < 0.01$ vs control

12.5 Role of Amino Acid Metabolic Signals

12.5.1 *Transgenerational Response of Leucine Metabolism–Related Signals to Pollutants*

After pollutant exposure, leucine metabolism–related signals could exhibit certain transgenerational responses. Exposure to pollutants, such as 6-PPDQ, could disrupt metabolisms of some amino acids, such as glutamate, methionine, and leucine [27–29]. At P0-G, 6-PPDQ reduced leucine content, which was related to activation of branched-chain amino acid transferase BCAT-1 governing catabolism of leucine to α-ketoisocaproate [29]. Moreover, 6-PPDQ exposure induced transgenerational activation of BCAT-1, and transgenerational activation of BCAT-1 induced by 10 μg/L 6-PPDQ could be observed till to the F4-G (Fig. 12.12) (Y.-X. Wang, and D.-Y., persona communication).

12.5.2 *Function of Leucine Metabolism–Related Signals in Regulating Toxicity of Pollutants*

Exposure to 6-PPDQ inhibited DBT-1 and IVD-1, the enzymes governing catabolism of α-ketoisocaproate to acetyl-CoA [29]. The 6-PPDQ-induced reduction in NADH and $FADH_2$ contents and mitochondrial dysfunction could be inhibited by RNAi of *bcat-1* and exacerbated by RNAi of *dbt-1* and *ivd-1* [29]. Meanwhile, RNAi of *bcat-1* caused activation of DBT-1 and IVD-1 [29], suggesting that the activation of BCAT-1 mediated 6-PPDQ toxicity on mitochondrial function by causing the inhibition in DBT-1 and IVD-1. Moreover, 6-PPDQ-induced intestinal oxidative stress, neurotoxicity, and reproductive toxicity were also lowered by RNAi of *bcat-1* and strengthened by RNAi of *dbt-1* and *ivd-1* [29]. These observations implied that transgenerational activation of BCAT-1 and transgenerational inhibition in DBT-1 and IVD-1 induced by pollutants possibly mediated transgenerational toxicity induction, such as transgenerational mitochondrial dysfunction.

12.5.3 Pharmacological Effect of Leucine Treatment on Toxicity of Pollutants

With 6-PPDQ as example of pollutants, 6-PPDQ-induced toxicity could be inhibited by following leucine treatment. The 6-PPDQ-induced mitochondrial dysfunction and reduction in NADH and $FADH_2$ contents were inhibited by leucine treatment [29]. Similarly, 6-PPDQ-induced intestinal oxidative stress, neurotoxicity, and reproductive toxicity were inhibited by leucine treatment [29]. Moreover, decrease in *dbt-1* and *ivd-1* expressions caused by 6-PPDQ were also inhibited by leucine treatment [29], suggesting that leucine has beneficial effect against pollutant toxicity by activating DBT-1 and IVD-1. These observations implied the important role of metabolic molecules, such as leucine, in modulating transgenerational toxicity of pollutants.

12.6 Role of Retinoic Acid Synthesis–Related Signals

12.6.1 Transgenerational Response of Retinoic Acid Synthesis–Related Signals to Pollutants

Exposure to pollutants could induce transgenerational response of retinoic acid synthesis–related signals. With 6-PPDQ as example of pollutants, its exposure caused inhibition in retinoic acid content [30]. This reduction in retinoic acid content was associated with inhibition in ALH-3 (enzyme governing metabolism from retinol to retinal) and DHS-19 (enzyme governing metabolism from retinal to retinoic acid) [30]. Among potential retinoic acid receptors, 6-PPDQ exposure inhibited the SEX-1, and the SEX-1 was further inhibited by RNAi of *alh-3* and *dhs-19* [30]. Moreover, after exposure to 10 μg/L 6-PPDQ, transgenerational inhibition in ALH-3, DHS-19, and SEX-1 was observed till to the F4-G (Fig. 12.13) (Y.-X Wang and D.-Y. Wang, personal communication).

12.6.2 Function of Retinoic Acid Synthesis–Related Signals in Regulating Toxicity of Pollutants

ALH-3, DHS-19, and SEX-1 were involved in control of pollutant toxicity. The 6-PPDQ-induced intestinal oxidative stress, neurotoxicity, and reproductive toxicity were exacerbated by RNAi of *alh-3*, *dhs-19*, and *sex-1* [30]. During this process, PMK-1, BAR-1, and DAF-16 were identified as downstream targets of ALH-3, DHS-19, and SEX-1 [30]. That is, transgenerational inhibition in ALH-3, DHS-19, and SEX-1 possibly mediated induction of transgenerational pollutant toxicity.

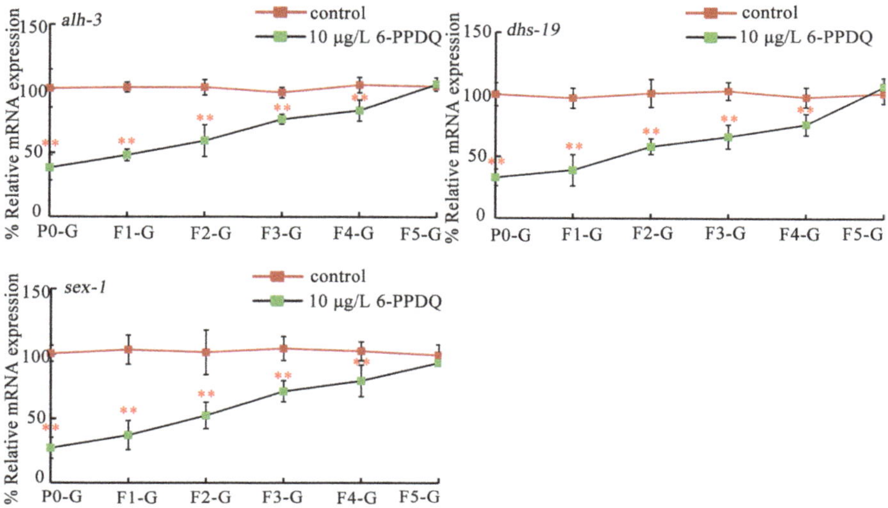

Fig. 12.13 Transgenerational response of retinoic acid metabolism–related signals to 6-PPDQ. $^{**}P < 0.01$ vs control

12.6.3 Pharmacological Effect of Retinoic Acid Treatment on Toxicity of Pollutants

The retinoic acid treatment could suppress toxicity of pollutants. The intestinal oxidative stress, neurotoxicity, and reproductive toxicity caused by 6-PPDQ could be inhibited by following retinoic acid treatment [30]. Therefore, retinoic acid is another metabolic molecule with the potential function in modulating transgenerational pollutant toxicity.

12.7 Perspectives

First of all, exposure to pollutants could induce certain forms of transgenerational response of metabolic molecules. For example, transgenerational elevation of glucose content and lipid accumulation were caused by 6-PPDQ. The 6-PPDQ-induced transgenerational response of glucose and lipid metabolism–related signals further supported the observed transgenerational elevation of glucose content and lipid accumulation. Besides these, transgenerational response of leucine and retinoic acid metabolisms and CAC-related signals could be induced by pollutant exposure, suggesting the possible transgenerational response of amino acids, retinoic acid, and CAC metabolites. Therefore, transgenerational response of metabolic molecules and corresponding metabolism-related signals may be the very conserved and commonly observed phenotype after pollutant exposure at the P0-G.

Accompanied with transgenerational response of metabolism-related signals induced by pollutants, function of some metabolism related signals in regulating pollutants toxicity could be confirmed. This suggested the crucial role of metabolism-related signals in modulating transgenerational pollutant toxicity. That is, these metabolism-related signals may function together with germline ligands and germline epigenetic regulation–related signals to regulate the transgenerational pollutant toxicity. Moreover, pharmacological rescue analysis indicated that treatment with some metabolic molecules, such as leucine and retinoic acid, could reverse the toxicity induced by pollutants. This further suggested the possibility that these metabolic molecules potentially directly participate in modulating the transgenerational pollutant toxicity after the formation of transgenerational response of these metabolic molecules.

References

1. Wang D-Y (2018) Nanotoxicology in *Caenorhabditis elegans*. Springer Nature Singapore Pte Ltd., Singapore
2. Wang D-Y (2020) Exposure toxicology in *Caenorhabditis elegans*. Springer Nature Singapore Pte Ltd., Singapore
3. Wang D-Y (2022) Toxicology at environmentally relevant concentrations in *Caenorhabditis elegans*. Springer Nature Singapore Pte Ltd., Singapore
4. Wang D-Y (2019) Molecular toxicology in *Caenorhabditis elegans*. Springer Nature Singapore Pte Ltd., Singapore
5. Wang D-Y (2019) Target organ toxicology in *Caenorhabditis elegans*. Springer Nature Singapore Pte Ltd., Singapore
6. Hua X, Feng X, Liang G-Y, Chao J, Wang D-Y (2023) Long-term exposure to tire-derived 6-PPD quinone causes intestinal toxicity by affecting functional state of intestinal barrier in *Caenorhabditis elegans*. Sci Total Environ 861:160591
7. Hua X, Feng X, Liang G-Y, Chao J, Wang D-Y (2023) Exposure to 6-PPD quinone at environmentally relevant concentrations causes abnormal locomotion behaviors and neurodegeneration in *Caenorhabditis elegans*. Environ Sci Technol 57:4940–4950
8. Hua X, Feng X, Liang G-Y, Chao J, Wang D-Y (2023) Long-term exposure to 6-PPD quinone reduces reproductive capacity by enhancing germline apoptosis associated with activation of both DNA damage and cell corpse engulfment in *Caenorhabditis elegans*. J Hazard Mater 454:131495
9. Hua X, Wang D-Y (2023) Exposure to 6-PPD quinone at environmentally relevant concentrations inhibits both lifespan and healthspan in *C. elegans*. Environ Sci Technol 57:19295–19303
10. Liu Z-Y, Bian Q, Wang D-Y (2024) Exposure to 6-PPD quinone causes ferroptosis activation associated with induction of reproductive toxicity in *Caenorhabditis elegans*. J Hazard Mater 471:134356
11. Wu J-W, Shen S-H, Wang D-Y (2024) 6-PPD quinone at environmentally relevant concentrations induces immunosenescence by causing immunosuppression during the aging process. Chemosphere 368:143719
12. Wang Y-X, Liang G-Y, Chao J, Wang D-Y (2024) Comparison of intestinal toxicity in enhancing intestinal permeability and in causing ROS production of six PPD quinones in *Caenorhabditis elegans*. Sci Total Environ 927:172306
13. Hua X, Liang G-Y, Chao J, Wang D-Y (2024) Exposure to 6-PPD quinone causes damage on mitochondrial complex I/II associated with lifespan reduction in *Caenorhabditis elegans*. J Hazard Mater 472:134598

14. Hua X, Wang D-Y (2025) 6-PPD quinone causes alteration in ubiquinone-mediated complex III associated with toxicity on mitochondrial function and longevity in *Caenorhabditis elegans*. J Environ Chem Eng 13:116571
15. Hua X, Wang D-Y (2025) 6-PPD quinone at environmentally relevant concentrations activates feedback response of electron transport chain to mediate damage on mitochondrial function and longevity in *Caenorhabditis elegans*. Environ Chem Ecotoxicol 7:2356–2365
16. Hua X, Wang D-Y (2024) 6-PPD quinone at environmentally relevant concentrations induced damage on longevity in *C. elegans*: mechanistic insight from inhibition in mitochondrial UPR response. Sci Total Environ 954:176275
17. Hua X, Wang D-Y (2025) Environmentally relevant concentration of 6-PPD quinone inhibits two types of mitophagy to cause mitochondrial dysfunction and lifespan reduction in *Caenorhabditis elegans*. Environ Sci Processes Impacts 27:1928–1940
18. Wang Y-X, Wang D-Y (2024) Transgenerational intestinal toxicity of 6-PPD quinone in causing ROS production, enhancement in intestinal permeability and suppression in innate immunity in *C. elegans*. Environ Pollut 363:125208
19. Liu Z-J, Li Y-H, Wang D-Y (2024) Exposure to 6-PPD quinone disrupts glucose metabolism associated with lifespan reduction by affecting insulin and AMPK signals in *Caenorhabditis elegans*. Chemosphere 363:142975
20. Wang Y-X, Wang D-Y (2024) Exposure to 6-PPD quinone enhances glycogen accumulation in *Caenorhabditis elegans*. Environ Pollut 359:124600
21. Liu Z-J, Li Y-H, Wang D-Y (2024) Transgenerational response of glucose metabolism in *Caenorhabditis elegans* exposed to 6-PPD quinone. Chemosphere 367:143653
22. Wan X, Liang G-Y, Wang D-Y (2025) 6-PPD quinone at environmentally relevant concentrations disrupts citric acid cycle in *Caenorhabditis elegans*: role of reduction in acetyl CoA and pyruvate contents. Environ. Chem. Ecotoxicol. 7:1119–1129
23. Li Z, Yu Z, Yin D (2021) Multi- and trans-generational disturbances of perfluorobutane sulfonate and perfluorohexane sulfonate on lipid metabolism in *Caenorhabditis elegans*. Chemosphere 280:130666
24. Wang Y-X, Wu J-W, Wang D-Y (2025) 6-PPD quinone causes lipid accumulation across multiple generations differentially affected by metabolic sensors and components of COMPASS complex in *Caenorhabditis elegans*. Environ Pollut 366:125539
25. Wang Y-X, Hua X, Wang D-Y (2023) Exposure to 6-PPD quinone enhances lipid accumulation through activating metabolic sensors of SBP-1 and MDT-15 in *Caenorhabditis elegans*. Environ Pollut 333:121937
26. Han S, Schroeder EA, Silva-Garcia CG, Hebestreit K, Mair WB, Brunet A (2017) Monounsaturated fatty acids link H3K4me3 modifiers to *C. elegans* lifespan. Nature 544:185–190
27. Wang W, Li Y-H, Wang D-Y (2025) Long-term exposure to 6-PPD quinone inhibits glutamate synthesis and glutamate receptor function associated with its toxicity induction in *Caenorhabditis elegans*. Toxics 13:434
28. Wang Y-X, Hu G-C, Wang D-Y (2025) Increased S-adenosyl methionine strengthens the suppression in mitochondrial unfolded protein response induced by 6-PPD quinone at environmentally relevant concentrations in *Caenorhabditis elegans*. Environ Pollut 386:127231
29. Wang W, Li Y-H, Wang D-Y (2025) Exposure to 6-PPD quinone disrupts adsorption and catabolism of leucine to cause mitochondrial dysfunction in *Caenorhabditis elegans*. Toxics 13:544
30. Hu D-Y, Wang Y-X, Hu G-C, Liu R, Wang D-Y (2025) 6-PPD quinone inhibited retinoic acid synthesis mediates toxicity through feedback loop between ALH-3/DHS-19-SEX-1 axis and intestinal signals in *Caenorhabditis elegans*. J Environ Expo Assess 4:40

Chapter 13
Regulation of Transgenerational Toxicity by Germline Small RNAs and RNAi-Associated Signaling

Abstract Regulatory RNAs represent an important form of epigenetic regulation of gene expression. In *Caenorhabditis elegans*, certain germline microRNAs (miRNAs) and germline long non-coding RNAs (lncRNAs) showed transgenerational response to pollutant exposure. In the germline, transgenerational response of miRNAs (such as *mir-38* and *mir-240*) regulated transgenerational pollutant toxicity by targeting certain downstream targets, such as NDK-1, NHL-3, WRT-3, and RAB-5. Transgenerational response of germline lncRNAs (such as *linc-7*) controlled transgenerational pollutant toxicity by targeting certain downstream targets, such as transcriptional factor DAF-12. Moreover, exposure to stresses could reset the transgenerational endogenously derived heritable small RNA silencing, which was under the control of Argonaute protein HRDE-1.

Keywords Transgenerational toxicology · MicroRNA · lncRNA · RNAi-associated signaling · *Caenorhabditis elegans*

13.1 Introduction

Caenorhabditis elegans is not only suitable for evaluating toxicity of stresses and pollutants at low doses and concentrations [1–3], but it also provides a useful platform for toxicological study, especially the study of molecular toxicology [4, 5]. Among the molecular signals involved in control of toxicity of stresses and pollutants, regulatory RNAs represent a form of epigenetic regulation.

Among the regulatory RNAs, as key post-transcriptional regulators, microRNAs (miRNAs) are small, single-stranded, and non-coding RNAs with typically ~22 nucleotides long. The miRNAs can silence gene expression by targeting certain mRNAs to cause the translational repression or mRNA degradation. In *C. elegans*, some miRNAs showed the response to exposure to stresses and pollutants and participated in controlling toxicity of stresses and pollutants [6–22]. The long non-coding RNAs (lncRNAs) are RNAs with longer than 200 nucleotides and do not code for functional proteins. In the cytoplasm, the lncRNAs regulate the gene

D. Wang, *Transgenerational Toxicology in Caenorhabditis elegans*,
https://doi.org/10.1007/978-981-92-1389-4_13

expression by influencing mRNA stability, translation, or degradation and competing with mRNAs for miRNA binding. In the nucleus, the lncRNAs are involved in controlling chromatin remodeling. Certain *C. elegans* lncRNAs exhibited the response to exposure to stresses and pollutants and were required for the control of toxicity of stresses and pollutants [23–26]. The circular RNAs (circRNAs) are circular non-coding RNAs generated through back-splicing and act as key post-transcriptional regulators by sponging miRNAs and proteins and scaffolding complexes. In *C. elegans*, certain circRNAs, such as *circ_0000115*, were also involved in regulating toxicity of pollutants, such as graphene oxide, a carbon-based nanomaterial [27].

In this chapter, we first introduced the transgenerational responses of germline miRNAs and germline lncRNAs to pollutant exposure and discussed the role of germline miRNAs and germline lncRNAs in regulating the induction of transgenerational pollutant toxicity. In addition, considering that the RNAi-associated signaling forms a crucial bridge between environmental stress and transgenerational inheritance, the possible role of RNAi-associated signaling in modulating transgenerational toxicity of stresses and pollutants was further discussed.

13.2 Role of Germline miRNAs

13.2.1 Involvement of Germline Argonaute Protein in Controlling Transgenerational Pollutant Toxicity

The *C. elegans* miRNA biogenesis is governed by Argonaute proteins [28, 29]. ALG-1, ALG-2, ALG-3, and ALG-5 are germline-expressed Argonaute proteins, and their transgenerational response implies the potential formation of transgenerational response of miRNAs. Nanoplastics is a family of emergent contaminants receiving the increasing attention for their environmental exposure risk [30]. Nanoplastic exposure caused multiple toxicities on nematodes, including lifespan reduction, metabolic disruption, and organ toxicity [31–42]. Nanoplastic exposure could further result in the severe transgenerational toxicity, such as transgenerational neurotoxicity and transgenerational reproductive toxicity [43–46]. Polystyrene nanoparticle (PS-NP) is a widely examined member of nanoplastics and selected as example of pollutants. Among these germline Argonaute proteins, activation of germline ALG-2 and inhibition in germline ALG-3 were induced by 1–100 μg/L PS-NPs (Fig. 13.1) [47]. After exposure to 1 μg/L PS-NPs, transgenerational activation of germline ALG-2 and transgenerational inhibition in germline ALG-3 were also induced (Fig. 13.1) [47]. These observations suggested that transgenerational change of germline ALG-2 and germline ALG-3 induced by pollutants mediated different transgenerational responses of germline miRNAs. Moreover, transgenerational neurotoxicity

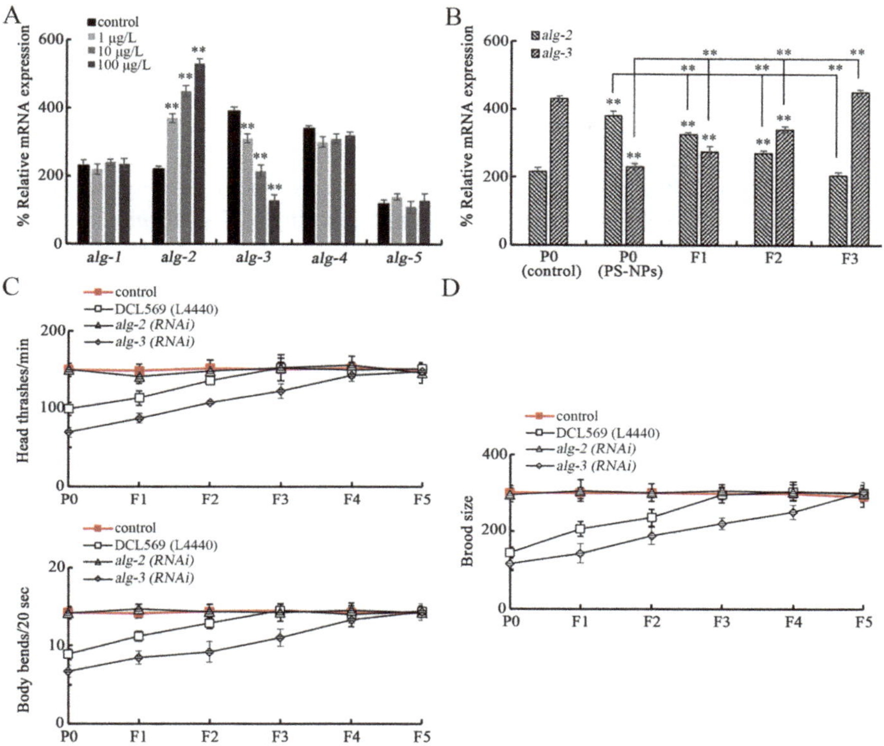

Fig. 13.1 Requirement of germline Argonaute genes for the control of transgenerational PS-NP toxicity [47]. (**a**) Effect of PS-NP exposure on expression of genes encoding Argonaute proteins. Bars represent means ± SD. $^{**}P < 0.01$ vs Control. (**b**) Transgenerational expressions of germline *alg-2* and *alg-3* after PS-NP exposure. A total of 30 intact gonads were used for the qRT-PCR assay. Exposure concentration of PS-NPs was 1 μg/L. Bars represent means ± SD. $^{**}P < 0.01$ vs P0 (control) (if not specially indicated). (**c**) Effect of germline RNAi knockdown of *alg-2* or *alg-3* on transgenerational PS-NP toxicity in decreasing locomotion behavior. Exposure concentration of PS-NPs was 1 μg/L. Bars represent means ± SD. The curves of DCL569(L4440) ($P < 0.01$) were significantly different from those of Control. The curves of *alg-2 (RNAi)* ($P < 0.01$) and *alg-3(RNAi)* ($P < 0.01$) were significantly different from those of DCL569(L4440). (**d**) Effect of germline RNAi knockdown of *alg-2* or *alg-3* on transgenerational PS-NP toxicity in reducing brood size. Exposure concentration of PS-NPs was 1 μg/L. Bars represent means ± SD. The curve of DCL569(L4440) ($P < 0.01$) was significantly different from that of Control. The curves of *alg-2(RNAi)* ($P < 0.01$) and *alg-3(RNAi)* ($P < 0.01$) were significantly different from that of DCL569(L4440)

and transgenerational reproductive toxicity of PS-NPs could be suppressed by germline RNAi of *alg-2* and increased by germline RNAi of *alg-3* (Fig. 13.1) [47]. This further suggested that, in the germline, ALG-2 and ALG-3 have opposite function in modulating the transgenerational pollutant toxicity.

13.2.2 *Screen of Germline miRNAs Involved in the Control of Transgenerational Pollutant Toxicity*

Among the miRNAs, at least 12 miRNAs can be expressed in the germline (Fig. 13.2) [48]. With the PS-NPs as the example, inhibition in germline *mir-38* and activation of *mir-240* were caused by 1–100 μg/L PS-NP exposure (Fig. 13.3) [47, 48]. Among the 12 germline miRNAs, although germline RNAi of *mir-36* caused the resistance to reproductive toxicity and neurotoxicity of PS-NPs (Fig. 13.2) [48], germline *mir-36* expression was not changed by 1–100 μg/L PS-NPs (Fig. 13.3) [48]. Therefore, so far, germline *mir-38* and germline *mir-240* were identified to have the potential role in regulating transgenerational pollutant toxicity, and they showed the opposite responses to pollutant exposure.

13.2.3 *Role of Germline* mir-240

13.2.3.1 Transgenerational Response of Germline *mir-240* to Pollutants

Transgenerational activation of germline *mir-240* could be induced by pollutant exposure at the parental generation (P0-G). After exposure to 1 μg/L PS-NPs, transgenerational activation of germline *mir-240* was induced, and the activation of germline *mir-240* could be observed till to the F2-G (Fig. 13.3) [48].

13.2.3.2 Function of Germline *mir-240* in Regulating Transgenerational Toxicity of Pollutants

Germline RNAi of *mir-240* conferred the resistance to transgenerational pollutant toxicity. PS-NP was used as example of pollutants. The transgenerational neurotoxicity and transgenerational reproductive toxicity of PS-NPs were inhibited by germline RNAi of *mir-240* (Fig. 13.4) [48]. That is, transgenerational activation of germline *mir-240* mediated induction of transgenerational pollutant toxicity.

13.2.3.3 Identification of Direct Downstream Targets for Germline *mir-240* in Regulating Transgenerational Toxicity of Pollutants

In the germline, RAB-5 was identified as the direct target of germline *mir-240* to regulate the transgenerational pollutant toxicity. Among predicted targeted genes of *mir-240*, ten genes can be expressed in the germline (Fig. 13.5) [48]. Among these predicted ten targets, inhibition in germline RAB-5 and RAB-62 was caused by 1–100 μg/L PS-NPs, and transgenerational inhibition in germline RAB-5 and RAB-6.2 could be further induced by PS-NP exposure (Fig. 13.5) [48]. Nevertheless, after PS-NP exposure, germline RNAi of *mir-240* could activate the germline

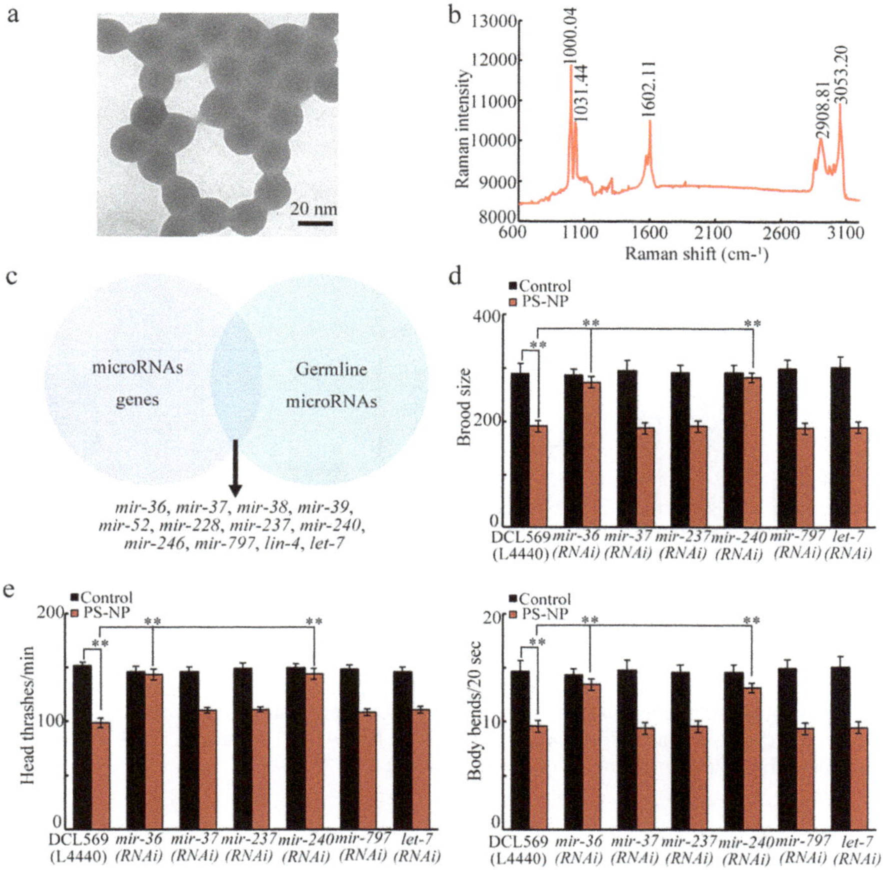

Fig. 13.2 Identification of germline microRNAs involved in regulating PS-NP toxicity [48]. (**a**) TEM image of PS-NP before sonication. (**b**) Raman spectrum of PS-NP. (**c**) microRNAs expressed in the germline of nematodes. (**d**) Effect of germline RNAi of candidate microRNAs on the brood size in PS-NP exposed nematodes. (**e**) Effect of germline RNAi of candidate microRNAs on the locomotion behavior in PS-NP-exposed nematodes. Exposure concentration of PS-NP was 1 μg L^{-1}. $^{**}P < 0.01$

RAB-6, but not the germline RAB-6.2 (Fig. 13.5) [48], suggesting that the RAB-5, an endosomal Rab-type GTPase, is the potential target of *mir-240* in the germline.

Functional analysis further demonstrated the role of RAB-5 as the direct target of germline *mir-240* in regulating transgenerational pollutant toxicity. Germline RNAi of *rab-5* caused the susceptibility to transgenerational neurotoxicity and transgenerational reproductive toxicity of PS-NPs (Fig. 13.6) [48]. Moreover, the resistance to transgenerational neurotoxicity and transgenerational reproductive toxicity of PS-NPs induced by germline RNAi of *mir-240* was inhibited by germline RNAi of *rab-5*, and the *rab-5(RNAi);mir-240(RNAi)* showed the susceptibility to transgenerational neurotoxicity and transgenerational reproductive toxicity of PS-NPs (Fig. 13.7) [48].

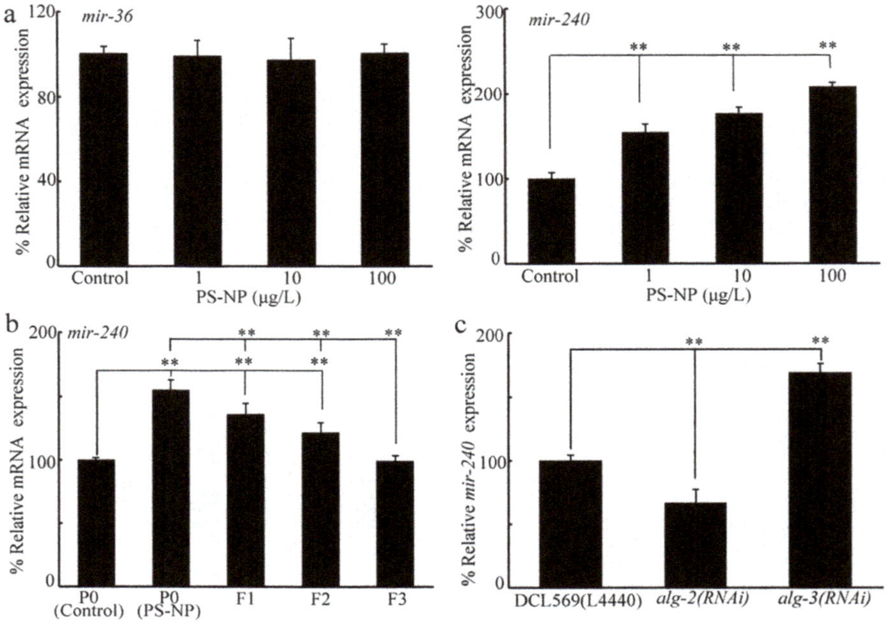

Fig. 13.3 Effect of PS-NP exposure on the expressions of *mir-36* and *mir-240* [48]. (**a**) Effect of PS-NP exposure on the germline expressions of *mir-36* and *mir-240* at P0-G. (**b**) Transgenerational expression of germline *mir-240* after PS-NP exposure. Exposure concentration of PS-NP was 1 μg L^{-1}. (**c**) Effect of germline RNAi of *alg-2* and *alg-3* on the *mir-240* expression in PS-NP-exposed nematodes. Exposure concentration of PS-NP was 1 μg L^{-1}. $^{**}P < 0.01$

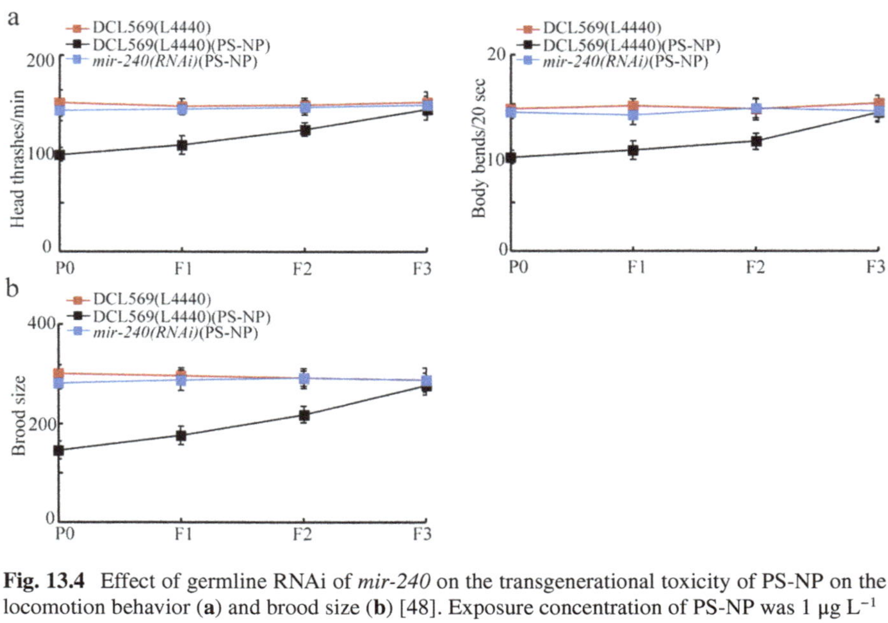

Fig. 13.4 Effect of germline RNAi of *mir-240* on the transgenerational toxicity of PS-NP on the locomotion behavior (**a**) and brood size (**b**) [48]. Exposure concentration of PS-NP was 1 μg L^{-1}

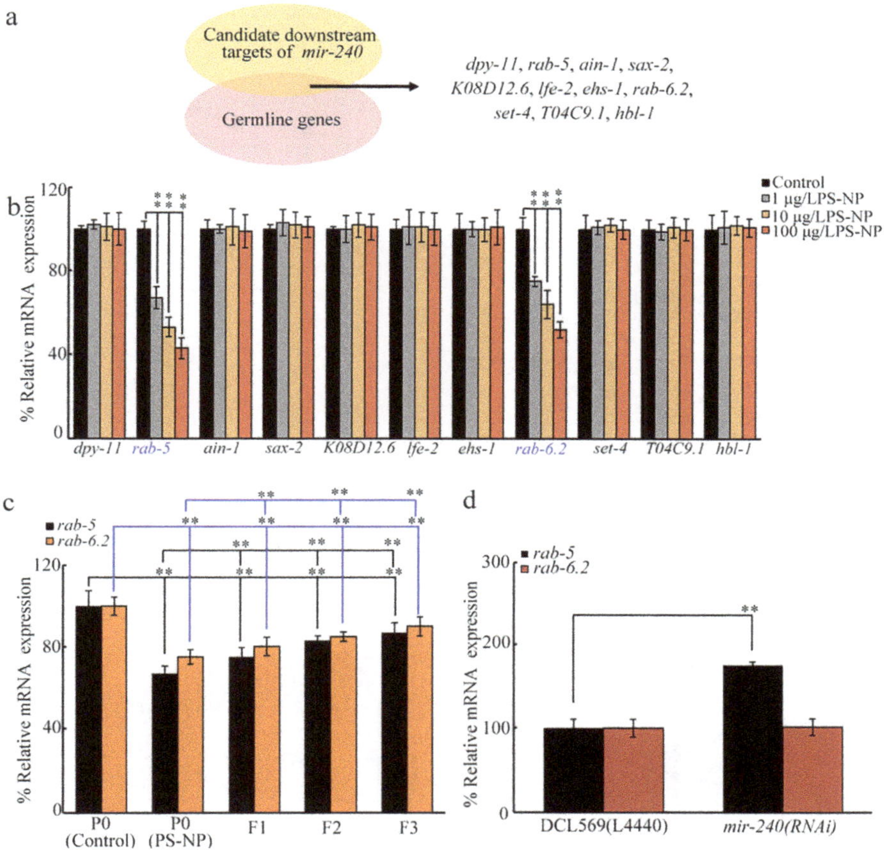

Fig. 13.5 Identification of potential targets of germline *mir-240* in regulating transgenerational PS-NP toxicity [48]. (**a**) Predicted germline targets of *mir-240*. (**b**) Effect of PS-NP exposure on the germline expressions of predicted germline target genes. $^{**}P < 0.01$ vs. control. (**c**) Transgenerational expressions of germline *rab-5* and *rab-6.2* after PS-NP exposure. Exposure concentration of PS-NP was 1 μg L^{-1}. $^{**}P < 0.01$. (**d**) Effect of the germline RNAi of *mir-240* on the expressions of *rab-5* and *rab-6.2* in PS-NP-exposed nematodes. Exposure concentration of PS-NP was 1 μg L^{-1}. $^{**}P < 0.01$

13.2.3.4 Identification of Secreted Ligands as Downstream Targets for Germline RAB-5 in Regulating Transgenerational Toxicity of Pollutants

Some secreted ligands including INS-3, INS-29, DAF-28, EGL-17, and EFN-3 were identified as downstream targets of RAB-5 in the germline to regulate transgenerational pollutant toxicity. In *C. elegans*, some secreted ligands functioned together with their receptors to regulate the transgenerational pollutant toxicity [49–53]. Among them, after PS-NP exposure, activation of insulin ligands (INS-3,

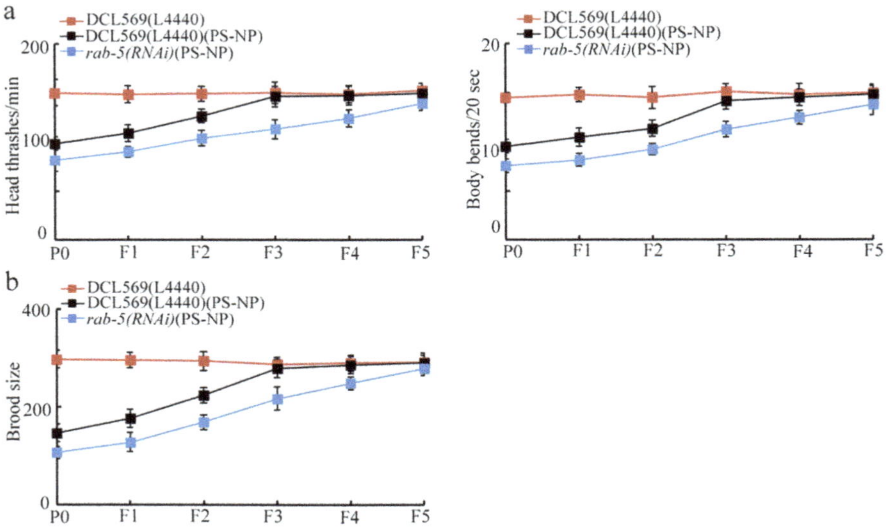

Fig. 13.6 Effect of the germline RNAi of *rab-5* on transgenerational PS-NP toxicity on the locomotion behavior (**a**) and brood size (**b**) [48]. Exposure concentration of PS-NP was 1 μg L^{-1}

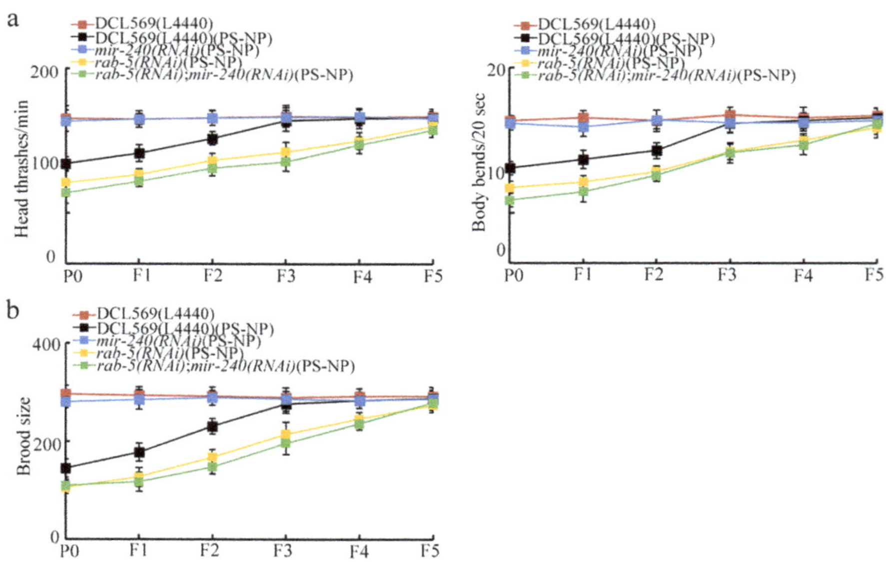

Fig. 13.7 Genetic interaction between *mir-240* and RAB-5 for regulating the transgenerational PS-NP toxicity on the locomotion behavior (**a**) and brood size (**b**) [48]. Exposure concentration of PS-NP was 1 μg L^{-1}

INS-29, DAF-28), FGF ligand (EGL-17), and Ephrin ligand (EFN-3) was caused by germline RNAi of *rab-5* (Fig. 13.8) [48]. Additionally, germline RNAi of *rab-5* resulted in activation of their corresponding receptors (DAF-2, EGL-15, and VAB-1) (Fig. 13.8) [48], which further supports the observation above. Therefore, germline RAB-5 regulates transgenerational pollutant toxicity by modulating generation and/ or secretion of certain ligands.

13.2.3.5 RNAi of *alg-2* and *alg-3* Had Opposite Function in Affecting Germline *mir-240* Expression

In the germline, ALG-2 and ALG-3 functioned as core upregulators for activation or inhibition of miRNAs. After PS-NP exposure, germline *mir-240* expression was activated by germline RNAi of *alg-3* and inhibited by germline RNAi of *alg-2* (Fig. 13.3) [48]. Therefore, function of germline *mir-240* in regulating transgenerational pollutant toxicity was under the control of its upregulators ALG-2 and ALG-3.

13.2.4 Role of Germline mir-38

13.2.4.1 Transgenerational Response of Germline *mir-38* to Pollutants

At the P0-G, *mir-38* could act in the germline to regulate the pollutant toxicity [54]. Exposure to PS-NPs caused inhibition in germline *mir-38* (Fig. 13.9) [47]. After exposure to 1 μg/L PS-NPs, transgenerational inhibition in germline *mir-38* was observed, and the inhibition in germline *mir-38* could be detected till to the F2-G (Fig. 13.9) [47].

13.2.4.2 Function of Germline *mir-38* in Regulating Transgenerational Toxicity of Pollutants

Germline overexpression of *mir-38* conferred the resistance to transgenerational pollutant toxicity. The transgenerational neurotoxicity and transgenerational reproductive toxicity of PS-NPs were inhibited by germline overexpression of *mir-38* (Fig. 13.10) [47]. That is, transgenerational inhibition in germline *mir-38* mediated induction of transgenerational pollutant toxicity (Fig. 13.10) [47].

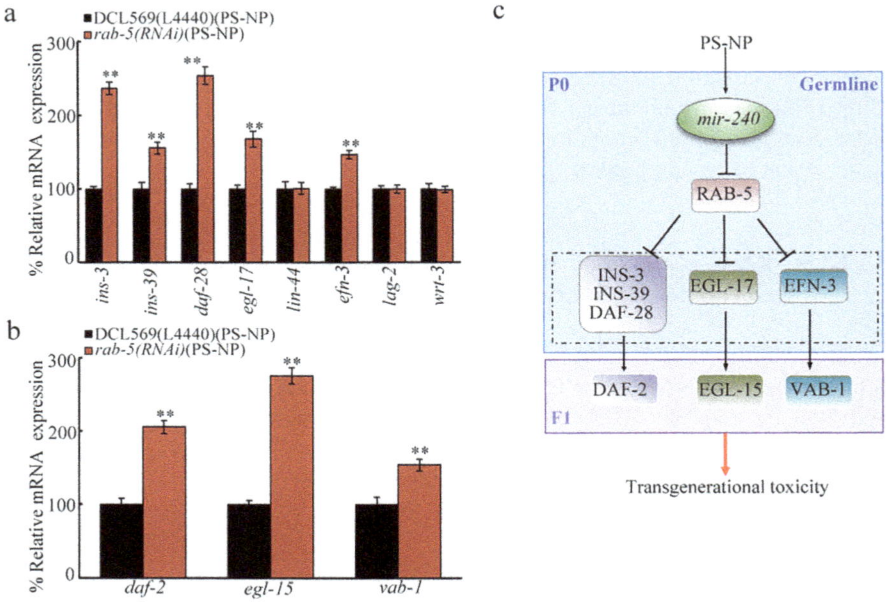

Fig. 13.8 Identification of targets of germline RAB-5 in regulating transgenerational PS-NP toxicity [48]. (**a**) Effect of germline RNAi of *rab-5* on expressions of *ins-3*, *ins-39*, *daf-28*, *egl-17*, *lin-44*, *efn-3*, *lag-2*, and *wrt-3* in PS-NP-exposed nematodes. Exposure concentration of PS-NP was 1 μg L^{-1}. $^{**}P < 0.01$ vs. DCL569(L4440). (**b**) Effect of germline RNAi of *rab-5* on the expressions of *daf-2*, *egl-15*, and *vab-1* in the F1-G of PS-NP-exposed nematodes. Exposure concentration of PS-NP was 1 μg L^{-1}. $^{**}P < 0.01$ vs. DCL569(L4440). (**c**) A diagram showing the molecular basis for *mir-240* and RAB-5 in the germline to regulate transgenerational PS-NP toxicity

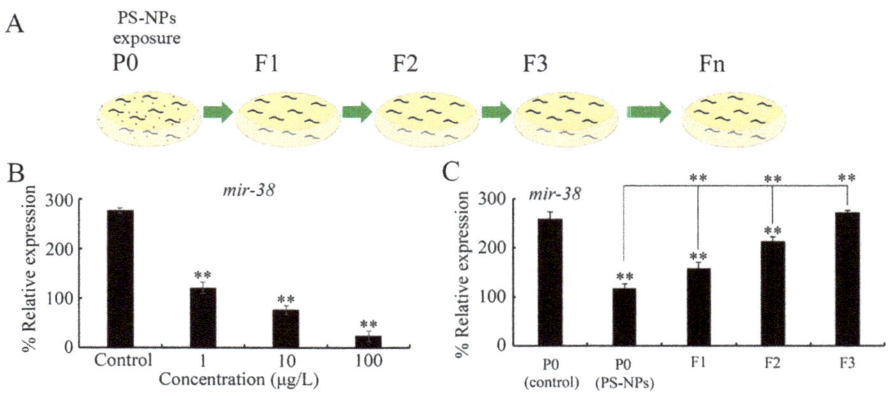

Fig. 13.9 Expression of germline *mir-38* in PS-NP-exposed nematodes [47]. (**a**) Assay system for assessment of transgenerational PS-NP toxicity. (**b**) Effect of 1–100 μg/L PS-NP exposure on germline *mir-38* expression. A total of 30 intact gonads were used for the qRT-PCR assay. Bars represent means ± SD. $^{**}P < 0.01$ vs Control. (**c**) Transgenerational expressions of germline *mir-38* after PS-NP exposure. A total of 30 intact gonads were used for the qRT-PCR assay. Exposure concentration of PS-NPs was 1 μg/L. Bars represent means ± SD. $^{**}P < 0.01$ vs P0 (control) (if not specially indicated)

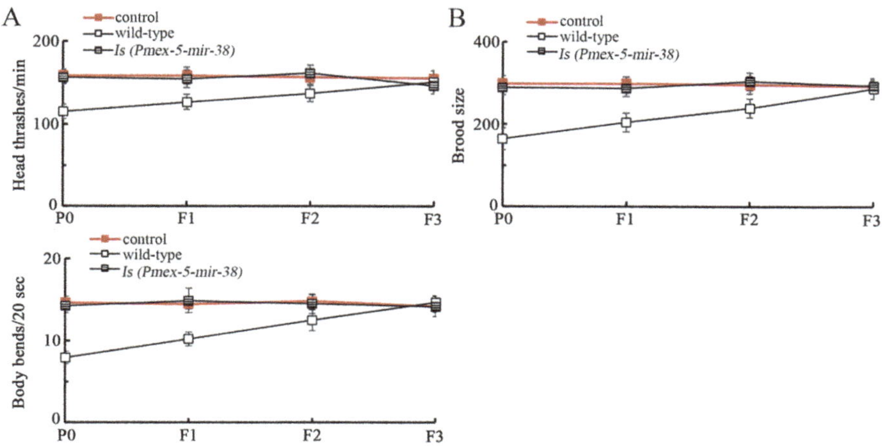

Fig. 13.10 Effect of germline overexpression of *mir-38* on transgenerational PS-NP toxicity [47]. (**a**) Effect of germline overexpression of *mir-38* on transgenerational PS-NP toxicity in decreasing locomotion behavior. Bars represent means ± SD. The curves of wild-type ($P < 0.01$) were significantly different from those of Control. The curves of *Is(Pmex-5-mir-38)* ($P < 0.01$) were significantly different from those of wild type. (**b**) Effect of germline overexpression of *mir-38* on transgenerational PSNP toxicity in reducing brood size. Exposure concentration of PS-NPs was 1 μg/L. Bars represent means ± SD. The curve of wild-type ($P < 0.01$) was significantly different from that of Control. The curve of *Is(Pmex-5-mir-38)* ($P < 0.01$) was significantly different from that of wild type

13.2.4.3 Identification of Direct Downstream Targets for Germline *mir-38* in Regulating Transgenerational Toxicity of Pollutants

In the germline, NDK-1, NHL-3, and WRT-3 were identified as downstream targets of *mir-38* in regulating transgenerational pollutant toxicity. Exposure to PS-NPs caused transgenerational activation of germline NDK-1, NHL-3, and WRT-3, and germline RNAi of *ndk-1*, *nhl-3*, and *wrt-3* resulted in resistance to transgenerational neurotoxicity and transgenerational reproductive toxicity of PS-NPs (Fig. 13.11) [47]. These observations indicated that transgenerational activation of germline NDK-1, NHL-3, and WRT-3 mediated transgenerational PS-NP toxicity. After PS-NP exposure, germline overexpression of *mir-38* inhibited expressions of germline *ndk-1*, *nhl-3*, and *wrt-3* (Fig. 13.11) [47], which suggested the role of NDK-1, NHL-3, and WRT-3 as downstream targets of germline *mir-38* to regulate transgenerational PS-NP toxicity. Germline overexpression of *ndk-1* containing its 3′-UTR caused the susceptibility to transgenerational neurotoxicity and transgenerational reproductive toxicity of PS-NPs (Fig. 13.12) [47]. Moreover, in nematodes with germline overexpression of *ndk-1* containing its 3′-UTR, the further germline overexpression of *mir-38* resulted in the resistance to transgenerational neurotoxicity and transgenerational reproductive toxicity of PS-NPs (Fig. 13.12) [47], which suggested the role of NDK-1 as the direct target of germline *mir-38* during the control of this process.

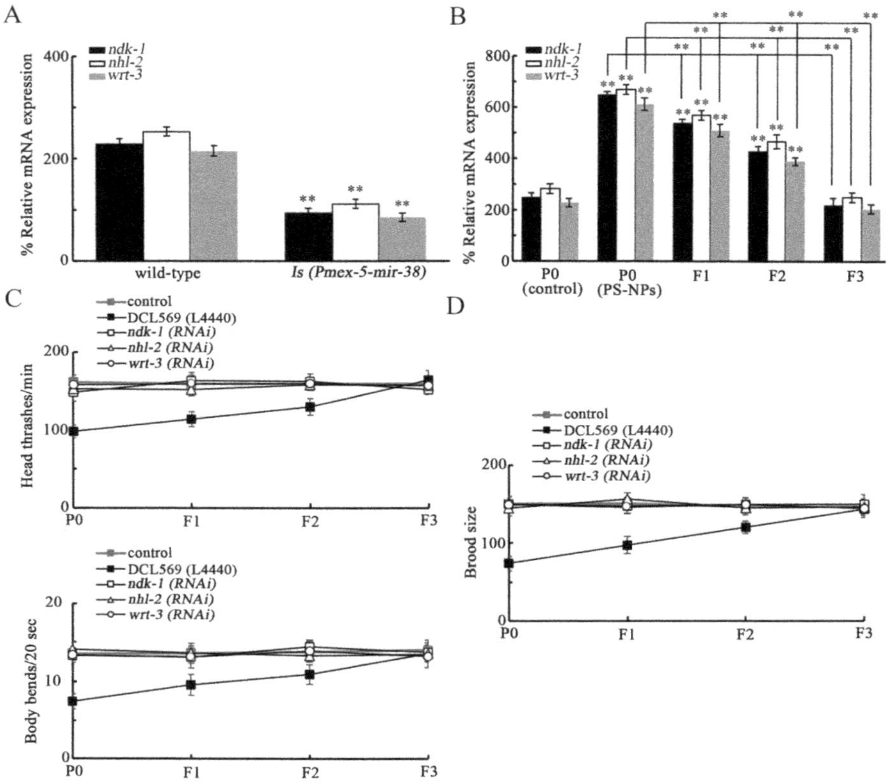

Fig. 13.11 Identification of potential targets of germline *mir-38* in controlling transgenerational PS-NP toxicity [47]. (**a**) Effect of germline overexpression *mir-38* on expressions of *ndk-1*, *nhl-2*, and *wrt-3* in PS-NP exposed wild-type nematodes. A total of 30 intact gonads were used for the qRT-PCR assay. Bars represent means ± SD. $^{**}P < 0.01$ vs wild type. (**b**) Transgenerational expressions of germline *ndk-1*, *nhl-2*, and *wrt-3* after PS-NP exposure. A total of 30 intact gonads were used for the qRT-PCR assay. Bars represent means ± SD. $^{**}P < 0.01$ vs P0 (control) (if not specially indicated). (**c**) Effect of germline RNAi knockdown of *ndk-1*, *nhl-2*, and *wrt-3* on transgenerational PS-NP toxicity in decreasing locomotion behavior. L4440, empty vector. Bars represent means ± SD. The curves of DCL569(L4440) ($P < 0.01$) were significantly different from those of Control. The curves of *ndk-1(RNAi)* ($P < 0.01$), *nhl-2(RNAi)* ($P < 0.01$), and *wrt-3(RNAi)* ($P < 0.01$) were significantly different from those of DCL569(L4440). (**d**) Effect of germline RNAi knockdown of *ndk-1*, *nhl-2*, and *wrt-3* on transgenerational PS-NP toxicity in reducing brood size. L4440, empty vector. Bars represent means ± SD. The curve of DCL569(L4440) ($P < 0.01$) was significantly different from that of Control. The curves of *ndk-1(RNAi)* ($P < 0.01$), *nhl-2(RNAi)* ($P < 0.01$), and *wrt-3(RNAi)* ($P < 0.01$) were significantly different from that of DCL569(L4440). Exposure concentration of PS-NPs was 1 μg/L

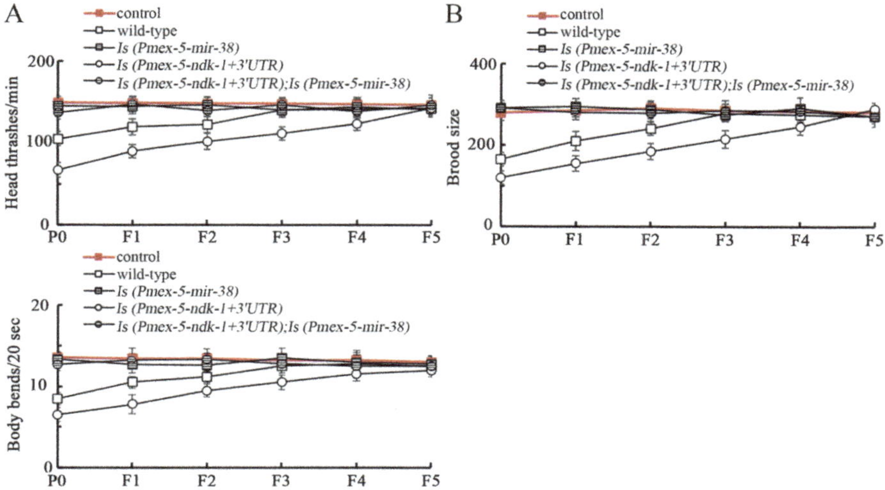

Fig. 13.12 Effects of germline overexpression of *mir-38* on transgenerational PS-NP toxicity in decreasing locomotion behavior (**a**) and reducing brood size (**b**) in nematodes overexpressing germline *ndk-1* containing 3′ UTR [47]. Exposure concentration was 1 μg/L. Bars represent means ± SD. The curves of wild-type ($P < 0.01$) were significantly different from those of Control. The curves of *Is(Pmex-5-mir-38)* ($P < 0.01$) and *Is(Pmex-5-ndk-1+3′UTR)* ($P < 0.01$) were significantly different from those of wild type. The curves *Is(Pmex-5-ndk-1+3′UTR);Is(Pmex-5-mir-38)* ($P < 0.01$) were significantly different from those of *Is(Pmex-5-ndk-1+3′UTR)*

13.2.4.4 Identification of KSR-1/2 as Downstream Targets for Germline NDK-1 in Regulating Transgenerational Toxicity of Pollutants

KSR-1 and KSR-2, two kinase suppressors of Ras, were identified as downstream targets for germline NDK-1 in regulating transgenerational pollutant toxicity. Exposure to PS-NPs induced transgenerational inhibition in KSR-1 and KSR-2, and germline RNAi of *ksr-1* and *ksr-2* led to the susceptibility to transgenerational neurotoxicity and transgenerational reproductive toxicity of PS-NPs (Fig. 13.13) [47], which suggested the role of transgenerational inhibition in KSR-1 and KSR-2 to mediate transgenerational pollutant toxicity. The first evidence to indicate the role of KSR-1 and KSR-2 as downstream targets of germline NDK-1 in regulating transgenerational pollutant toxicity is that germline RNAi of *ndk-1* caused activation of germline of KSR-1 and KSR-2 in PS-NP-exposed nematodes (Fig. 13.13) [47]. Second, the resistance to transgenerational neurotoxicity and transgenerational reproductive toxicity of PS-NPs induced by germline RNAi of *ndk-1* was inhibited by germline RNAi of *ksr-1* and *ksr-2*, and the *ndk-1(RNAi);ksr-1(RNAi)* and *ndk-1(RNAi);ksr-2(RNAi)* showed the susceptibility to transgenerational neurotoxicity and transgenerational reproductive toxicity of PS-NPs (Fig. 13.13) [47].

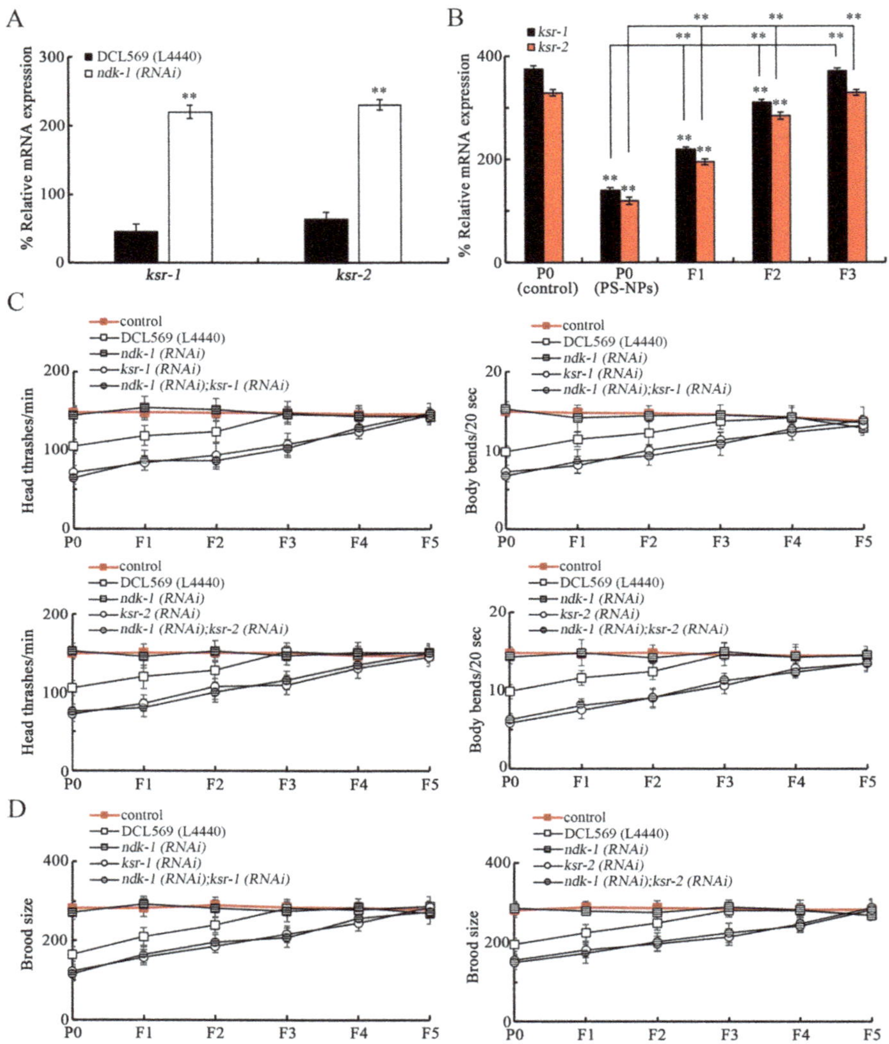

Fig. 13.13 Genetic interaction between germline NDK-1 and KSR-1 or KSR-2 in controlling transgenerational PS-NP toxicity [47]. (**a**) Effect of germline RNAi knockdown of *ndk-1* on expressions of germline *ksr-1* and *ksr-2* in PS-NP-exposed nematodes. A total of 30 intact gonads were used for the qRT-PCR assay. Exposure concentration was 1 μg/L. Bars represent means ± SD. $^{**}P < 0.01$ vs DCL569 (L4440). (**b**) Transgenerational expressions of germline *ksr-1* and *ksr-2* after PS-NP exposure. A total of 30 intact gonads were used for the qRT-PCR assay. Bars represent means ± SD. $^{**}P < 0.01$ vs P0 (control) (if not specially indicated). (**c**) Genetic interaction between germline NDK-1 and KSR-1 or KSR-2 in controlling transgenerational PS-NP toxicity to decrease locomotion behavior. Exposure concentration of PS-NPs was 1 μg/L. Bars represent means ± SD. The curves of DCL569(L4440) ($P < 0.01$) were significantly different from those of Control. The curves of *ndk-1(RNAi)* ($P < 0.01$), *ksr-1(RNAi)* ($P < 0.01$), and *ksr-2(RNAi)* ($P < 0.01$) were significantly different from those of DCL569(L4440). The curves of *ndk-1(RNAi);ksr-1(RNAi)* ($P < 0.01$) and *ndk-1(RNAi);ksr-2(RNAi)* ($P < 0.01$) were significantly different from those of *ndk-1(RNAi)*. (**d**) Genetic interaction between germline NDK-1 and KSR-1 or KSR-2 in controlling

13.2.4.5 RNAi of *alg-3* Affected Germline *mir-38* Expression

After PS-NP exposure, germline RNAi of *alg-2* did not influence the germline *mir-38* expression (Fig. 13.14) [47]. Nevertheless, germline RNAi of *alg-3* caused the inhibition in germline *mir-38* (Fig. 13.14) [47]. This suggested that the transgenerational response of germline *mir-38* to pollutant exposure was mediated by the transgenerational response of ALG-3.

13.3 Role of Germline lncRNAs

13.3.1 Screen of Germline lncRNAs Involved in Control of Pollutant Toxicity

In *C. elegans*, certain lncRNAs were involved in the control of pollutant toxicity. Multi-walled carbon nanotubes (MWCNTs) are a member of carbon-based nanomaterials. Exposure to MWCNTs could induce multiple aspects of damage on *C. elegans*, including organ toxicity and immunosuppression [7, 11, 14, 15, 55–61]. MWCNT was selected as example of pollutants. Among the known lncRNAs expressed in the germline based on the available information, exposure to MWCNTs induced activation of germline *linc-7* and germline *linc-32* and inhibition in germline *linc-2*, germline *linc-9*, and germline *linc-50* (Fig. 13.15) [62]. Nevertheless, MWCNT-induced neurotoxicity and reproductive toxicity were not affected by germline RNAi of *linc-2*, *linc-9*, *linc-32*, and *linc-50* (Fig. 13.16) [62]. In contrast, MWCNT-induced neurotoxicity and reproductive toxicity could be inhibited by germline RNAi of *linc-7* (Fig. 13.16) [62], suggesting the involvement of germline *linc-7* in regulating pollutant toxicity.

13.3.2 Role of Germline linc-7

13.3.2.1 Transgenerational Response of Germline *linc-7* to Pollutants

Exposure to pollutants could induce transgenerational activation of germline *linc-7*. With the MWCNT as example of pollutants, MWCNT exposure activated the germline *linc-7* (Fig. 13.17) [62]. Moreover, transgenerational activation of germline

Fig. 13.13 (continued) transgenerational PS-NP toxicity to reduce brood size. Exposure concentration of PS-NPs was 1 μg/L. Bars represent means ± SD. The curves of DCL569 (L4440) ($P < 0.01$) were significantly different from those of Control. The curves of *ndk-1(RNAi)* ($P < 0.01$), *ksr-1(RNAi)* ($P < 0.01$), and *ksr-2(RNAi)* ($P < 0.01$) were significantly different from those of DCL569(L4440). The curves of *ndk-1(RNAi);ksr-1(RNAi)* ($P < 0.01$) and *ndk-1(RNAi);ksr-2(RNAi)* ($P < 0.01$) were significantly different from those of *ndk-1(RNAi)*. L4440, empty vector

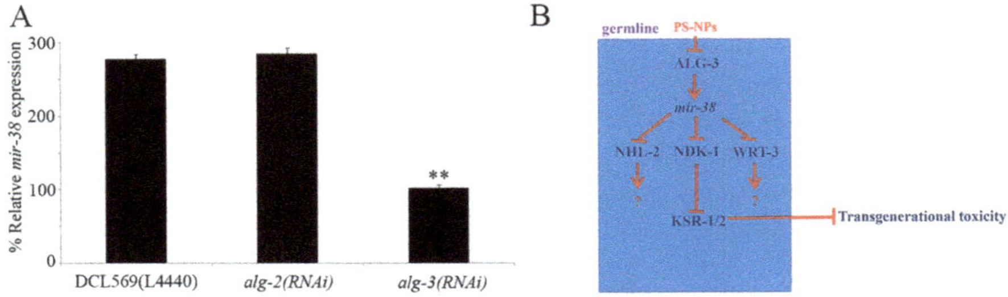

Fig. 13.14 Germline RNAi knockdown of *alg-3* affected *mir-38* expression in PS-NP-exposed nematodes [47]. (**a**) Effect of germline RNAi knockdown of *alg-2* or *alg-3* on *mir-38* expression after PS-NP exposure. A total of 30 intact gonads were used for the qRT-PCR assay. L4440, empty vector. Exposure concentration of PS-NPs was 1 μg/L. Bars represent means ± SD. $^{**}P < 0.01$ vs DCL569(L4440). (**b**) A diagram showing molecular basis for germline *mir-38* in controlling transgenerational PS-NP toxicity

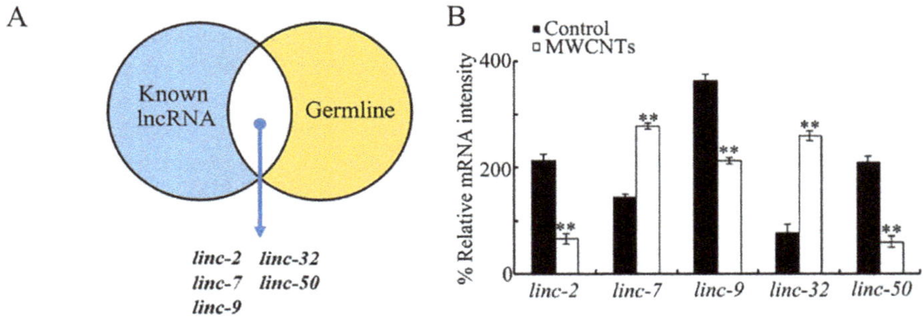

Fig. 13.15 Effect of MWCNT exposure on lncRNA expressions in wild-type nematodes [62]. (**a**) Known lncRNAs expressed in the germline. (**b**) Effect of MWCNT exposure on expressions of *linc-2*, *linc-7*, *linc-9*, *linc-32*, and *linc-50* in wild-type nematodes. MWCNT exposure concentration was 0.1 μg/L. Bars represent means ± SD. $^{**}P < 0.01$ vs Control

linc-7 could be induced by MWCNT exposure, and transgenerational activation of germline *linc-7* was observed till to the F2-G after exposure to 0.1 μg/L MWCNTs (Fig. 13.17) [62].

13.3.2.2 Function of Germline *linc-7* in Regulating Transgenerational Toxicity of Pollutants

Germline RNAi of *linc-7* caused the resistance to transgenerational pollutants toxicity. In *C. elegans*, exposure to MWCNTs caused transgenerational neurotoxicity and transgenerational reproductive toxicity (Fig. 13.17) [62]. After germline RNAi of *linc-7*, transgenerational neurotoxicity and transgenerational reproductive toxicity caused by MWCNTs were inhibited (Fig. 13.17) [62]. Therefore, transgenerational activation of germline *linc-7* mediated the induction of transgenerational pollutant toxicity.

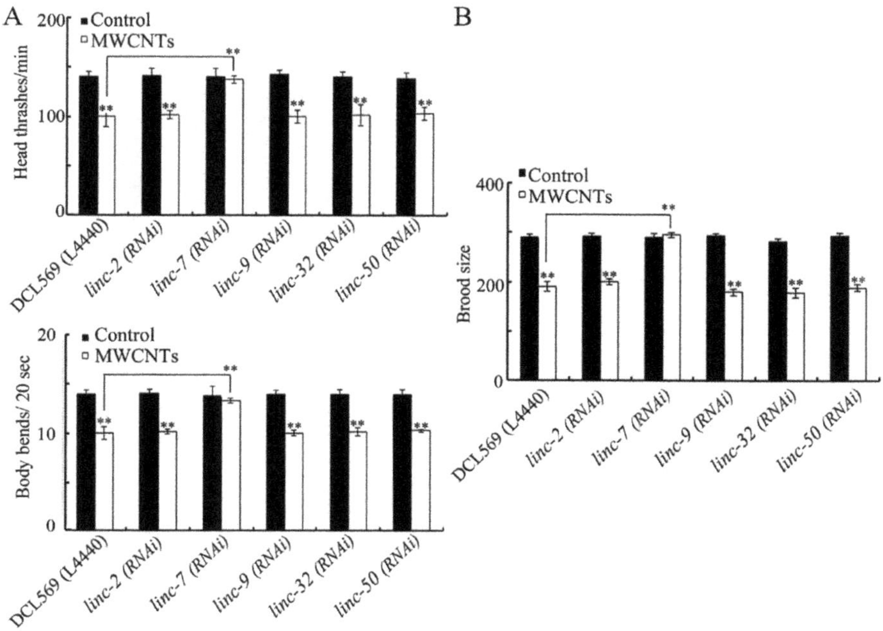

Fig. 13.16 Identification of germline lncRNAs involved in the control of MWCNT toxicity [62]. (**a**) Effect of germline RNAi knockdown of *linc-2*, *linc-7*, *linc-9*, *linc-32*, and *linc-50* on MWCNT toxicity in decreasing locomotion behavior. (**b**) Effect of germline RNAi knockdown of *linc-2*, *linc-7*, *linc-9*, *linc-32*, and *linc-50* on MWCNT toxicity in reducing brood size. L4440, empty vector. MWCNT exposure concentration was 0.1 μg/L. Bars represent means ± SD. $^{**}P < 0.01$ vs Control (if not specially indicated)

13.3.2.3 Identification of Direct Downstream Targets for Germline *linc-7* in Regulating Transgenerational Toxicity of Pollutants

Certain transcriptional factors (TFs) could be identified as downstream targets for *linc-7* in the germline to regulate transgenerational pollutant toxicity. In organisms, TFs play pivotal functions by regulating expressions of multiple downstream targeted genes. Among the TFs with the potential binding ability with *linc-7*, exposure to MWCNTs induced inhibition in DAF-16, DVE-1, CBP-1, and SKN-1 and activation of DAF-12 (Fig. 13.18) [62]. Among these five candidate TFs, DVE-1, DAF-16, and SKN-1 functioned in the intestine to regulate the pollutant toxicity [41, 63, 64]. After MWCNT exposure, although germline *linc-7* did not alter germline CBP-1 expression, inhibition in germline DAF-12 was caused by germline RNAi of *linc-7* (Fig. 13.18) [62]. Meanwhile, transgenerational activation of germline DAF-12 could be induced by MWCNT exposure (Fig. 13.18) [62], suggesting the possible involvement of germline DAF-12 in regulating transgenerational pollutant toxicity.

The functional analysis further demonstrated the role of DAF-12 as the downstream target of germline *linc-7* to regulate the transgenerational pollutant

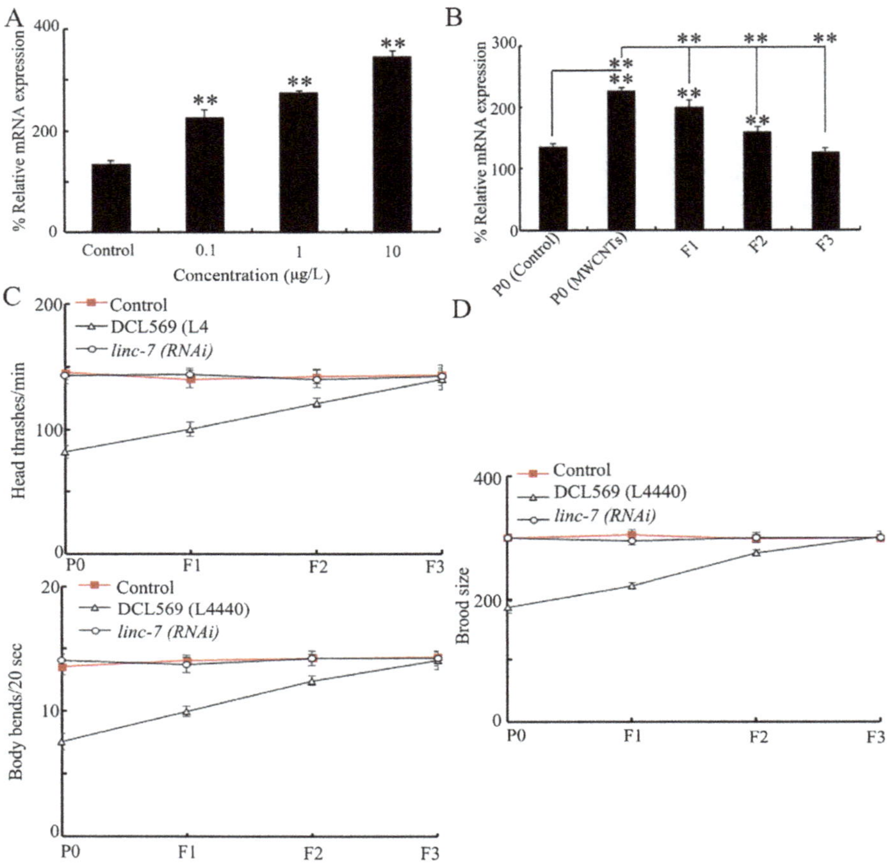

Fig. 13.17 Germline RNAi knockdown of *linc-7* suppressed transgenerational MWCNT toxicity [62]. (**a**) Effect of MWCNT (0.1–10 μg/L) exposure on germline *linc-7* expression. A total of 30 intact gonads were used for the qRT-PCR assay. Bars represent means ± SD. $^{**}P < 0.01$ vs Control. (**b**) Transgenerational expression of germline *linc-7* after MWCNT exposure. A total of 30 intact gonads were used for the qRT-PCR assay. MWCNT exposure concentration was 0.1 μg/L. Bars represent means ± SD. $^{**}P < 0.01$ vs Control (if not specially indicated). (**c**) Effect of germline knockdown of *linc-7* on transgenerational toxicity of MWCNT in decreasing locomotion behavior. L4440, empty vector. MWCNT exposure concentration was 0.1 μg/L. Bars represent means ± SD. (**d**) Effect of germline knockdown of *linc-7* on transgenerational toxicity of MWCNT in reducing brood size. L4440, empty vector. MWCNT exposure concentration was 0.1 μg/L. Bars represent means ± SD

toxicity. Germline RNAi of *daf-12* led to the resistance to transgenerational neurotoxicity and transgenerational reproductive toxicity of MWCNTs (Fig. 13.19) [62]. In contrast, germline *linc-7* overexpression caused the susceptibility to the transgenerational neurotoxicity and transgenerational reproductive toxicity of MWCNTs (Fig. 13.19) [62]. More importantly, the susceptibility to the transgenerational neurotoxicity and transgenerational reproductive toxicity of MWCNTs

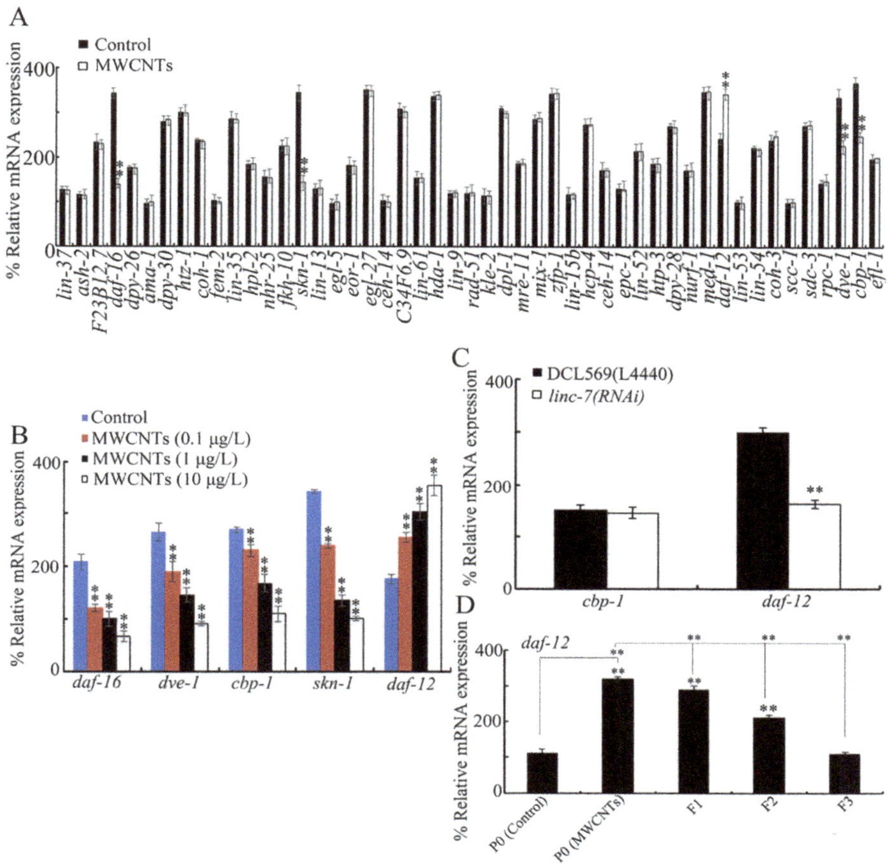

Fig. 13.18 Identification of potential downstream transcriptional factors for germline *linc-7* in regulating transgenerational MWCNT toxicity [62]. (**a**) Effect of MWCNT exposure on expressions of genes encoding transcriptional factors. MWCNT exposure concentration was 0.1 μg/L. Bars represent means ± SD. $^{**}P < 0.01$ vs Control. (**b**) Effect of exposure to MWCNT (0.1–10 μg/L) on expressions of *daf-16*, *dve-1*, *cbp-1*, *skn-1*, and *daf-12*. Bars represent means ± SD. $^{**}P < 0.01$ vs Control. (**c**) Effect of germline RNAi knockdown of *linc-7* on expressions of germline *cbp-1* and *daf-12* in MWCNT-exposed nematodes. L4440, empty vector. A total of 30 intact gonads were used for the qRT-PCR assay. MWCNT exposure concentration was 0.1 μg/L. Bars represent means ± SD. $^{**}P < 0.01$ vs DCL569. (**d**) Transgenerational expression of germline *daf-12* after MWCMT exposure. A total of 30 intact gonads were used for the qRT-PCR assay. MWCNT exposure concentration was 0.1 μg/L. Bars represent means ± SD. $^{**}P < 0.01$ vs Control (if not specially indicated)

induced by germline *linc-7* overexpression could be inhibited by RNAi of *daf-12*, and the *Is(Pmex-5-linc-7)daf-12(RNAi)* showed the resistance to transgenerational neurotoxicity and transgenerational reproductive toxicity of MWCNTs (Fig. 13.19) [62].

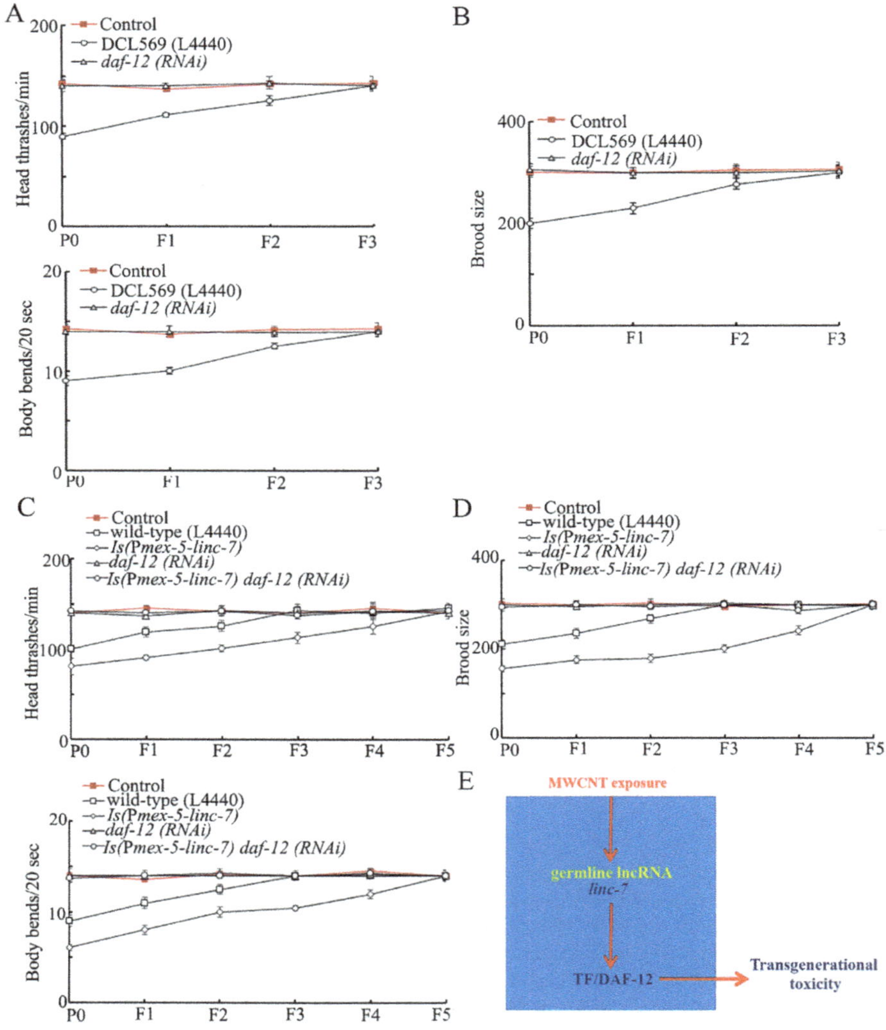

Fig. 13.19 Genetic interaction between DAF-12 and germline *linc-7* in regulating transgenerational MWCNT toxicity [62]. (**a**) Effect of germline RNAi knockdown of *daf-12* on transgenerational toxicity of MWCNT in decreasing locomotion behavior. (**b**) Effect of germline RNAi knockdown of *daf-12* on transgenerational toxicity of MWCNT in reducing brood size. (**c**) Genetic interaction between DAF-12 and germline *linc-7* in regulating transgenerational MWCNT toxicity in decreasing locomotion behavior. (**d**) Genetic interaction between DAF-12 and germline *linc-7* in regulating transgenerational MWCNT toxicity in reducing brood size. L4440, empty vector. MWCNT exposure concentration was 0.1 μg/L. Bars represent means ± SD. (**e**) A diagram showing molecular basis for germline *linc-7* in regulating transgenerational MWCNT toxicity

13.4 Role of RNAi Signaling

13.4.1 Transgenerational Endogenously Derived Heritable Small RNA Silencing Can Be Reset by Stresses

In *C. elegans*, after exposure to stresses, such as heat stress, osmotic stress, and starvation, transgenerational heritable small RNA silencing was found to be reset [65]. Among them, transgenerational endo-siRNA- and piRNA-induced silencing could be reset by heat stress [65]. After exposure to stresses, genome-wide small RNA changes were formed [65]. During this process, PMK-1/p38 MAPK and SKN-1/TF regulated the transgenerational small RNAs resetting in response to stresses [65]. Additionally, stress-induced transgenerational small RNA resetting also depended on the function of H3K9 methyltransferase MET-2 [65].

13.4.2 Role of HRDE-1

HRDE-1 is an Argonaute protein. In *C. elegans*, mutation of *hrde-1* strengthened the transgenerational reproductive toxicity induced by temperature shift [66]. Mutation of *hrde-1* enhanced the activation in mRNA expression of a subset of heat-induced genes, which suggested that the germline HRDE-1 prevents the heat activation in mRNA expression of target genes [66]. Many of the affected genes were in or near long-terminal repeat retrotransposons, and the heat stress-induced activation of some of these genes was intensified in *hrde-1* mutant nematodes [66]. Additionally, mutation of *hrde-1* led to siRNA expression changes of some of these genes, which was affected by both examined temperatures and generations [66]. These observations implied that germline nuclear RNAi may antagonize the transgenerational response induced by heat stress.

Besides this, germline HRDE-1 also regulated the chemotaxis behavior under the stress condition by governing generation of some RDE-4-dependent small RNAs in the neurons [67]. These germline small RNAs could result in the amplified endogenous small interfering RNAs (siRNAs) and the following transgenerational response of germline genes [67]. For the molecular control, SAEG-2 was identified as the downstream target of candidate small RNAs [67]. The transgenerational inhibition in SAEG-2 was induced, and silencing of *saeg-2* following neuronal small RNA biogenesis was required for chemotaxis behavior under the stress condition [67], which suggested the small-RNA-based mechanism for communication of neuronal processes transgenerationally.

Moreover, the HRDE-1 regulated transgenerational inheritance of sexual attractiveness via affecting small RNAs under the heat stress condition [68]. Under heat stress condition, hermaphrodites secreted male-attracting pheromone, which can be transmitted transgenerationally under the control of heritable small RNAs and HRDE-1 [68]. Response of endo-siRNAs that target sperm genes was identified to transgenerationally regulate the sexual attraction [68]. That is, reproductive organs or cells, such as the sperms, can act as the "stress-sensor" to show the transgenerational response to stresses.

13.5 Perspectives

Among *C. elegans* regulatory RNAs, so far, only two germline miRNAs (*mir-38* and *mir-240*) and one germline lncRNA (*linc-7*) were identified to regulate the transgenerational pollutant toxicity. In contrast, the germline circRNAs possibly involved in regulating the transgenerational pollutant toxicity still remain unclear. In addition, the other germline miRNAs and lncRNAs with transgenerational response to other pollutants with more severe toxicity, as well as their functions, need to be further determined.

For the direct downstream targets of regulatory RNAs during the control of transgenerational pollutant toxicity, even for germline *mir-38* and *mir-240*, other direct targets may still exist in regulating the transgenerational toxicity of other pollutants with more severe toxicity. The forms of downstream targets for lncRNAs are more complex than those of miRNAs. For example, besides TFs, miRNAs can also act as the downstream targets of germline lncRNAs in regulating transgenerational pollutant toxicity. Accompanied with the future identification of germline circRNAs having the function to regulate transgenerational pollutant toxicity, their different forms of potential downstream targets in the germline need to be further identified.

Besides regulatory RNAs, certain RNAi-associated signals, such as HRDE-1, have been identified to regulate transgenerational toxicity of stresses. Nevertheless, the detailed molecular network involved in regulating transgenerational toxicity of stresses and pollutants remains unclear. Recently, three rules governing transgenerational small RNA inheritance have been described [69]. Additionally, certain signals (such as HSF-1) regulated small RNA inheritance [69], and some signals (such as RSD-2 and RSD-3) were required for maintenance of small RNA inheritance [70]. These observations provide important clues for further elucidation of related molecular network regulating transgenerational toxicity of stresses and pollutants.

References

1. Wang D-Y (2018) Nanotoxicology in *Caenorhabditis elegans*. Springer Nature Singapore Pte Ltd., Singapore
2. Wang D-Y (2020) Exposure toxicology in *Caenorhabditis elegans*. Springer Nature Singapore Pte Ltd., Singapore
3. Wang D-Y (2022) Toxicology at environmentally relevant concentrations in *Caenorhabditis elegans*. Springer Nature Singapore Pte Ltd., Singapore
4. Wang D-Y (2019) Molecular toxicology in *Caenorhabditis elegans*. Springer Nature Singapore Pte Ltd., Singapore

5. Wang D-Y (2019) Target organ toxicology in *Caenorhabditis elegans*. Springer Nature Singapore Pte Ltd., Singapore
6. Wu Q-L, Zhao Y-L, Zhao G, Wang D-Y (2014) microRNAs control of *in vivo* toxicity from graphene oxide in *Caenorhabditis elegans*. Nanomed Nanotechnol Biol Med 10:1401–1410
7. Zhao Y-L, Wu Q-L, Li Y-P, Nouara A, Jia R-H, Wang D-Y (2014) *In vivo* translocation and toxicity of multi-walled carbon nanotubes are regulated by microRNAs. Nanoscale 6:4275–4284
8. Zhao Y-L, Wu Q-L, Wang D-Y (2015) A microRNAs-mRNAs network involved in the control of graphene oxide toxicity in *Caenorhabditis elegans*. RSC Adv 5:92394–92405
9. Zhao Y-L, Wu Q-L, Wang D-Y (2016) An epigenetic signal encoded protection mechanism is activated by graphene oxide to inhibit its induced reproductive toxicity in *Caenorhabditis elegans*. Biomaterials 79:15–24
10. Yang R-L, Ren M-X, Rui Q, Wang D-Y (2016) A *mir-231*-regulated protection mechanism against the toxicity of graphene oxide in nematode *Caenorhabditis elegans*. Sci Rep 6:32214
11. Zhao Y-L, Yang J-N, Wang D-Y (2016) A microRNA-mediated insulin signaling pathway regulates the toxicity of multi-walled carbon nanotubes in nematode *Caenorhabditis elegans*. Sci Rep 6:23234
12. Zhi L-T, Yu Y-L, Li X-Y, Wang D-Y (2017) Molecular control of innate immune response to *Pseudomonas aeruginosa* infection by intestinal *let-7* in *Caenorhabditis elegans*. PLoS Pathog 13:e1006152
13. Zhi L-T, Yu Y-L, Jiang Z-X, Wang D-Y (2017) *mir-355* functions as an important link between p38 MAPK signaling and insulin signaling in the regulation of innate immunity. Sci Rep 7:14560
14. Zhao L, Wan H-X, Liu Q-Z, Wang D-Y (2017) Multi-walled carbon nanotubes-induced alterations in microRNA *let-7* and its targets activate a protection mechanism by conferring a developmental timing control. Part Fibre Toxicol 14:27
15. Zhao Y-L, Jin L, Wang Y, Kong Y, Wang D-Y (2019) Prolonged exposure to multi-walled carbon nanotubes dysregulates intestinal *mir-35* and its direct target MAB-3 in nematode *Caenorhabditis elegans*. Sci Rep 9:12144
16. Zhao L, Dong S-S, Zhao Y-L, Shao H-M, Krasteva N, Wu Q-L, Wang D-Y (2019) Dysregulation of *let-7* by PEG modified graphene oxide in nematodes with deficit in epidermal barrier. Ecotoxicol Environ Saf 169:1–7
17. Qu M, Luo L-B, Yang Y-H, Kong Y, Wang D-Y (2019) Nanopolystyrene-induced microRNAs response in *Caenorhabditis elegans* after long-term and lose-dose exposure. Sci Total Environ 697:134131
18. Sun L-M, Li W-J, Li D, Wang D-Y (2020) microRNAs involved in the control of toxicity on locomotion behavior induced by simulated microgravity stress in *Caenorhabditis elegans*. Sci Rep 10:17510
19. Wang S-T, Liu H-L, Zhao Y-Y, Rui Q, Wang D-Y (2020) Dysregulated *mir-354* enhanced the protective response to nanopolystyrene by affecting the activity of TGF-β signaling pathway in nematode *Caenorhabditis elegans*. NanoImpact 20:100256
20. Qiu Y-X, Liu Y-Q, Li Y-H, Wang D-Y (2020) Intestinal *mir-794* responds to nanopolystyrene by linking insulin and p38 MAPK signaling pathways in nematode *Caenorhabditis elegans*. Ecotoxicol Environ Saf 201:110857
21. Li D, Yuan Y-J, Wang D-Y (2020) Regulation of response to nanopolystyrene by intestinal microRNA *mir-35* in nematode *Caenorhabditis elegans*. Sci Total Environ 736:139677
22. Liu H-L, Zhao Y-Y, Bi Q, Rui Q, Wang D-Y (2021) Dysregulated *mir-76* mediated a protective response to nanopolystyrene by modulating heme homeostasis related molecular signaling in nematode *Caenorhabditis elegans*. Ecotoxicol Environ Saf 212:112018
23. Wu Q-L, Zhou X-F, Han X-X, Zhuo Y-Z, Zhu S-T, Zhao Y-L, Wang D-Y (2016) Genome-wide identification and functional analysis of long noncoding RNAs involved in the response to graphene oxide. Biomaterials 102:277–291
24. Qu M, Zhao Y-L, Zhao Y-Y, Rui Q, Kong Y, Wang D-Y (2019) Identification of long noncoding RNAs in response to nanopolystyrene in *Caenorhabditis elegans* after long-term and low-dose exposure. Environ Pollut 255:113137

25. Sun L-M, Li D, Yuan Y-J, Wang D-Y (2021) Intestinal long non-coding RNAs in response to simulated microgravity stress in *Caenorhabditis elegans*. Sci Rep 11:1997
26. Zhao Y-Y, Xu R-R, Chen X, Wang J, Rui Q, Wang D-Y (2021) Induction of protective response to polystyrene nanoparticles associated with dysregulation of intestinal long non-coding RNAs in *Caenorhabditis elegans*. Ecotoxicol Environ Saf 212:111976
27. Shi L-F, Jia X-H, Guo T-T, Cheng L, Han X-X, Wu Q-L, Wang D-Y (2019) A circular RNA *circ_0000115* in response to graphene oxide in nematodes. RSC Adv 9:13722–13735
28. Bouasker S, Simard MJ (2012) The slicing activity of miRNA-specific Argonautes is essential for the miRNA pathway in *C. elegans*. Nucleic Acids Res 40:10452–10462
29. Zisoulis DG, Kai ZS, Chang RK, Pasquinelli AE (2012) Auto-regulation of miRNA biogenesis by *let-7* and Argonaute. Nature 486:541–544
30. Hua X, Wang D-Y (2022) Cellular uptake, transport, and organelle response after exposure to microplastics and nanoplastics: current knowledge and perspectives for environmental and health risks. Rev Environ Contam Toxicol 260:12
31. Qu M, Kong Y, Yuan Y-J, Wang D-Y (2019) Neuronal damage induced by nanopolystyrene particles in nematode *Caenorhabditis elegans*. Environ Sci Nano 6:2591–2601
32. Qu M, Qiu Y-X, Kong Y, Wang D-Y (2019) Amino modification enhances reproductive toxicity of nanopolystyrene on gonad development and reproductive capacity in nematode *Caenorhabditis elegans*. Environ Pollut 254:112978
33. Qu M, Nida A, Kong Y, Du H-H, Xiao G-S, Wang D-Y (2019) Nanopolystyrene at predicted environmental concentration enhances microcystin-LR toxicity by inducing intestinal damage in *Caenorhabditis elegans*. Ecotoxicol Environ Saf 183:109568
34. Qu M, Xu K-N, Li Y-H, Wong G, Wang D-Y (2018) Using *acs-22* mutant *Caenorhabditis elegans* to detect the toxicity of nanopolystyrene particles. Sci Total Environ 643:119–126
35. Qiu Y-X, Liu Y-Q, Li Y-H, Li G-J, Wang D-Y (2020) Effect of chronic exposure to nanopolystyrene on nematode *Caenorhabditis elegans*. Chemosphere 256:127172
36. Qiu Y-X, Luo L-B, Yang Y-H, Kong Y, Li Y-H, Wang D-Y (2020) Potential toxicity of nanopolystyrene on lifespan and aging process of nematode *Caenorhabditis elegans*. Sci Total Environ 705:135918
37. Li D, Deng Y-J, Wang S-T, Du H-H, Xiao G-S, Wang D-Y (2020) Assessment of nanopolystyrene toxicity under fungal infection condition in *Caenorhabditis elegans*. Ecotoxicol Environ Saf 197:110625
38. Liu H-L, Shao H-M, Guo Z-J, Wang D-Y (2020) Nanopolystyrene exposure activates a fat metabolism related signaling-mediated protective response in *Caenorhabditis elegans*. NanoImpact 17:100204
39. Dong S-S, Qu M, Rui Q, Wang D-Y (2018) Combinational effect of titanium dioxide nanoparticles and nanopolystyrene particles at environmentally relevant concentrations on nematodes *Caenorhabditis elegans*. Ecotoxicol Environ Saf 161:444–450
40. Zhao Y-Y, Li D, Rui Q, Wang D-Y (2020) Toxicity induction of nanopolystyrene under microgravity stress condition in *Caenorhabditis elegans*. Sci Total Environ 703:135623
41. Liu H-L, Wang D-Y (2021) Intestinal mitochondrial unfolded protein response induced by nanoplastic particles in *Caenorhabditis elegans*. Chemosphere 267:128917
42. Tang M, Ding G, Lu X, Huang Q, Du H, Xiao G, Wang D (2022) Exposure to nanoplastic particles enhances *Acinetobacter* survival, biofilm formation, and serum resistance. Nanomaterials 12:4222
43. Zhao L, Qu M, Wong G, Wang D-Y (2017) Transgenerational toxicity of nanopolystyrene particles in the range of μg/L in nematode *Caenorhabditis elegans*. Environ Sci Nano 4:2356–2366
44. Sun L-M, Liao K, Wang D-Y (2021) Comparison of transgenerational reproductive toxicity induced by pristine and amino modified nanoplastics in *Caenorhabditis elegans*. Sci Total Environ 768:144362
45. Liu H-L, Tian L-J, Wang S-T, Wang D-Y (2021) Size-dependent transgenerational toxicity induced by nanoplastics in nematode *Caenorhabditis elegans*. Sci Total Environ 790:148217

46. Zhao Y, Chen J, Wang R, Pu X, Wang D (2023) A review of transgenerational and multigenerational toxicology in the *in vivo* model animal *Caenorhabditis elegans*. J Appl Toxicol 43:122–145
47. Hua X, Zhao Y, Yuan Y-J, Zhang L, Bian Q, Wang D-Y (2022) Nanoplastics cause transgenerational toxicity through inhibiting germline microRNA *mir-38* in *C. elegans*. J Hazard Mater 437:129302
48. Hua X, Zhang L, Wang D-Y (2024) Nanoplastic at environmentally relevant concentrations activates germline *mir-240-rab-5* signaling cascade to affect secreted ligands associated with transgenerational toxicity induction in *C. elegans*. Environ Sci Nano 11:3615
49. Zhao Y, Hua X, Bian Q, Wang D-Y (2022) Nanoplastic exposure at predicted environmental concentrations induces activation of germline Ephrin signal associated with toxicity formation in the *Caenorhabditis elegans* offspring. Toxics 10:699
50. Xu R-R, Hua X, Rui Q, Wang D-Y (2022) Alteration in Wnt signaling mediates induction of transgenerational toxicity of polystyrene nanoplastics in *C. elegans*. NanoImpact 28:100425
51. Liu H-L, Zhao Y-L, Hua X, Wang D-Y (2022) Induction of transgenerational toxicity is associated with the activated germline insulin signals in nematodes exposed to nanoplastic at predicted environmental concentrations. Ecotoxicol Environ Saf 243:114022
52. He W-M, Gu A-H, Wang D-Y (2023) Sulfonate-modified polystyrene nanoparticle at precited environmental concentrations induces transgenerational toxicity associated with increase in germline Notch signal of *Caenorhabditis elegans*. Toxics 11:511
53. Hua X, Cao C, Zhang L, Wang D-Y (2023) Activation of FGF signal in germline mediates transgenerational toxicity of polystyrene nanoparticles at predicted environmental concentrations in *Caenorhabditis elegans*. J Hazard Mater 451:131174
54. Yang Y-H, Wu Q-L, Wang D-Y (2020) Epigenetic response to nanopolystyrene in germline of nematode *Caenorhabditis elegans*. Ecotoxicol Environ Saf 206:111404
55. Wu Q-L, Li Y-X, Li Y-P, Zhao Y-L, Ge L, Wang H-F, Wang D-Y (2013) Crucial role of biological barrier at the primary targeted organs in controlling translocation and toxicity of multi-walled carbon nanotubes in nematode *Caenorhabditis elegans*. Nanoscale 5:11166–11178
56. Nouara A, Wu Q-L, Li Y-X, Tang M, Wang H-F, Zhao Y-L, Wang D-Y (2013) Carboxylic acid functionalization prevents the translocation of multi-walled carbon nanotubes at predicted environmental relevant concentrations into targeted organs of nematode *Caenorhabditis elegans*. Nanoscale 5:6088–6096
57. Zhao Y-L, Wu Q-L, Li Y-P, Wang D-Y (2013) Translocation, transfer, and in vivo safety evaluation of engineered nanomaterials in the non-mammalian alternative toxicity assay model of nematode *Caenorhabditis elegans*. RSC Adv 3:5741–5757
58. Shakoor S, Sun L-M, Wang D-Y (2016) Multi-walled carbon nanotubes enhanced fungal colonization and suppressed innate immune response to fungal infection in nematodes. Toxicol Res 5:492–499
59. Zhi L-T, Fu W, Wang X, Wang D-Y (2016) ACS-22, a protein homologous to mammalian fatty acid transport protein 4, is essential for the control of toxicity and translocation of multi-walled carbon nanotubes in *Caenorhabditis elegans*. RSC Adv 6:4151–4159
60. Zhuang Z-H, Li M, Liu H, Luo L-B, Gu W-D, Wu Q-L, Wang D-Y (2016) Function of RSKS-1-AAK-2-DAF-16 signaling cascade in enhancing toxicity of multi-walled carbon nanotubes can be suppressed by *mir-259* activation in *Caenorhabditis elegans*. Sci Rep 6:32409
61. Zhao Y-Y, Hua X, Rui Q, Wang D-Y (2023) Exposure to multi-walled carbon nanotubes causes suppression in octopamine signal associated with transgenerational toxicity induction in *C. elegans*. Chemosphere 318:137986
62. Zhao Y-Y, Xu R-R, Hua X, Rui Q, Wang D-Y (2022) Multi-walled carbon nanotubes induce transgenerational toxicity associated with activation of germline long non-coding RNA *linc-7* in *C. elegans*. Chemosphere 301:134687
63. Shao H-M, Han Z-Y, Krasteva N, Wang D-Y (2019) Identification of signaling cascade in the insulin signaling pathway in response to nanopolystyrene particles. Nanotoxicology 13:174–188

64. Qu M, Liu Y-Q, Xu K-N, Wang D-Y (2019) Activation of p38 MAPK signaling-mediated endoplasmic reticulum unfolded protein response by nanopolystyrene particles. Adv Biosyst 3:1800325
65. Houri-Zeevi L, Teichman G, Gingold H, Rechavi O (2021) Stress resets ancestral heritable small RNA responses. elife 10:e65797
66. Ni JZ, Kalinava N, Chen E, Huang A, Trinh T, Gu SG (2016) A transgenerational role of the germline nuclear RNAi pathway in repressing heat stress-induced transcriptional activation in *C. elegans*. Epigenetics Chromatin 9:3
67. Posner R, Toker IA, Antonova O, Star E, Anava S, Azmon E, Hendricks M, Bracha S, Gingold H, Rechavi O (2019) Neuronal small RNAs control behavior transgenerationally. Cell 177:1814–1826
68. Toker IA, Lev I, Mor Y, Gurevich Y, Fisher D, Houri-Zeevi L, Antonova O, Doron H, Anava S, Gingold H, Hadany L, Shaham S, Oded Rechavi O (2022) Transgenerational inheritance of sexual attractiveness via small RNAs enhances evolvability in *C. elegans*. Dev Cell 57:298–309
69. Houri-Zeevi L, Kohanim YK, Antonova O, Rechavi O (2020) Three rules explain transgenerational small RNA inheritance in *C. elegans*. Cell 182:1186–1197
70. Sakaguchi A, Sarkies P, Simon M, Doebley A, Goldstein LD, Hedges A, Ikegami K, Alvares SM, Yang L, LaRocque JR, Hall J, Miska EA, Ahmed S (2014) *Caenorhabditis elegans* RSD-2 and RSD-6 promote germ cell immortality by maintaining small interfering RNA populations. Proc Natl Acad Sci USA 25:E4323–E4331

Chapter 14
Regulation of Transgenerational Toxicity by Germline Methylation and Acetylation–Related Signals

Abstract Both histone/DNA methylation and histone acetylation play important functions in regulating transgenerational toxicity of stresses and pollutants. Pollutant exposure could cause transgenerational response of certain forms of histone methylation. In the germline, histone methyltransferases (HMTs) and histone demethylases (HDMs) could show transgenerational response to pollutants, and some germline HMTs regulated transgenerational pollutant toxicity by modulating expression and function of their downstream targets. The germline histone acetyltransferases (HATs) and germline histone deacetylases (HDACs) could also exhibit transgenerational response to pollutants, and some germline HATs and germline HDACs regulated transgenerational pollutants toxicity by affecting expressions of their downstream targets, such as certain secreted germline ligands. Exposure to stresses and pollutants could also induce transgenerational change of DNA 6mA level and transgenerational response of DNA methyltransferase DAMT-1 and DNA demethylase NMAD-1. DAMT-1 played the important function in regulating transgenerational pollutant toxicity in the germline.

Keywords Transgenerational toxicology · Methylation · Acetylation · *Caenorhabditis elegans*

14.1 Introduction

Based on high sensitivity to environmental exposure [1] and established system of exposure toxicology [2], *Caenorhabditis elegans* has been widely used to assess toxicity of different pollutants [3–8]. More importantly, *C. elegans* can provide useful research platform for the study of molecular toxicology and target organ toxicology [9, 10]. In Chaps. 5, 6, 7, 8, 9, 10 and 11, we have introduced the role of different germline secreted ligands including neurotransmitters and their receptors in regulating transgenerational pollutants toxicity. In Chap. 12, we further discussed the role of germline G protein–coupled receptors (GPCRs) and germline nuclear hormone receptors (NHRs) in the control of transgenerational pollutant toxicity.

D. Wang, *Transgenerational Toxicology in Caenorhabditis elegans*,
https://doi.org/10.1007/978-981-92-1389-4_14

Besides these molecular signals, some forms of epigenetic regulation also participated in controlling transgenerational pollutant toxicity, such as the role of germline microRNAs, germline long non-coding RNAs, and germline RNAi-associated signals, which were introduced in Chap. 14. Nevertheless, at the most cases, the described transgenerational inheritance is associated with the regulation by histone and/or DNA methylation and histone acetylation–related signals. Histone methylation and acetylation are covalent epigenetic modifications in which methyl groups ($-CH_3$) are added to lysine and arginine residues or an acetyl group is added to lysine residues. DNA methylation is an epigenetic mechanism controlling gene expression by adding methyl group to cytosine bases, primarily at CpG sites. Histone and/or DNA methylation and histone acetylation are epigenetic marks that fine-tune chromatin function and gene expression without altering the underlying DNA sequence, and they form "histone codes" regulating the chromatin structure and gene expression. The histone methylation process is catalyzed by histone methyltransferases (HMTs) and removed by histone demethylases (HDMs), and the methyl groups are donated by S-adenosyl methionine (SAM). The histone acetylation process is catalyzed by histone acetyltransferases (HATs) and removed by histone deacetylases (HDACs). The DNA methylation process is catalyzed by DNA methyltransferases (DNMTs) and removed by TET enzymes (Ten-eleven translocation) to oxidize 5mC to 5-hydroxymethylcytosine (5hmC). At the pollutant-exposed parental generation (P0-G), some histone methylation and histone acetylation–related signals were involved in control of pollutant toxicity [11, 12]. In this chapter, we first introduced the role of histone methylation and histone acetylation–related signals in regulating transgenerational pollutant toxicity. Moreover, the potential function of DNA methylation–related signals in modulating transgenerational toxicity of pollutants and stresses was discussed.

14.2 Role of Germline Histone Methylation–Related Signals

14.2.1 Transgenerational Response of Histone Methylation to Pollutants

After exposure to pollutants at the P0-G, transgenerational response of certain forms of histone methylation can be detected. With arsenite as example of pollutants, accompanied with induction of transgenerational increase in glucose content, arsenite exposure caused transgenerational elevation of H3K4me2 level (Fig. 14.1) [13]. After exposure to 50 and 100 μM arsenite, transgenerational elevation of H3K4me2 level could persist till to the F5-G (Figure 14.1) [13]. Besides this, exposure to arsenite also caused transgenerational elevation in dimethylation of global H3K4 [14]. In addition, accompanied with transgenerational reproductive decline induced by aged nanoplastics, transgenerational increase in H3K4 and H3K9 methylation levels could be observed till to the F2-G [15].

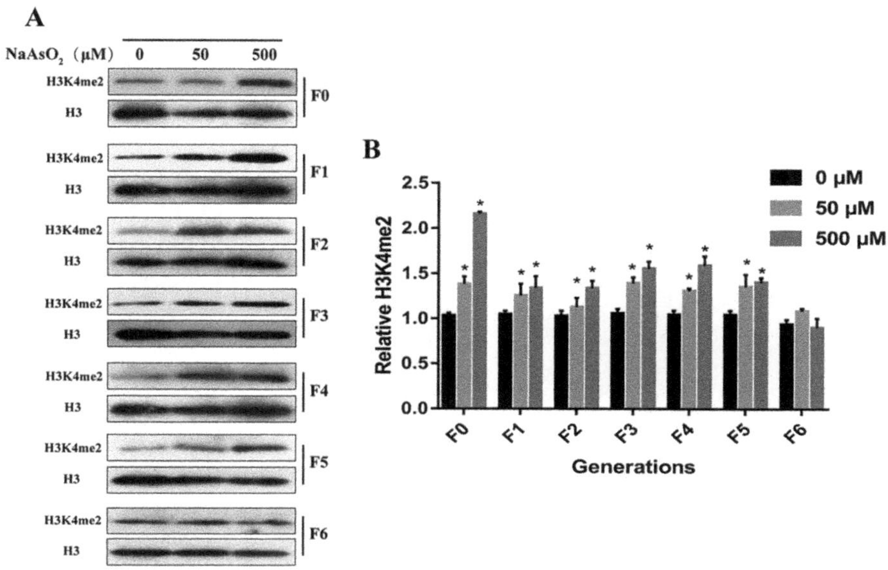

Fig. 14.1 Arsenite-induced decreases of H3K4me2 levels were transgenerated in *C. elegans* [13]. After *C. elegans* of F0 generation were exposed to 0, 50, or 500 μM arsenite for 72 h. For F1 generation *C. elegans*, pregnant F0 adults were bleached to obtain synchronized eggs, which were cultured in media containing normal diet (arsenite-free). This protocol was repeated until F6 generation. (**a**) Western blots were performed, and (**b**) relative protein levels of H3K4me2 in adult worms (F0-F6) were determined. Data are expressed as mean ± SD for three independent experiments. $^{*}P < 0.05$ significant differences from control group of the same generation

14.2.2 Transgenerational Response of Germline Histone Methylation–Related Signals to Pollutants

14.2.2.1 Transgenerational Response of Germline Histone Methylation–Related Signals to Nanoplastics

Nanoplastic exposure can induce transgenerational response of both HMTs and HDMs. In *C. elegans*, nanoplastic exposure resulted in different aspects of toxicity, including organ toxicity, metabolic disruption, and lifespan reduction [16–28]. At the P0-G, nanoplastic could affect expression of genes encoding both HMTs and HDMs [29, 30]. After exposure at P0-G, exposure to nanoplastics such as polystyrene nanoparticles (PS-NPs) further caused transgenerational toxicity on reproduction and locomotion [31–33]. PS-NP was selected as example of nanoplastics. PS-NPs exposure induced transgenerational response of certain germline HMTs. Among germline HMTs, activation of MET-2, TTLL-2, and SET-6 and inhibition in SET-5, SET-12, and SET-23 were induced by 1 μg/L PS-NPs (Fig. 14.2) [34]. The activation of germline MET-2, TTLL-2, and SET-6 and inhibition in germline SET-5, SET-12, and SET-23 were concentration dependent in 1–100 μg/L PS-NPs

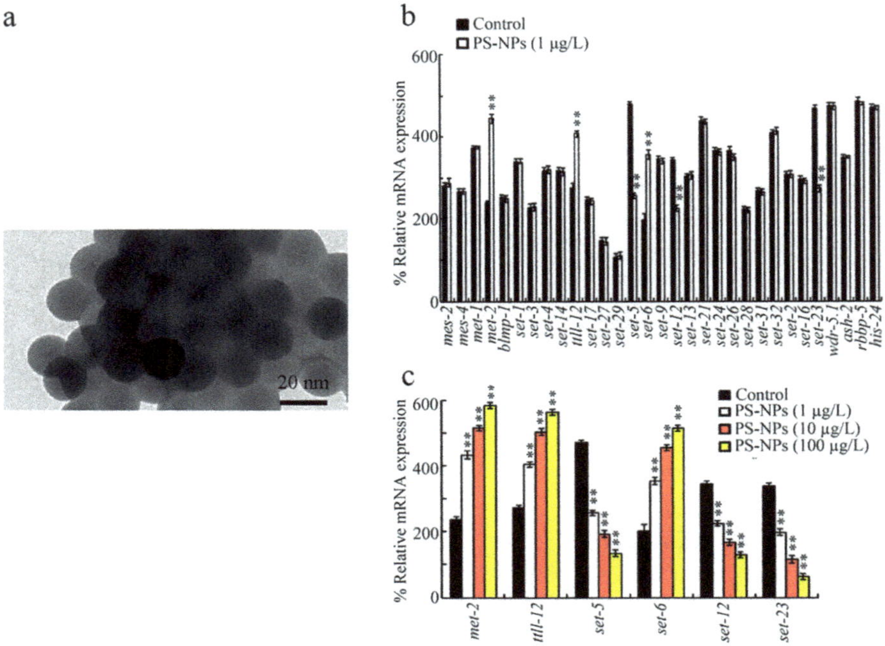

Fig. 14.2 Effect of PS-NP exposure on expressions of genes encoding histone methyltransferases (HMTs) in the germline [34]. (**a**) TEM image of PS-NPs before the sonication. (**b**) Effect of exposure to 1 μg L^{-1} PS-NPs on expressions of genes encoding HMTs in the germline. (**c**) Effect of exposure to PS-NPs (1–100 μg L^{-1}) on expressions of germline *met-2*, *ttll-12*, *set-5*, *set-6*, *set-12*, and *set-23*. A total of 30 intact gonads were used for the qRT-PCR assay. PS-NPs, Polystyrene nanoparticles. Bars represent means ± SD. $^{**}P < 0.01$ vs. control

(Fig. 14.2) [34]. Moreover, transgenerational activation of germline MET-2 and germline SET-6 was induced by 1 μg/L PS-NPs, which lasted till to the F2-G (Fig. 14.3) [34].

Besides transgenerational response of germline HMTs, nanoplastic exposure could also induce transgenerational response of HDMs. After exposure to aged PS-NPs, transgenerational inhibition in SPR-5, a H3K4me2 demethylase, was induced [15]. Similarly, transgenerational inhibition in SPR-5 could also be caused by arsenite exposure [14].

14.2.2.2 Transgenerational Response of Germline Histone Methylation–Related Signals to Microplastics

In *C. elegans*, exposure to microplastics (such as polylactic acid microparticles, PLA-MPs) also resulted in transgenerational response of certain germline HMTs. PLA-MPs exposure induced the severe reproductive toxicity [35, 36]. After exposure at P0-G, both pristine and aged PLA-MPs caused the transgenerational

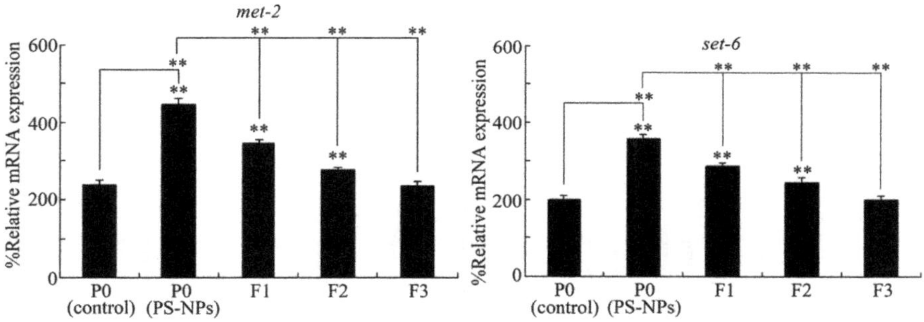

Fig. 14.3 Transgenerational expressions of germline MET-2 and SET-6 after PS-NP exposure [34]. A total of 30 intact gonads were used for the qRT-PCR assay. PS-NPs, Polystyrene nanoparticles. The exposure concentration of PS-NPs was 1 μg L^{-1}. Bars represent means ± SD. $^{**}P < 0.01$ vs. P0 (control) (if not specifically indicated)

reproductive toxicity [37, 38]. Among 31 germline HMTs, 1–100 μg/L photo-aged PLA-MPs induced activation of MET-2, SET-3, SET-4, SET-17, and SET-6 and inhibition in MES-2, BLMP-1, SET-27, SET-13, SET-24, SET-28, SET-31, SET-32, SET-2, SET-16, SET-23, and RBBP-5 at the P0-G (Fig. 14.4) [38]. Moreover, among these candidate germline HMTs, transgenerational activation of MET-2, SET-3, and SET-6 and transgenerational inhibition in MES-2, SET-27, SET-24, SET-28, SET-31, and SET-2 could be induced by 100 μg/L photo-aged PLA-MPs (Fig. 14.5) [38].

14.2.3 Role of Germline MET-2 in Regulating Transgenerational Toxicity of Pollutants

In *C. elegans*, transgenerational activation of germline MET-2 mediated induction of transgenerational pollutant toxicity. PS-NP was used as example of pollutants. Germline RNAi of *met-2* led to the resistance to transgenerational neurotoxicity of PS-NPs (Fig. 14.6) [34]. Besides this, transgenerational reproductive toxicity induced by crude oil was also inhibited by mutation of *met-2* [39]. Therefore, transgenerational pollutant toxicity can be mediated by transgenerational activation of germline MET-2.

During the control of transgenerational pollutant toxicity, PAT-12 and WRT-3 were first identified as downstream targets of germline MET-2. PS-NP exposure induced transgenerational activation of germline PAT-12 (a component of hemidesmosomes) and germline WRT-3 (a Hedgehog ligand) (Fig. 14.7) [34]. After PS-NP exposure, germline RNAi of *met-2* could cause inhibition in germline PAT-12 and germline WRT-3 (Fig. 14.7) [34]. Similar to the effect of germline RNAi of *met-2*, germline RNAi of *pat-12* and *wrt-3* also caused resistance to transgenerational neurotoxicity and transgenerational reproductive toxicity of PS-NPs (Fig. 14.7) [34]. In contrast, germline MET-2 overexpression resulted in the susceptibility to

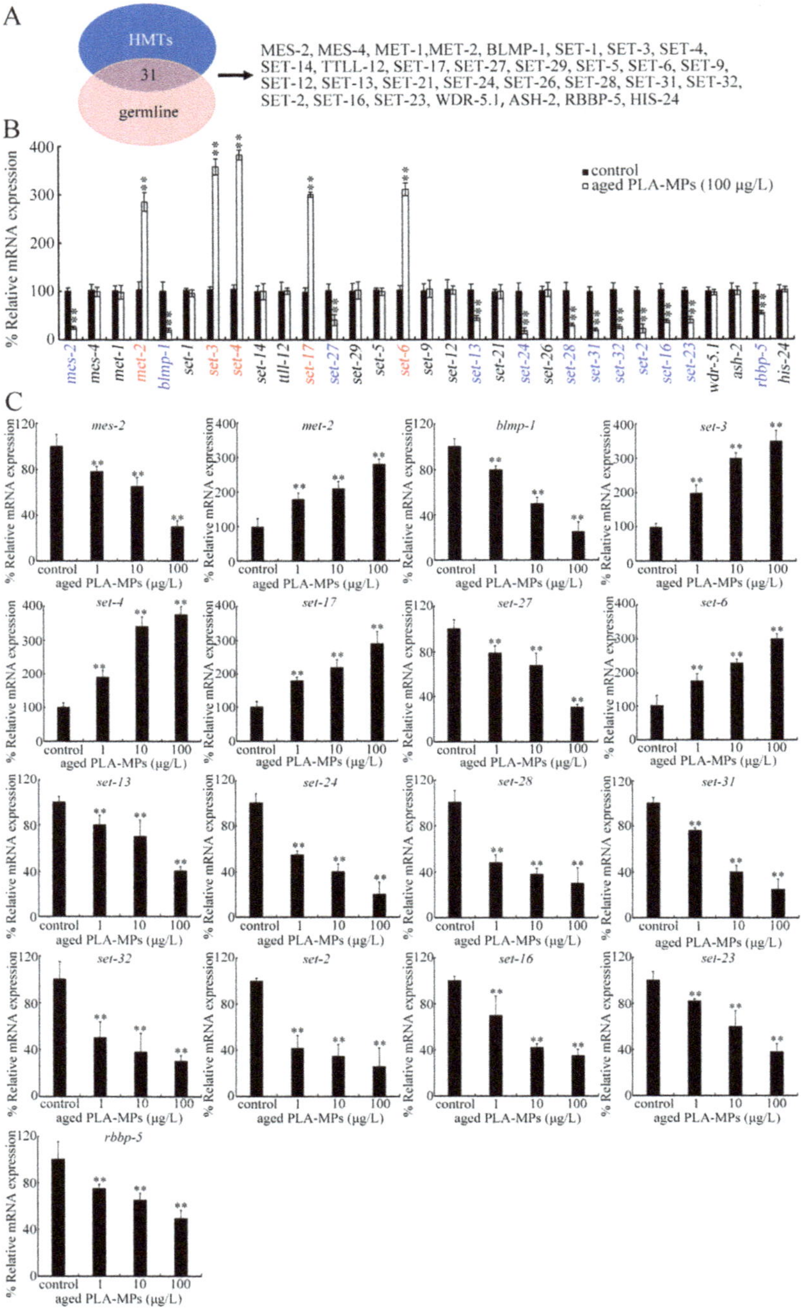

Fig. 14.4 Identification of candidate germline histone methylation transferase genes in regulating reproductive toxicity of photo-aged PLA-MPs [38]. (**a**) Histone methylation transferases expressed in the germline. (**b**) Effect of photo-aged PLA-MPs (100 μg/L) on expressions of germline histone methylation transferase genes. Three independent experiments were performed. (**c**) Expression of candidate germline histone methylation transferase genes in 1–100 μg/L photo-aged PLA-MP-exposed nematodes. A total of 30 intact gonads were used for the qRT-PCR assay. Three independent experiments were performed. $^{**}P < 0.01$ vs. control

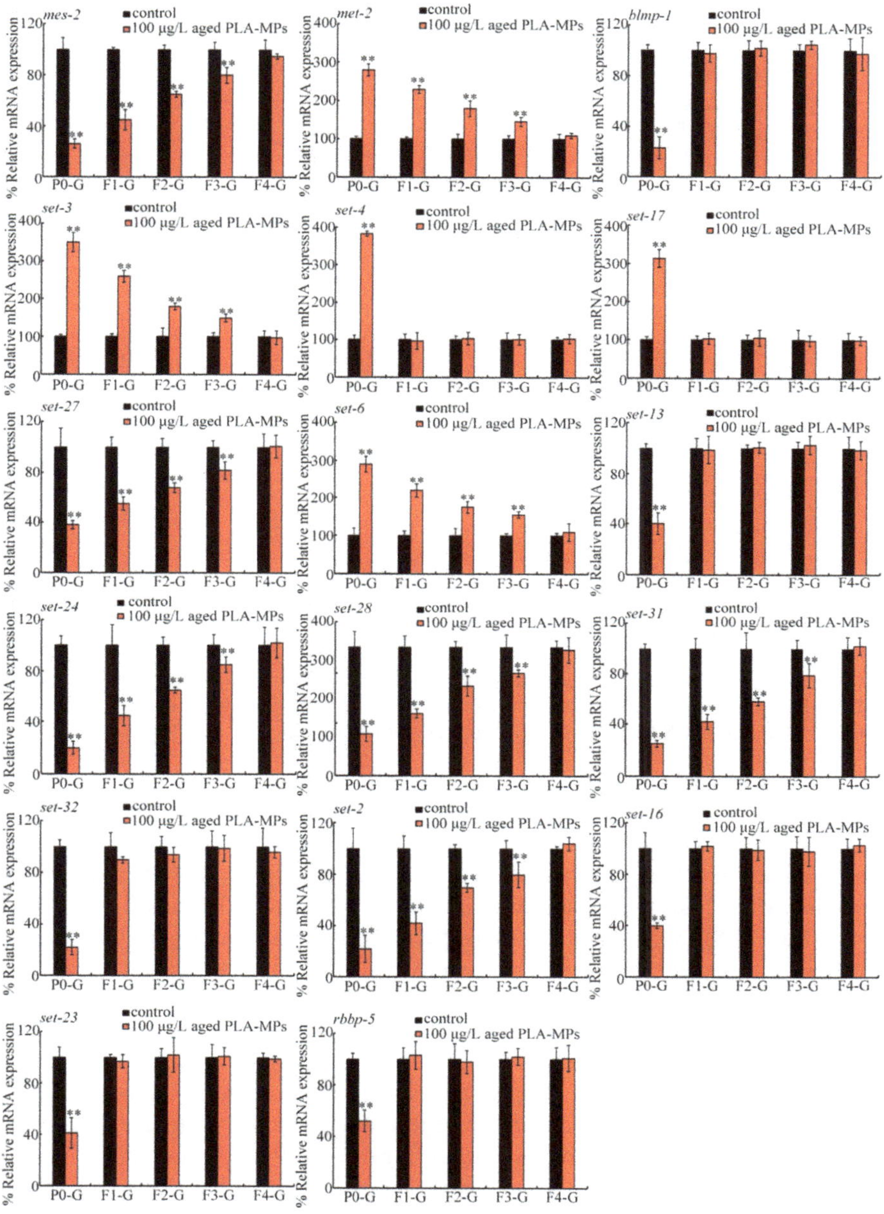

Fig. 14.5 Transgenerational expression of germline histone methylation transferase genes after exposure to photo-aged PLA-MPs [38]. A total of 30 intact gonads were used for the qRT-PCR assay. Exposure concentration of photo-aged PLA-MPs was 100 μg/L. Three independent experiments were performed. $^{**}P < 0.01$ vs. control

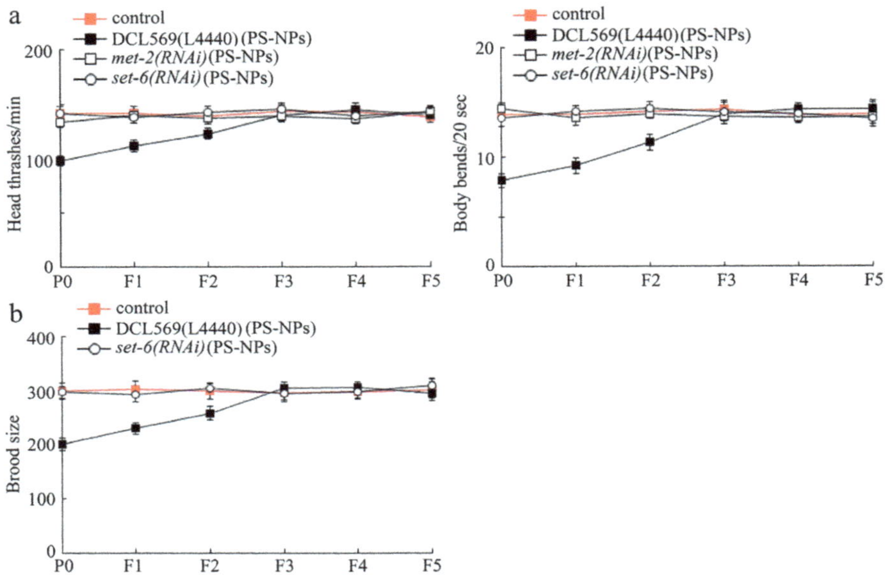

Fig. 14.6 Effect of germline RNAi knockdown of *met-2* or *set-6* on transgenerational PS-NP toxicity [34]. (**a**) Effect of germline RNAi knockdown of *met-2* or *set-6* on transgenerational PS-NP toxicity that decreases locomotion behavior. (**b**) Effect of germline RNAi knockdown of *set-6* on transgenerational PS-NP toxicity that reduces brood size. Control, DCL569 nematodes without PS-NP exposure. L4440, empty vector. PS-NPs, Polystyrene nanoparticles. The exposure concentration of PS-NPs was 1 μg L^{-1}. Bars represent means ± SD

transgenerational neurotoxicity and transgenerational reproductive toxicity of PS-NPs (Fig. 14.8) [34]. More importantly, the susceptibility to transgenerational neurotoxicity and transgenerational reproductive toxicity of PS-NPs induced by germline MET-2 overexpression was inhibited by germline RNAi of *pat-12* and *wrt-3* (Fig. 14.8) [34]. Therefore, transgenerational activation of germline MET-2 mediated transgenerational pollutant toxicity by causing the transgenerational activation of its downstream targets of PAT-12 and WRT-3.

Besides germline WRT-3, some other germline-secreted ligands were further identified as downstream targets of germline MET-2 in regulating transgenerational toxicity of pollutants [40]. In *C. elegans*, some germline-secreted ligands function with their receptors to regulate the transgenerational pollutant toxicity [41–46]. Among these germline ligands, inhibition in germline INS-39 (an insulin ligand) and its receptor DAF-2 could be further caused by germline RNAi of *met-2* (Fig. 14.9) [40], suggesting the role of INS-39 as another downstream target of germline MET-2 to regulate the transgenerational pollutant toxicity.

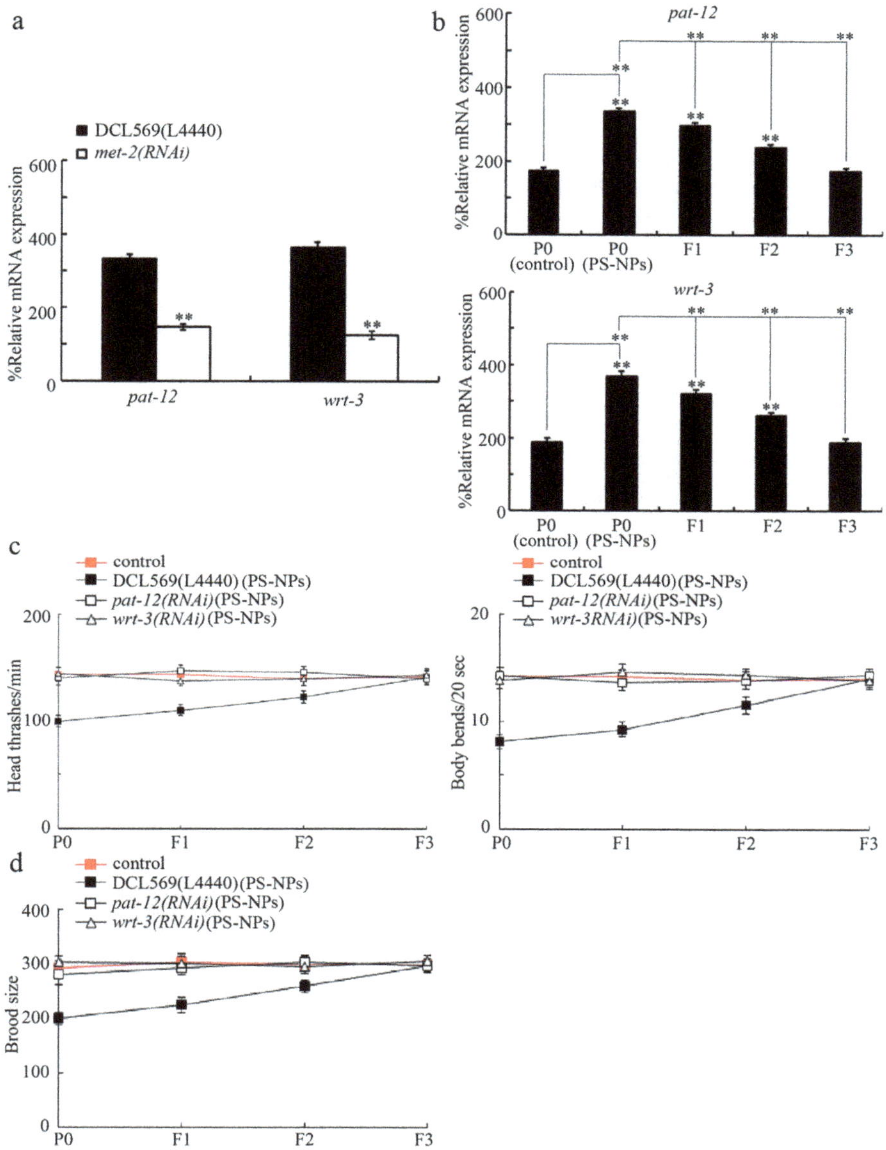

Fig. 14.7 Identification of PAT-12 and WRT-3 as potential targets of germline MET-2 in regulating transgenerational PS-NP toxicity [34]. (**a**) Effect of germline RNAi knockdown of *met-2* on expressions of *pat-12* and *wrt-3* in PS-NP-exposed DCL569 nematodes. Bars represent means ± SD. $^{**}P < 0.01$ vs. DCL569(L4440). (**b**) Transgenerational expressions of germline PAT-12 and WRT-3 after PS-NP exposure. A total of 30 intact gonads were used for the qRT-PCR assay. Bars represent means ± SD. $^{**}P < 0.01$ vs. P0 (control) (if not specially indicated). (**c**) Effect of germline RNAi knockdown of *pat-12* or *wrt-3* on transgenerational PS-NP toxicity that decreases locomotion behavior. Control, DCL569 nematodes without PSNP exposure. Bars represent means ± SD. (**d**) Effect of germline RNAi knockdown of *pat-12* or *wrt-3* on transgenerational PS-NP toxicity that reduces brood size. Control, DCL569 nematodes without PS-NP exposure. Bars represent means ± SD. PS-NPs, Polystyrene nanoparticles. L4440, empty vector. The exposure concentration of PS-NPs was 1 μg L^{-1}

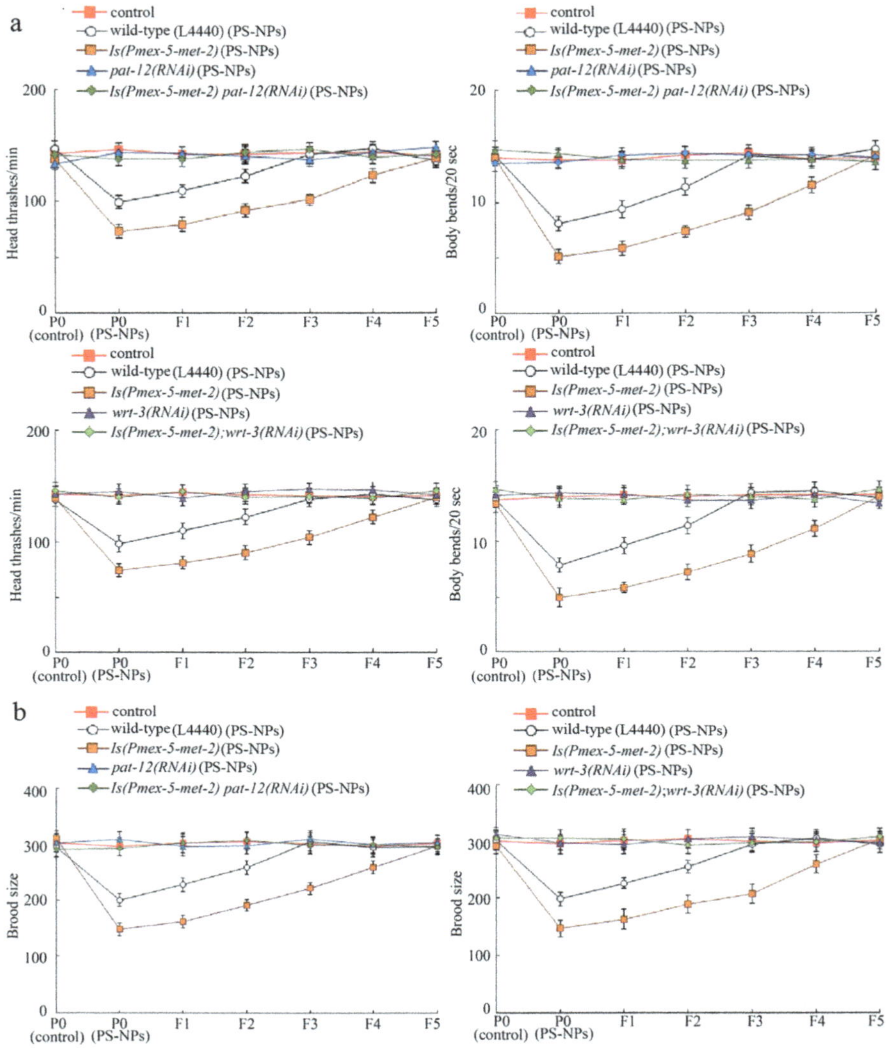

Fig. 14.8 Genetic interaction between germline MET-2 and PAT-12 or WRT-3 in regulating transgenerational PS-NP toxicity [34]. (**a**) Genetic interaction between germline MET-2 and PAT-12 or WRT-3 to regulate transgenerational PS-NP toxicity that decreases locomotion behavior. (**b**) Genetic interaction between germline MET-2 and PAT-12 or WRT-3 to regulate transgenerational PS-NP toxicity that reduces brood size. L4440, empty vector. Control, wild-type nematodes without PS-NP exposure. PS-NPs, Polystyrene nanoparticles. The exposure concentration of PS-NPs was 1 μg L^{-1}. Bars represent means ± SD

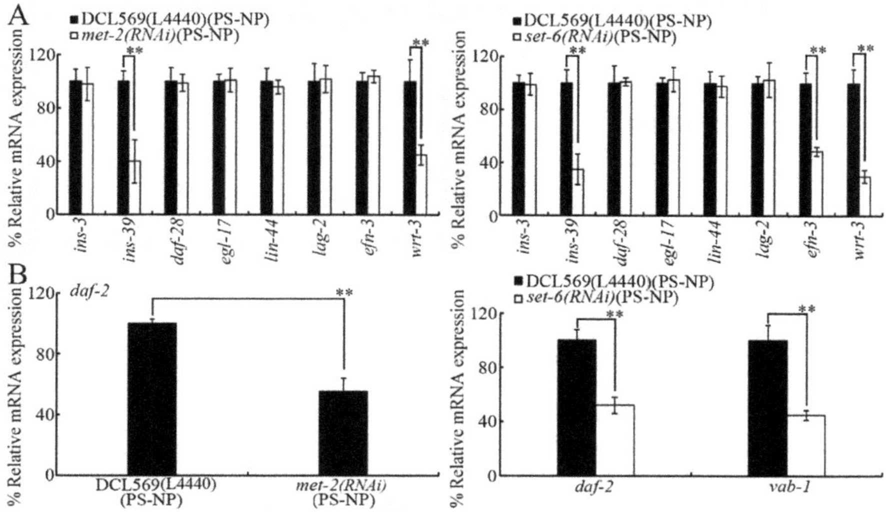

Fig. 14.9 Effect of germline RNAi of *met-2* and *set-6* on expressions of germline *ins-3*, *ins-39*, *daf-28*, *egl-17*, *lin-44*, *lag-2*, *efn-3*, and *wrt-3* in PS-NP-exposed nematodes (**a**) and expressions of *daf-2* and *vab-1* at F1-G of PS-NP-exposed nematodes (**b**) [40]. For analysis of germline ligand genes, approximately 40 intact gonads were isolated for the extraction of total RNAs and qRT-PCR assay. PS-NP exposure concentration was 1 μg/L. $^{**}P < 0.01$

14.2.4 *Role of Germline SET-6 in Regulating Transgenerational Toxicity of Pollutants*

Transgenerational activation of germline SET-6 also mediated the transgenerational pollutant toxicity. PS-NP was selected as example of pollutants. Germline RNAi of *set-6* caused the resistance to transgenerational neurotoxicity and transgenerational reproductive toxicity of PS-NPs (Fig. 14.6) [34]. That is, transgenerational pollutant toxicity can be further caused by transgenerational activation of germline SET-6.

During the control of transgenerational pollutant toxicity, some secreted ligands acted as downstream targets of germline SET-6. After PS-NP exposure, germline RNAi of *set-6* induced inhibition in germline INS-39, EFN-3 (an Ephrin ligand), and WRT-3, as well as receptors of INS-39 and EFN-3 (Fig. 14.9) [40]. That is, activation of germline SET-6 could mediate transgenerational pollutant toxicity by causing transgenerational activation of certain germline ligands, such as insulin, Ephrin, and Hedgehog ligands.

14.2.5 Role of Germline MES-2, SET-27, SET-24, SET-28, SET-31, SET-2, MET-2, SET-3, and SET-6 in Regulating Transgenerational Reproductive Toxicity of Pollutants

Accompanied with the transgenerational response of germline MES-2, SET-27, SET-24, SET-28, SET-31, SET-2, MET-2, SET-3, and SET-6, these germline HMTs participated in controlling transgenerational reproductive toxicity of pollutants. The photo-aged PLA-MP-induced transgenerational germline apoptosis induction, activation of apoptosis-promoting genes (*ced-3*, *ced-4*, and *egl-1*), and inhibition in anti-apoptotic gene *ced-9* could be strengthened by RNAi of *mes-2*, *set-2*, *set-24*, *set-27*, *set-28*, and *set-31* and inhibited by RNAi of *met-2*, *set-3*, and *set-6* (Fig. 14.10) [38]. The DNA damage checkpoints mediate the induction of germline apoptosis in pollutant-exposed nematodes [18, 47–50]. During this process, DNA damage checkpoints (HUS-1, CLK-2, and MRT-2) were identified as downstream targets of these germline HMTs. The photo-aged PLA-MP-induced transgenerational activation of germline HUS-1, CLK-2, and MRT-2 was increased by RNAi of *mes-2*, *set-2*, *set-24*, *set-27*, *set-28*, and *set-31* and suppressed by RNAi of *met-2*, *set-3*, and *set-6* (Fig. 14.10) [38].

14.2.6 Role of SET-2 and RBR-2 in Regulating Transgenerational Lipid Accumulation Induced by Pollutants

Certain histone methylation–related signals further regulated the transgenerational lipid accumulation induced by pollutants. At the P0-G, lipid accumulation could be induced by some pollutants, such as 6-PPD quinone (6-PPDQ) [24, 51–53]. Exposure to 6-PPDQ further caused transgenerational lipid accumulation, which was associated with transgenerational activation of enzyme genes governing fatty acid and monounsaturated fatty acyl-CoA synthesis and metabolic sensors (MDT-15 and SBP-1) and inhibition in enzyme genes governing fatty acid β-oxidation [54]. Accompanied with these, 0.1–10 μg/L 6-PPDQ caused transgenerational activation of HMT/SET-2 and transgenerational inhibition in HDM/RBR-2 (Fig. 14.11) [54]. The 6-PPDQ-induced transgenerational lipid accumulation and increase in triglyceride content could be increased by RNAi of *set-2* and inhibited by RNAi of *rbr-2* (Fig. 14.11) [54], suggesting that SET-2 and RBR-2 had opposite function in modulating 6-PPDQ-induced transgenerational lipid accumulation. Moreover, SBP-1 and MDT-15 were identified as downstream targets of SET-2 and RBR-2 in regulating transgenerational lipid accumulation caused by 6-PPDQ. After 6-PPDQ exposure, RNAi of *set-2* caused activation of SBP-1 and MDT-15, whereas RNAi of *rbr-2* resulted in the inhibition of SBP-1 and MDT-15 (Fig. 14.11) [54].

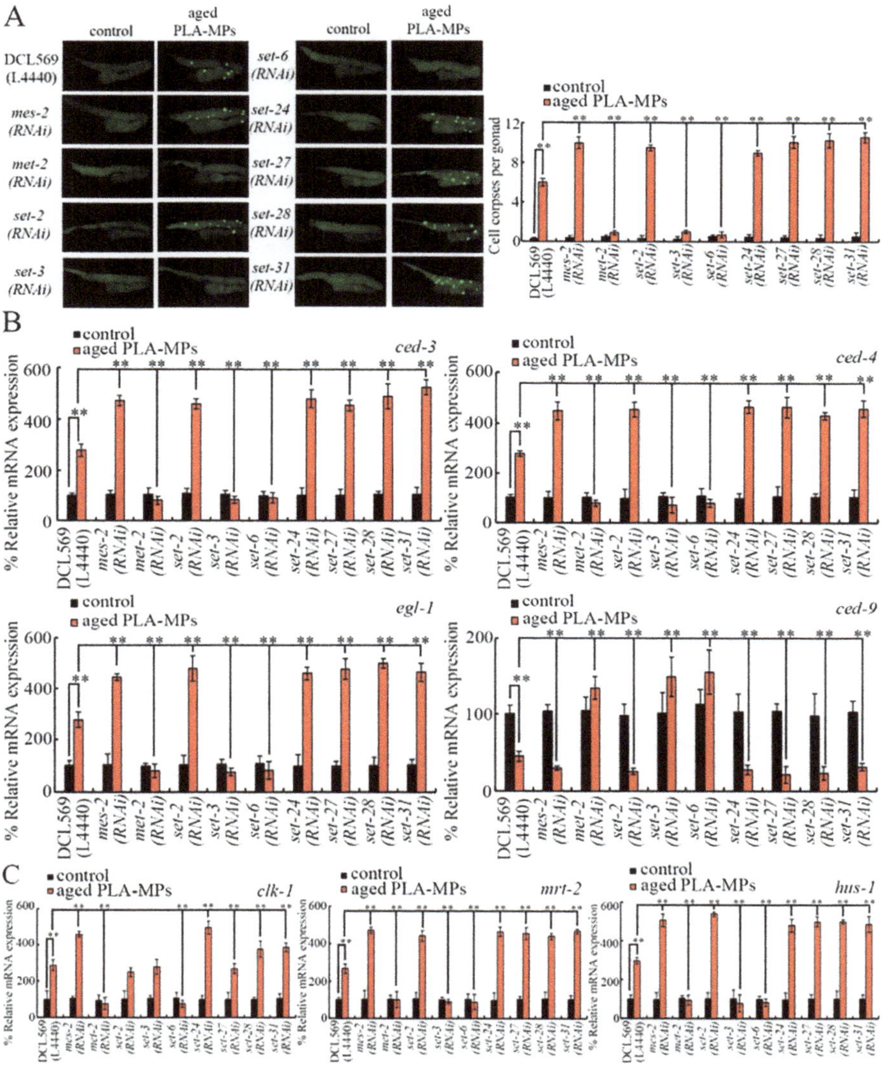

Fig. 14.10 Effect of germline RNAi of *mes-2*, *set-27*, *set-24*, *set-28*, *set-31*, *set-2*, *met-2*, *set-3*, and *set-6* on reproductive toxicity of photo-aged PLA-MPs [38]. (**a**) Effect of germline RNAi of *mes-2*, *set-27*, *set-24*, *set-28*, *set-31*, *set-2*, *met-2*, *set-3*, and *set-6* on toxicity of photo-aged PLA-MPs in inducing germline apoptosis. Thirty adult animals (adult day 1) were assessed for each treatment, and experiments were repeated three times. (**b**) Effect of RNAi of *mes-2*, *set-27*, *set-24*, *set-28*, *set-31*, *set-2*, *met-2*, *set-3*, and *set-6* on germline expression of genes governing germline apoptosis in photo-aged PLA-MP-exposed nematodes. A total of 30 intact gonads were used for the qRT-PCR assay. Three independent experiments were performed. (**c**) Effect of RNAi of *mes-2*, *set-27*, *set-24*, *set-28*, *set-31*, *set-2*, *met-2*, *set-3*, and *set-6* on germline expression of DNA damage checkpoint genes in photo-aged PLA-MP-exposed nematodes. A total of 30 intact gonads were used for the qRT-PCR assay. Three independent experiments were performed. Exposure concentration of photo-aged PLA-MPs was 100 μg/L. $^{**}P < 0.01$ vs. control (if not specially indicated)

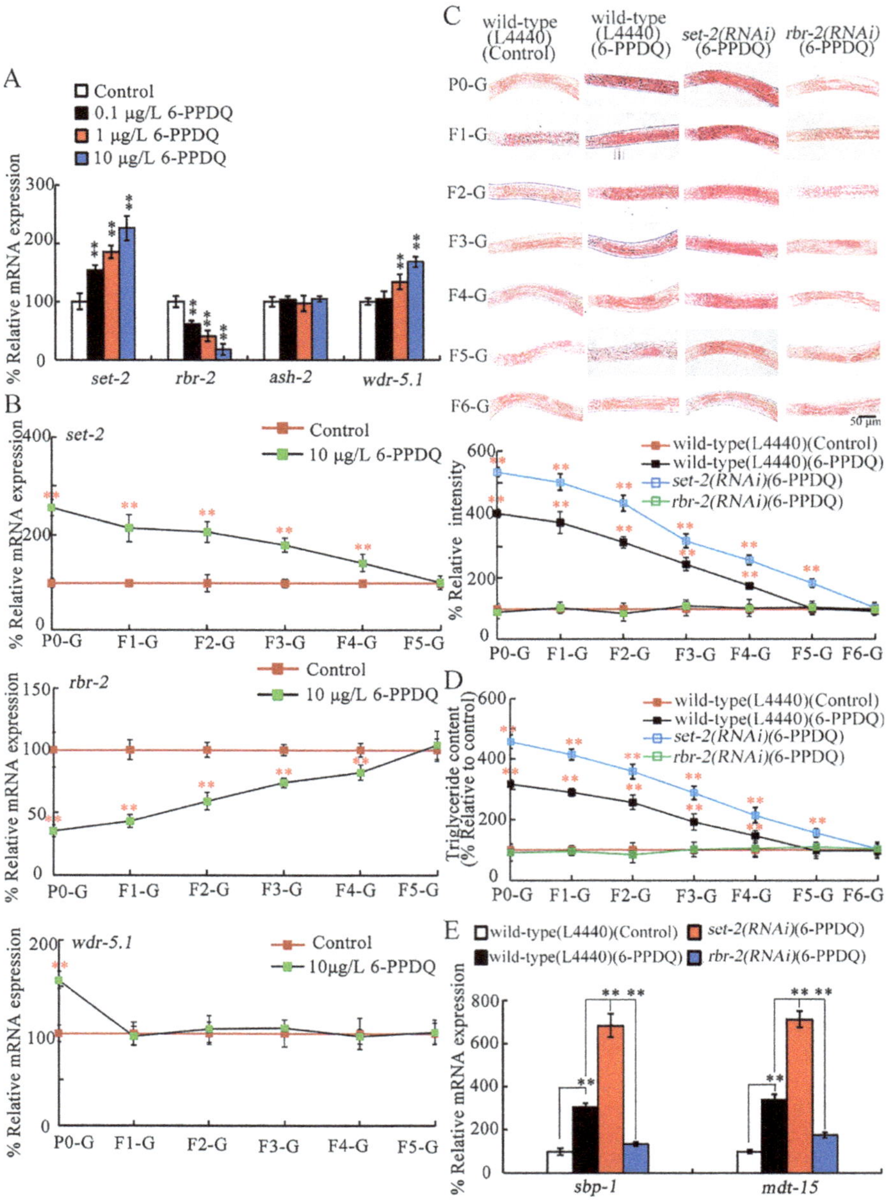

Fig. 14.11 SET-2 and RBR-2 regulated transgenerational lipid accumulation in 6-PPDQ-exposed nematodes [54]. (**a**) Effect of 6-PPDQ exposure on expressions of *set-2*, *rbr-2*, *ash-2*, and *wdr-5.1* at P0-G. $^{**}P < 0.01$ vs. control. (**b**) Transgenerational expression of *set-2*, *rbr-2*, and *wdr-5.1* after 6-PPDQ exposure at P0-G. $^{**}P < 0.01$ vs. control. (**c**) Effect of RNAi of *set-2* and *rbr-2* on transgenerational lipid accumulation induced by 6-PPDQ (10 μg/L). $^{**}P < 0.01$ vs. wild-type (L4440) (Control). (**d**) Effect of RNAi of *set-2* and *rbr-2* on transgenerational change of triglyceride content after 6-PPDQ (10 μg/L) exposure at P0-G. $^{**}P < 0.01$ vs. wild-type (L4440)(control). (**e**) Effect of RNAi of *set-2* and *rbr-2* on expressions of *sbp-1* and *mdt-15* in 6-PPDQ (10 μg/L)-exposed nematodes. $^{**}P < 0.01$

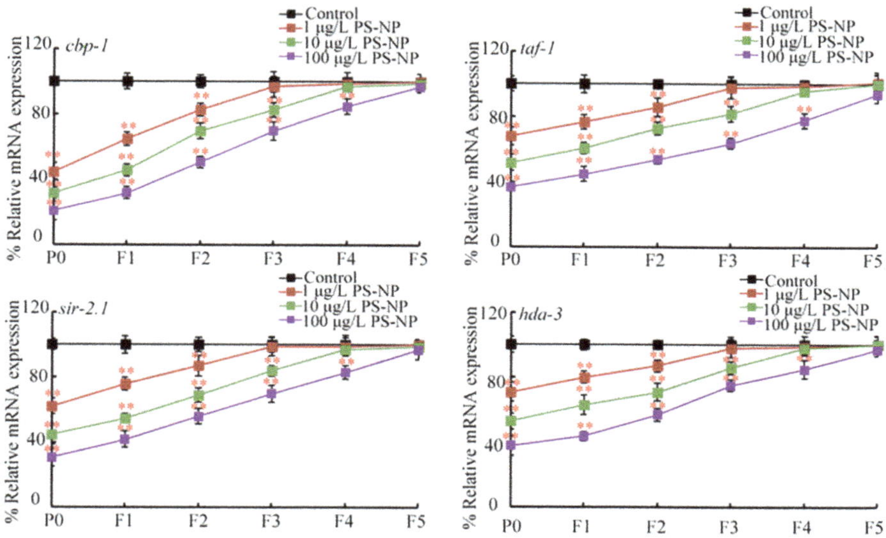

Fig. 14.12 Transgenerational expressions of germline *cbp-1*, *taf-1*, *sir-2.1*, and *hda-3* after exposure to 1–100 μg/L PS-NP at P0-G (30 intact gonads per group, 3 independent experiments) [55]. $^{**}P < 0.01$ vs. control

14.3 Role of Germline Histone Acetylation–Related Signals

14.3.1 Transgenerational Response of Germline Histone Acetylation–Related Signals to Pollutants

In *C. elegans*, certain germline HATs and germline HDACs showed the transgenerational response to pollutant exposure. PS-NP was selected as example of pollutants. At the P0-G, PS-NP exposure caused inhibition in germline CBP-1 and germline TAF-1 among germline HATs and inhibition in germline SIR-2.1 and germline HDA-3 among germline HDACs [55]. Moreover, PS-NPs induced transgenerational inhibition in germline CBP-1, TAF-1, SIR-2.1, and HDA-3 (Fig. 14.12) [55]. After exposure to 100 μg/L PS-NPs, the transgenerational inhibition in germline CBP-1, TAF-1, SIR-2.1, and HDA-3 could be detected till to the F4-G (Fig. 14.12) [55].

14.3.2 Role of Germline HATs and HDACs in Regulating Transgenerational Toxicity of Pollutants

Germline RNAi of *cbp-1*, *taf-1*, *sir-2.1*, and *hda-3* caused the susceptibility to transgenerational pollutant toxicity. PS-NP-induced transgenerational neurotoxicity and transgenerational reproductive toxicity were increased by germline RNAi of *HAT*

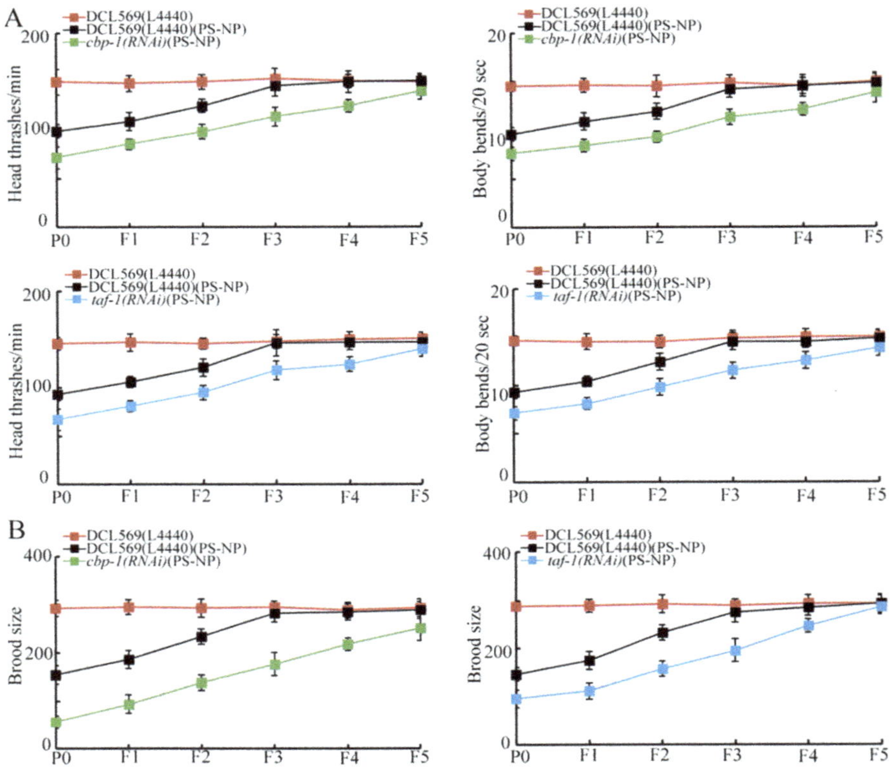

Fig. 14.13 Effect of germline RNAi of *cbp-1* and *taf-1* on transgenerational PS-NP toxicity [55]. (**a**) Effect of RNAi of *cbp-1* and *taf-1* on transgenerational PS-NP toxicity in decreasing locomotion behavior. (**b**) Effect of RNAi of *cbp-1* and *taf-1* on transgenerational PS-NP toxicity in reducing brood size. Nematodes were exposed to 1 μg/L PS-NP. The curves of DCL569(L4440)(PS-NP) were significantly ($P < 0.01$) different from that of DCL569(L4440). The curves of *cbp-1(RNAi)*(PS-NP) and *taf-1(RNAi)*(PS-NP) were significantly ($P < 0.01$) different from that of DCL569(L4440)(PS-NP)

genes, such as *cbp-1* and *taf-1* (Fig. 14.13) [55]. Similarly, transgenerational neurotoxicity and transgenerational reproductive toxicity of PS-NPs could be strengthened by germline RNAi of *HDAC* genes, such as *sir-2.1* and *hda-3* (Fig. 14.14) [55]. Therefore, transgenerational inhibition in germline CBP-1, TAF-1, SIR-2.1, and HDA-3 mediated the formation of transgenerational pollutant toxicity.

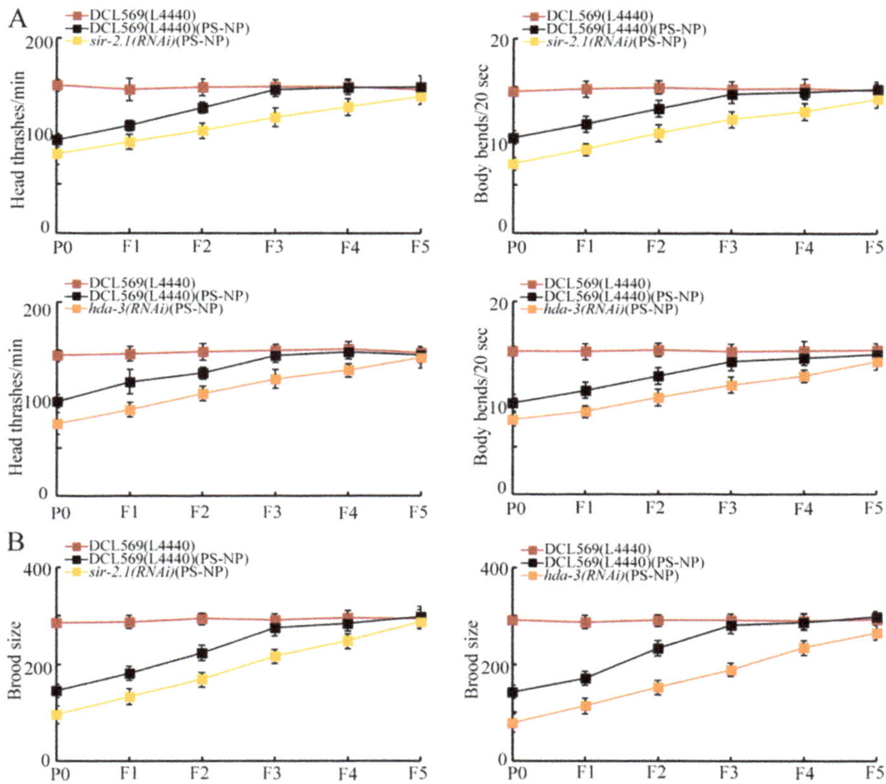

Fig. 14.14 Effect of germline RNAi of *sir-2.1* and *hda-3* on transgenerational PS-NP toxicity [55]. (**a**) Effect of RNAi of *sir-2.1* and *hda-3* on transgenerational PS-NP toxicity in decreasing locomotion behavior. (**b**) Effect of RNAi of *sir-2.1* and *hda-3* on transgenerational PS-NP toxicity in reducing brood size. Nematodes were exposed to 1 μg/L PS-NP. The curves of DCL569(L4440)(PS-NP) were significantly ($P < 0.01$) different from that of DCL569(L4440). The curves of *sir-2.1(RNAi)*(PS-NP) and *hda-3(RNAi)*(PS-NP) were significantly ($P < 0.01$) different from that of DCL569(L4440)(PS-NP)

14.3.3 Identification of Downstream Targets for Germline CBP-1, TAF-1, SIR-2.1, and HDA-3 in Regulating Transgenerational Toxicity of Pollutants

Certain secreted ligands were identified as downstream targets for germline CBP-1, TAF-1, SIR-2.1, and HDA-3 to regulate transgenerational pollutant toxicity. After PS-NP exposure, germline RNAi of *cbp-1* caused activation of germline DAF-28 (an insulin ligand) and germline LIN-44 (Wnt ligand), and germline RNAi of *taf-1* induced activation of germline INS-39, germline EGL-17 (FGF ligand), germline LIN-44, and germline LAG-2 (Notch ligand) (Fig. 14.15) [55]. More importantly, the susceptibility to transgenerational neurotoxicity of PS-NPs induced by germline

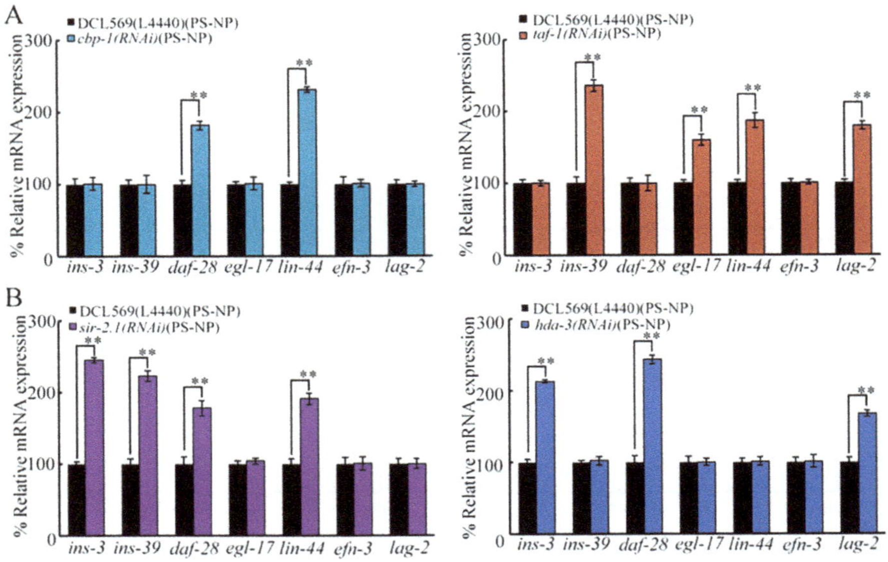

Fig. 14.15 Identification of downstream targets for CBP-1, TAF-1, SIR-2.1, HDA-3 in 6-PPDQ-exposed nematodes [55]. (**a**) Effect of germline RNAi of *cbp-1* and *taf-1* on expressions of secreted germline ligand genes in PS-NP-exposed nematodes. (**b**) Effect of germline RNAi of *sir-2.1* and *hda-3* on expressions of secreted germline ligand genes in PS-NP-exposed nematodes. Exposure concentration of PS-NP was 1 μg/L. $^{**}P < 0.01$

RNAi of *cbp-1* was inhibited by germline RNAi of *daf-28* and *lin-44* (Fig. 14.16) [55]. Additionally, the susceptibility to transgenerational neurotoxicity of PS-NPs caused by germline RNAi of *taf-1* was inhibited by germline RNAi of *ins-39, egl-17*, *lin-44*, and *lag-2* (Fig. 14.16) [55]. Therefore, inhibition in germline CBP-1 mediated transgenerational pollutant toxicity by causing transgenerational activation of germline DAF-28 and LIN-44, and inhibition in germline TAF-1 mediated transgenerational pollutant toxicity by resulting in transgenerational activation of germline INS-39, EGL-17, LIN-44, and LAG-2.

After PS-NP exposure, germline RNAi of *sir-2.1* caused activation of germline insulin ligands (INS-3, INS-39, and DAF-28) and germline LIN-44, and germline RNAi of *hda-3* induced activation of germline INS-3, germline DAF-28, and germline LAG-2 (Fig. 14.15) [55]. Moreover, the susceptibility to transgenerational neurotoxicity of PS-NPs induced by germline RNAi of *sir-2.1* was inhibited by germline RNAi of *ins-3*, *ins-39*, *daf-28*, and *lin-44* (Fig. 14.16) [55]. In addition, the susceptibility to transgenerational neurotoxicity of PS-NPs caused by germline RNAi of *hda-3* was inhibited by germline RNAi of *ins-3*, *daf-28*, and *lag-2* (Fig. 14.16) [55]. That is, inhibition in germline SIR-2.1 mediated transgenerational pollutant toxicity by causing transgenerational activation of germline insulin ligands and Wnt ligand LIN-44, and inhibition in germline HDA-3 mediated transgenerational pollutant toxicity by causing transgenerational activation of germline insulin ligands and Notch ligand LAG-2.

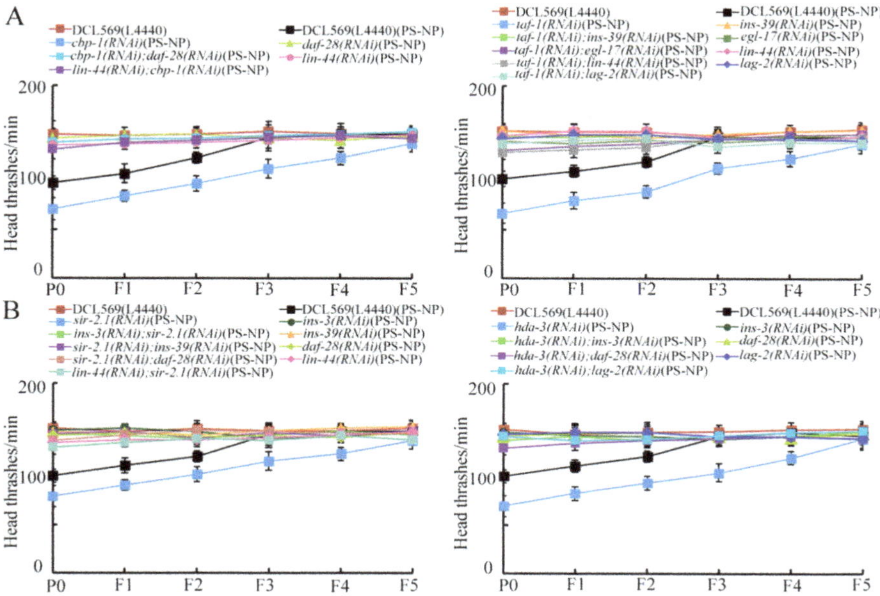

Fig. 14.16 Interaction between *cbp-1*, *taf-1*, *sir-2.1*, *hda-3* and their downstream targets in regulating transgenerational PS-NP toxicity on head thrash [55]. Exposure concentration of PS-NP was 1 μg/L. DCL569(L4440)(PS-NP) were significantly ($P < 0.01$) different from that of DCL569(L4440). The curves of *cbp-1(RNAi)*(PS-NP), *taf-1(RNAi)*(PS-NP), *sir-2.1(RNAi)*(PS-NP), *hda-3(RNAi)*(PS-NP), *daf-28(RNAi)*(PS-NP), *lin-44(RNAi)*(PS-NP), *ins-39(RNAi)*(PS-NP), *egl-17(RNAi)*(PS-NP), *lag-2(RNAi)*(PSNP), and *ins-3(RNAi)*(PS-NP) were significantly ($P < 0.01$) different from that of DCL569(L4440)(PS-NP). The curves of *cbp-1(RNAi);daf-28(RNAi)*(PS-NP) and *lin-44(RNAi);cbp-1(RNAi)*(PS-NP) were significantly ($P < 0.01$) different from that of *cbp-1(RNAi)*(PS-NP). The curves of *taf-1(RNAi);ins-39(RNAi)*(PS-NP), *taf-1(RNAi);egl-17(RNAi)*(PS-NP), *taf-1(RNAi);lin-44(RNAi)*(PS-NP), and *taf-1(RNAi);lag-2(RNAi)*(PS-NP) were significantly ($P < 0.01$) different from that of *taf-1(RNAi)*(PS-NP). The curves of *ins-3(RNAi);sir-2.1(RNAi)*(PS-NP), *sir-2.1(RNAi);ins-39(RNAi)*(PS-NP), *sir-2.1(RNAi);daf-28(RNAi)*(PS-NP), and *lin-44(RNAi);sir-2.1(RNAi)*(PS-NP) were significantly ($P < 0.01$) different from that of *sir-2.1(RNAi)*(PS-NP). The curves of *hda-3(RNAi);ins-3(RNAi)*(PS-NP), *hda-3(RNAi);daf-28(RNAi)*(PS-NP), and *hda-3(RNAi);lag-2(RNAi)*(PS-NP) were significantly ($P < 0.01$) different from that of *hda-3(RNAi)*(PS-NP)

14.4 Role of DNA Methylation–Related Signals

14.4.1 Transgenerational Change of DNA 6mA Level by Pollutants

Exposure to pollutants can cause the transgenerational change of DNA 6mA level. For example, exposure to antimycin caused the transgenerational elevation of DNA 6mA level till to the F4-G [56]. The antimycin could induce transgenerational mitochondrial stress adaptation, which suggested the association of change in global DNA 6mA level with antimycin-induced transgenerational mitochondrial stress adaptation [56].

14.4.2 *Transgenerational Response of DNA Methylation–Related Signals to Pollutants*

Certain DNA methylation–related signals exhibited transgenerational response to pollutants exposure. DNA methyltransferase DAMT-1 and DNA demethylase NMAD-1 regulate the 6mA levels [57]. Atrazine exposure induced transgenerational reproductive toxicity [58]. Accompanied with this caused transgenerational reproductive toxicity, transgenerational activation of DAMT-1 and NMAD-1 was induced by atrazine exposure at the P0-G [58]. Accompanied with transgenerational deficit in motor neurons caused by hexafluoropropylene oxide dimer acid (GenX), transgenerational response of DAMT-1 and NMAD-1 was induced [59]. Nevertheless, exposure to GenX induced the transgenerational inhibition in DAMT-1 and NMAD-1 [59].

14.4.3 *Function of Germline DAMT-1 in Regulating Transgenerational Toxicity of Pollutants and Stresses*

In *C. elegans*, germline DAMT-1 is required for the control of transgenerational pollutant toxicity. The antimycin was selected as the example of pollutants. Antimycin-induced transgenerational elevation in DNA 6mA level was inhibited by mutation of *damt-1* [56]. Mutation of *damt-1* further blocked the antimycin-induced transgenerational mitochondrial stress adaptation [56]. Meanwhile, the lacking ability to transmit mitochondrial stress adaptation caused by *damt-1* mutation was rescued by germline *damt-1* expression [56]. For the underlying molecular basis of DAMT-1 function in regulating this process, it was raised that the transcriptional factor ATFS-1- and ATFS-1-dependent stress response genes played an important role [56]. Moreover, the hormetic heat stress–induced transgenerational survival advantages could also be under the control of DAMT-1 [60], which implied that the N^6-methyladenine marked the transgenerational heat stress response.

14.5 Perspectives

In this chapter, we first introduced the transgenerational response of histone methylation and related HMTs and HDMs in the germline to pollutant exposure. Among the identified germline HMTs with transgenerational response to pollutants, their functions and corresponding downstream targets were further introduced. In contrast, the functions and possible downstream targets of germline HDMs in regulating transgenerational pollutant toxicity are relatively limited. So far, the identified downstream targets of germline HMTs in regulating transgenerational pollutants toxicity can be classified into two groups. One group contains the germline proteins with clues from the published references. Another group contains the germline-secreted ligands. More detailed identification of downstream targets for germline HMTs and germline HDMs in regulating transgenerational pollutant toxicity is suggested to be further performed in order to systematically understand the basis of germline HMTs and HDMs functions. In addition, the upregulators for germline HMTs and germline HDMs in regulating transgenerational pollutant toxicity are also needed to be further identified. The identified upregulators for HMTs during controlling biological processes, such as DAF-12 and histone chaperone SPT2, may provide some useful clues [61, 62].

In *C. elegans*, transgenerational response of germline histone acetylation–related signals, such as HATs and HDACs, has been described. However, still no direct evidence was provided to show the transgenerational response of histone acetylation to pollutant exposure. In the germline, some HATs and HDACs could act on their certain downstream targets (such as secreted germline ligands) to regulate the transgenerational pollutant toxicity. Nevertheless, during the control of transgenerational pollutant toxicity, information regarding both the downstream targets and the upregulators for germline HATs and germline HDACs is still limited.

The information regarding the role of DNA methylation in regulating transgenerational pollutant toxicity is even more limited. The transgenerational response of DNA 6mA level and expressions of DNA methyltransferase DAMT-1 and DNA demethylase NMAD-1 have been described. The function of DAMT-1 in regulating transgenerational pollutants toxicity has been indicated; however, the function of NMAD-1 remains unclear. Similarly, both the downstream targets and the upregulators for germline DAMT-1 and germline DAMT-1 in regulating transgenerational pollutants toxicity are still unknown. Besides these, m7G methyltransferase BUD-23/BUD23 and $m^{6,2}A$ DIMT-1/DIMT1 were recently described [63], and their possible function in regulating transgenerational pollutants toxicity is also needed to be considered.

References

1. Wang D-Y (2022) Toxicology at environmentally relevant concentrations in *Caenorhabditis elegans*. Springer Nature Singapore Pte Ltd., Singapore
2. Wang D-Y (2020) Exposure toxicology in *Caenorhabditis elegans*. Springer Nature Singapore Pte Ltd., Singapore
3. Wang D-Y (2018) Nanotoxicology in *Caenorhabditis elegans*. Springer Nature Singapore Pte Ltd., Singapore
4. Chen H, Wang C, Li H, Ma R, Yu Z, Li L, Xiang M, Chen X, Hua X, Yu Y (2019) A review of toxicity induced by persistent organic pollutants (POPs) and endocrine-disrupting chemicals (EDCs) in the nematode *Caenorhabditis elegans*. J Environ Manag 237:519–525
5. Weinhouse C, Truong L, Meyer JN, Allard P (2018) *Caenorhabditis elegans* as an emerging model system in environmental epigenetics. Environ Mol Mutagen 59:560–575
6. Tejeda-Benitez L, Olivero-Verbel J (2016) *Caenorhabditis elegans*, a biological model for research in toxicology. Rev Environ Contam Toxicol 237:1–35

7. Leung MC, Williams PL, Benedetto A, Au C, Helmcke KJ, Aschner M, Meyer JN (2008) *Caenorhabditis elegans*: an emerging model in biomedical and environmental toxicology. Toxicol Sci 106:5–28
8. Zhao Y, Chen J, Wang R, Pu X, Wang D (2023) A review of transgenerational and multi-generational toxicology in the *in vivo* model animal *Caenorhabditis elegans*. J Appl Toxicol 43:122–145
9. Wang D-Y (2019) Molecular toxicology in *Caenorhabditis elegans*. Springer Nature Singapore Pte Ltd., Singapore
10. Wang D-Y (2019) Target organ toxicology in *Caenorhabditis elegans*. Springer Nature Singapore Pte Ltd., Singapore
11. Wang S-T, Zhang R-J, Wang D-Y (2021) Induction of protective response to polystyrene nanoparticles associated with methylation regulation in *Caenorhabditis elegans*. Chemosphere 271:129589
12. Liu H-L, Tian L-J, Qu M, Wang D-Y (2021) Acetylation regulation associated with the induction of protective response to polystyrene nanoparticles in *Caenorhabditis elegans*. J Hazard Mater 411:125035
13. Gu C-X, Wen Y, Wu L, Wang Y-D, Wu Q-L, Wang D-Y, Wang Y-B, Liu Q-Z, Zhang J-S (2020) Arsenite-induced transgenerational glycometabolism is associated with up-regulation of H3K4me2 via inhibiting *spr-5* in *Caenorhabditis elegans*. Toxicol Lett 326:11–17
14. Yu C, Liao VH (2016) Transgenerational reproductive effects of arsenite are associated with H3K4 dimethylation and SPR-5 downregulation in *Caenorhabditis elegans*. Environ Sci Technol 50:10673–10681
15. Chen H, Gu Y, Jiang Y, Yu J, Chen C, Shi C, Li H (2023) Photoaged polystyrene nanoplastics result in transgenerational reproductive toxicity associated with the methylation of histone H3K4 and H3K9 in *Caenorhabditis elegans*. Environ Sci Technol 2023(57):19341–19351
16. Hua X, Wang D-Y (2022) Cellular uptake, transport, and organelle response after exposure to microplastics and nanoplastics: current knowledge and perspectives for environmental and health risks. Rev Environ Contam Toxicol 260:12
17. Qu M, Kong Y, Yuan Y-J, Wang D-Y (2019) Neuronal damage induced by nanopolystyrene particles in nematode *Caenorhabditis elegans*. Environ Sci Nano 6:2591–2601
18. Qu M, Qiu Y-X, Kong Y, Wang D-Y (2019) Amino modification enhances reproductive toxicity of nanopolystyrene on gonad development and reproductive capacity in nematode *Caenorhabditis elegans*. Environ Pollut 254:112978
19. Qu M, Nida A, Kong Y, Du H-H, Xiao G-S, Wang D-Y (2019) Nanopolystyrene at predicted environmental concentration enhances microcystin-LR toxicity by inducing intestinal damage in *Caenorhabditis elegans*. Ecotoxicol Environ Saf 183:109568
20. Qu M, Xu K-N, Li Y-H, Wong G, Wang D-Y (2018) Using *acs-22* mutant *Caenorhabditis elegans* to detect the toxicity of nanopolystyrene particles. Sci Total Environ 643:119–126
21. Qiu Y-X, Liu Y-Q, Li Y-H, Li G-J, Wang D-Y (2020) Effect of chronic exposure to nanopolystyrene on nematode *Caenorhabditis elegans*. Chemosphere 256:127172
22. Qiu Y-X, Luo L-B, Yang Y-H, Kong Y, Li Y-H, Wang D-Y (2020) Potential toxicity of nanopolystyrene on lifespan and aging process of nematode *Caenorhabditis elegans*. Sci Total Environ 705:135918
23. Li D, Deng Y-J, Wang S-T, Du H-H, Xiao G-S, Wang D-Y (2020) Assessment of nanopolystyrene toxicity under fungal infection condition in *Caenorhabditis elegans*. Ecotoxicol Environ Saf 197:110625
24. Liu H-L, Shao H-M, Guo Z-J, Wang D-Y (2020) Nanopolystyrene exposure activates a fat metabolism related signaling-mediated protective response in *Caenorhabditis elegans*. NanoImpact 17:100204
25. Dong S-S, Qu M, Rui Q, Wang D-Y (2018) Combinational effect of titanium dioxide nanoparticles and nanopolystyrene particles at environmentally relevant concentrations on nematodes *Caenorhabditis elegans*. Ecotoxicol Environ Saf 161:444–450

26. Zhao Y-Y, Li D, Rui Q, Wang D-Y (2020) Toxicity induction of nanopolystyrene under microgravity stress condition in *Caenorhabditis elegans*. Sci Total Environ 703:135623
27. Liu H-L, Wang D-Y (2021) Intestinal mitochondrial unfolded protein response induced by nanoplastic particles in *Caenorhabditis elegans*. Chemosphere 267:128917
28. Tang M, Ding G, Lu X, Huang Q, Du H, Xiao G, Wang D (2022) Exposure to nanoplastic particles enhances *Acinetobacter* survival, biofilm formation, and serum resistance. Nanomaterials 12:4222
29. Yu C, Luk TC, Liao VH (2021) Long-term nanoplastics exposure results in multi and transgenerational reproduction decline associated with germline toxicity and epigenetic regulation in *Caenorhabditis elegans*. J Hazard Mater 412:125173
30. Wei C, Yen P, Chaikritsadakarn A, Huang C, Chang C, Liao VH (2020) Parental CuO nanoparticles exposure results in transgenerational toxicity in *Caenorhabditis elegans* associated with possible epigenetic regulation. Ecotoxicol Environ Saf 203:111001
31. Zhao L, Qu M, Wong G, Wang D-Y (2017) Transgenerational toxicity of nanopolystyrene particles in the range of μg/L in nematode *Caenorhabditis elegans*. Environ Sci Nano 4:2356–2366
32. Sun L-M, Liao K, Wang D-Y (2021) Comparison of transgenerational reproductive toxicity induced by pristine and amino modified nanoplastics in *Caenorhabditis elegans*. Sci Total Environ 768:144362
33. Liu H-L, Tian L-J, Wang S-T, Wang D-Y (2021) Size-dependent transgenerational toxicity induced by nanoplastics in nematode *Caenorhabditis elegans*. Sci Total Environ 790:148217
34. Zhang L, Wang S-T, Zhao Y, Bi K, Wang D-Y (2022) Increase in germline methyltransferases governing methylation of histone H3K9 is associated with transgenerational nanoplastic toxicity in *Caenorhabditis elegans*. Environ Sci Nano 9:265–274
35. Shao Y-T, Wang Y-X, Hua X, Li Y-H, Wang D-Y (2023) Polylactic acid microparticles in the range of μg/L reduce reproductive capacity by affecting the gonad development and the germline apoptosis in *Caenorhabditis elegans*. Chemosphere 336:139193
36. Shao Y-T, Hua X, Li Y-H, Wang D-Y (2024) Comparison of reproductive toxicity between pristine and aged polylactic acid microplastics in *Caenorhabditis elegans*. J Hazard Mater 466:133545
37. Shao Y-T, Li Y-H, Wang D-Y (2024) Polylactic acid microplastics cause transgenerational reproductive toxicity associated with activation of insulin and hedgehog ligands in *C. elegans*. Sci Total Environ 942:173746
38. Wu J-W, Shao Y-T, Hua X, Li Y-H, Wang D-Y (2025) Photo-aged polylactic acid microplastics causes severe transgenerational decline in reproductive capacity in *C. elegans*: insight into activation of DNA damage checkpoints affected by multiple germline histone methyltransferases. Environ Pollut 382:126697
39. Yang J, Chatterjee N, Kim Y, Roh J, Kwon J, Park M, Choi J (2018) Histone methylation-associated transgenerational inheritance of reproductive defects in *Caenorhabditis elegans* exposed to crude oil under various exposure scenarios. Chemosphere 200:358–365
40. Wu J-W, Shao Y-T, Hua X, Wang Y-X, Wang D-Y (2024) Nanoplastic at environmentally relevant concentrations induces toxicity across multiple generations associated with inhibition in germline G protein-coupled receptor CED-1 in *Caenorhabditis elegans*. Chemosphere 364:143011
41. Zhao Y, Hua X, Bian Q, Wang D-Y (2022) Nanoplastic exposure at predicted environmental concentrations induces activation of germline Ephrin signal associated with toxicity formation in the *Caenorhabditis elegans* offspring. Toxics 10:699
42. Xu R-R, Hua X, Rui Q, Wang D-Y (2022) Alteration in Wnt signaling mediates induction of transgenerational toxicity of polystyrene nanoplastics in *C. elegans*. NanoImpact 28:100425
43. Liu H-L, Zhao Y-L, Hua X, Wang D-Y (2022) Induction of transgenerational toxicity is associated with the activated germline insulin signals in nematodes exposed to nanoplastic at predicted environmental concentrations. Ecotoxicol Environ Saf 243:114022

44. He W-M, Gu A-H, Wang D-Y (2023) Sulfonate-modified polystyrene nanoparticle at precited environmental concentrations induces transgenerational toxicity associated with increase in germline Notch signal of *Caenorhabditis elegans*. Toxics 11:511
45. Hua X, Cao C, Zhang L, Wang D-Y (2023) Activation of FGF signal in germline mediates transgenerational toxicity of polystyrene nanoparticles at predicted environmental concentrations in *Caenorhabditis elegans*. J Hazard Mater 451:131174
46. Wu J-W, Shao Y-T, Hua X, Wang D-Y (2024) Activated hedgehog and insulin ligands by decreased transcriptional factor DAF-16 mediate transgenerational nanoplastic toxicity in *Caenorhabditis elegans*. J Hazard Mater 480:135909
47. Liu Z-Y, Bian Q, Wang D-Y (2025) 6-PPD quinone induces response of nuclear hormone receptors in the germline associated with formation of reproductive toxicity in *Caenorhabditis elegans*. J Hazard Mater 495:138815
48. Liu Z-Y, Hua X, Zhao Y, Bian Q, Wang D-Y (2024) Polyethylene nanoplastics cause reproductive toxicity associated with activation of both estrogenic hormone receptor NHR-14 and DNA damage checkpoints in *C. elegans*. Sci Total Environ 906:167471
49. Hua X, Feng X, Liang G-Y, Chao J, Wang D-Y (2023) Long-term exposure to 6-PPD quinone reduces reproductive capacity by enhancing germline apoptosis associated with activation of both DNA damage and cell corpse engulfment in *Caenorhabditis elegans*. J Hazard Mater 454:131495
50. Zhao Y-L, Wu Q-L, Wang D-Y (2016) An epigenetic signal encoded protection mechanism is activated by graphene oxide to inhibit its induced reproductive toxicity in *Caenorhabditis elegans*. Biomaterials 79:15–24
51. Wang Y-X, Hua X, Wang D-Y (2023) Exposure to 6-PPD quinone enhances lipid accumulation through activating metabolic sensors of SBP-1 and MDT-15 in *Caenorhabditis elegans*. Environ Pollut 333:121937
52. Yang Y-H, Shao H-M, Wu Q-L, Wang D-Y (2020) Lipid metabolic response to polystyrene particles in nematode *Caenorhabditis elegans*. Environ Pollut 256:113439
53. Wu Q-L, Zhi L-T, Qu Y-Y, Wang D-Y (2016) Quantum dots increased fat storage in intestine of *Caenorhabditis elegans* by influencing molecular basis for fatty acid metabolism. Nanomed Nanotechnol Biol Med 12:1175–1184
54. Wang Y-X, Wu J-W, Wang D-Y (2025) 6-PPD quinone causes lipid accumulation across multiple generations differentially affected by metabolic sensors and components of COMPASS complex in *Caenorhabditis elegans*. Environ Pollut 366:125539
55. Hua X, Wang D-Y (2024) Transgenerational response of germline histone acetyltransferases and deacetylases to nanoplastics at predicted environmental doses in *Caenorhabditis elegans*. Sci Total Environ 952:175903
56. Ma C, Niu R, Huang T, Shao L, Peng Y, Ding W, Wang Y, Jia G, He C, Li C, He A, Liu Y (2019) N6-methyldeoxyadenine is a transgenerational epigenetic signal for mitochondrial stress adaptation. Nat Cell Biol 21:319–327
57. Greer EL, Blanco MA, Gu L, Sendinc E, Liu J, Aristizábal-Corrales D, Hsu CH, Aravind L, He C, Shim Y (2015) DNA methylation on N6-Adenine in *C. elegans*. Cell 161:868–878
58. Yin J-C, Hong X, Wang J, Li W-X, Shi Y-C, Wang D-Y, Liu R (2023) DNA methylation 6 mA and histone methylation involved in multi-/trans-generational reproductive effects in *Caenorhabditis elegans* induced by atrazine. Ecotoxicol Environ Saf 249:114348
59. Wei C, How C, Huang C, Huang W (2025) Long-term exposure to environmentally relevant concentrations of hexafluoropropylene oxide dimer acid (GenX) induces trans-generational motor deficits through H3K27 demethylase jmjd-1.2/PHF8 in *Caenorhabditis elegans*. Environ Pollut 384:126981
60. Wan Q, Meng X, Dai W, Luo Z, Wang C, Fu X, Yang J, Ye Q, Zhou Q (2021) N^6-methyldeoxyadenine and histone methylation mediate transgenerational survival advantages induced by hormetic heat stress. Sci Adv 7:eabc3026
61. Greer EL, Becker B, Latza C, Antebi A, Shi Y (2016) Mutation of *C. elegans* demethylase *spr-5* extends transgenerational longevity. Cell Res 26:229–238

62. Saredi G, Carelli FN, Rolland SGM, Furlan G, Piquet S, Appert A, Sanchez-Pulido L, Price JL, Alcon P, Lampersberger L, Déclais A, Ramakrishna NB, Toth R, Macartney T, Alabert C, Ponting CP, Polo SE, Miska EA, Gartner A, Ahringer J, Rouse J (2024) The histone chaperone SPT2 regulates chromatin structure and function in Metazoa. Nat Struct Mol Biol 31:523–535
63. Liberman N, Rothi MH, Gerashchenko MV, Zorbas C, Boulias K, MacWhinnie FG, Ying AK, Taylor AF, Haddad JA, Shibuya H, Roach L, Dong A, Dellacona S, Lafontaine DLJ, Gladyshev VN, Greer EL (2023) 18S rRNA methyltransferases DIMT1 and BUD23 drive intergenerational hormesis. Mol Cell 83:3268–3282

Chapter 15
Regulation of Transgenerational Toxicity by Germline G Protein–Coupled Receptors and Nuclear Hormone Receptors

Abstract Responses of G protein–coupled receptors (GPCRs) and nuclear hormone receptors (NHRs) trigger the stresses and pollutant-induced damage on *Caenorhabditis elegans*. In the germline, certain GPCRs (such as CED-1) and NHRs (such as NHR-12, NHR-14, NHR-47, and DAF-12) showed transgenerational response to pollutants, which mediated induction of transgenerational pollutant toxicity. Inhibition in germline CED-1 mediated transgenerational pollutant toxicity by causing transgenerational activation of MET-2 and SET-6. Change of germline NHRs expressions mediated transgenerational pollutant toxicity by resulting in transgenerational response of secreted ligands.

Keywords Transgenerational toxicology · G protein–coupled receptor · Nuclear hormone receptor · *Caenorhabditis elegans*

15.1 Introduction

Caenorhabditis elegans has been widely used in toxicity assessment of various stresses and pollutants, especially at low dose and concentrations [1–3]. As a classic model animal, *C. elegans* provides an important platform for the further study of toxicology, such as molecular toxicology and target organ toxicology [4, 5]. After exposure to stresses and pollutants, if they could cause the damage on organisms, first of all, they must attack the G protein–coupled proteins (GPCRs) and/or nuclear hormone receptors (NHRs).

GPCRs are seven-transmembrane-domain receptors located on the cell surface, and pollutants function as the ligands to bind to GPCRs and to activate intracellular G proteins, initiating a cascade of signaling events that lead to a cellular response. Exposure to stresses and pollutants could induce the response of different tissues of GPCRs, and GPCRs could function in different tissues, such as intestine, neurons, and germline, to regulate pollutant toxicity [6–9]. NHRs are a family of ligand-activated transcription factors located in the nucleus, and fat-soluble pollutants could bind to NHRs and affect expression of specific target genes of NHRs. Exposure to pollutants,

D. Wang, *Transgenerational Toxicology in Caenorhabditis elegans*,
https://doi.org/10.1007/978-981-92-1389-4_15

such as nanoplastics and 6-PPD quinone (6-PPDQ), could activate or inhibit certain NHRs, and the NHRs could regulate the toxicity induction of these pollutants [10, 11].

In this chapter, we first introduced the transgenerational responses of germline GPCRs and germline NHRs induced by pollutants and the roles of germline GPCRs and germline NHRs in regulating transgenerational pollutants toxicity. Moreover, underlying molecular mechanisms for germline GPCRs and germline NHRs in regulating transgenerational pollutant toxicity were further discussed.

15.2 Role of Germline GPCRs

15.2.1 Transgenerational Response of Germline GPCRs

Exposure to pollutants could induce transgenerational inhibition in germline CED-1. Nanoplastic exposure caused multiple toxicities on *C. elegans*, including lifespan reduction, disrupted metabolism, and organ toxicity [12–24]. The transgenerational toxicity on locomotion and reproduction could be further caused by nanoplastic exposure at the parental generation (P0-G) [25–28]. Polystyrene nanoparticle (PS-NP), a member of nanoplastics, was used as example of pollutants. CED-1 is a GPCR identified in the germline to regulate the nanoplastic toxicity at the P0-G [29]. Inhibition in CED-1 was caused by 1–100 μg/L PS-NPs (Fig. 15.1) [30]. Moreover, after exposure to 1 μg/L PS-NPs, inhibition in CED-1 could be observed till to the F2-G (Fig. 15.1) [30].

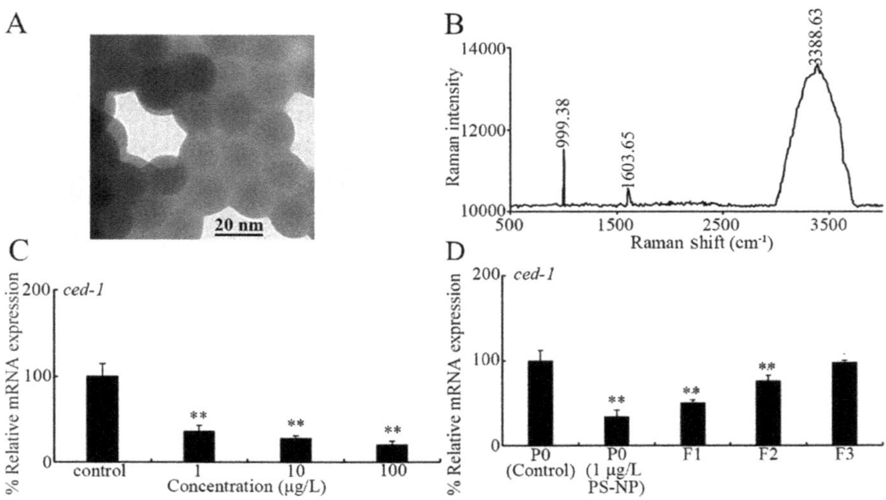

Fig. 15.1 Transgenerational decrease in expression of *ced-1* after PS-NP exposure at P0-G [30]. (**a**) TEM image of PS-NP before sonication. (**b**) Raman spectrum of PS-NPs. (**c**) Effect of PS-NP exposure on germline *ced-1* expression at P0-G. (**d**) Transgenerational expression of germline *ced-1* after PS-NP exposure at P0-G. Approximately 40 intact gonads were isolated for the extraction of total RNAs and qRT-PCR assay. $^{**}P < 0.01$ vs control

15.2.2 Function of Germline GPCRs in Regulating Transgenerational Toxicity of Pollutants

Functional analysis indicated that germline RNAi of *ced-1* caused susceptibility to transgenerational pollutant toxicity. Transgenerational neurotoxicity and transgenerational reproductive toxicity of PS-NPs could be increased by germline RNAi of *ced-1* (Fig. 15.2) [30]. That is, transgenerational inhibition in germline CED-1 mediates induction of transgenerational pollutant toxicity.

15.2.3 Downstream Targets of Germline CED-1 in Regulating Transgenerational Toxicity of Pollutants

In the germline, MET-2 and SET-6 were identified as downstream targets of CED-1 to regulate transgenerational pollutant toxicity. MET-2 and SET-6 are histone methylation transferases, which were involved in controlling transgenerational pollutant toxicity [31]. The first evidence was that, after PS-NP exposure, activation of germline MET-2 and SET-6 was induced by germline RNAi of *ced-1* (Fig. 15.3) [30]. In *C. elegans*, germline RNAi of *met-2* and *set-6* led to resistance to transgenerational neurotoxicity and transgenerational reproductive toxicity of PS-NPs (Fig. 15.4) [30]. The second evidence was that the susceptibility to transgenerational

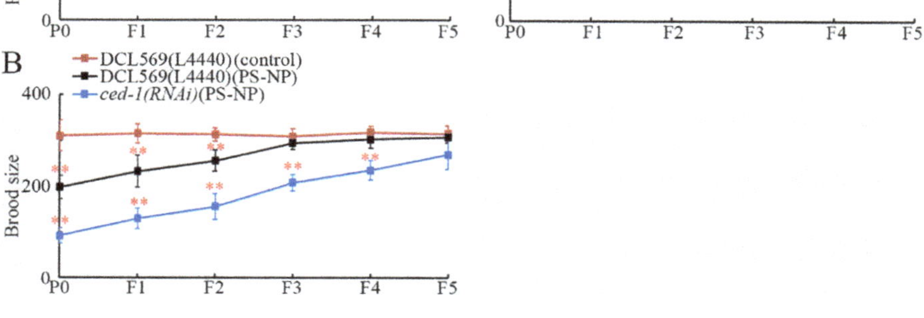

Fig. 15.2 Effect of germline RNAi of *ced-1* on transgenerational PS-NP toxicity in decreasing locomotion behavior (**a**) and in reducing brood size (**b**) [30]. PS-NP exposure concentration was 1 μg/L. The curve of DCL569(L4440)(PS-NP) was significantly ($P < 0.01$) different from that of DCL569(L4440)(control). The curve of *ced-1(RNAi)*(PS-NP) was significantly ($P < 0.01$) different from that of DCL569(L4440)(PS-NP). $^{**}P < 0.01$ vs DCL569(L4440)(control)

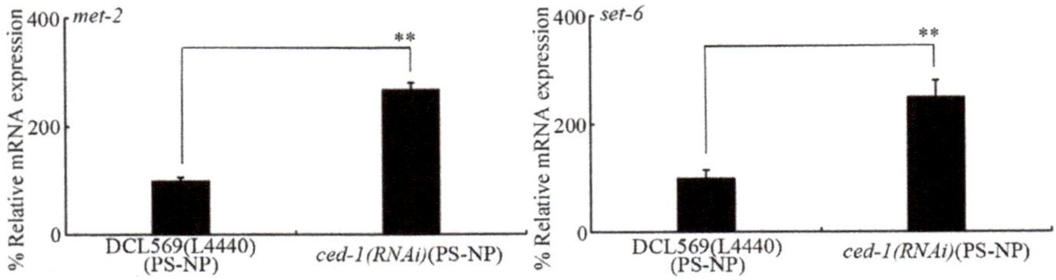

Fig. 15.3 Effect of germline RNAi of *ced-1* on germline expressions of *met-2* and *set-6* in PS-NP-exposed nematodes [30]. Approximately 40 intact gonads were isolated for the extraction of total RNAs and qRT-PCR assay. PS-NP exposure concentration was 1 μg/L $^{**}P < 0.01$

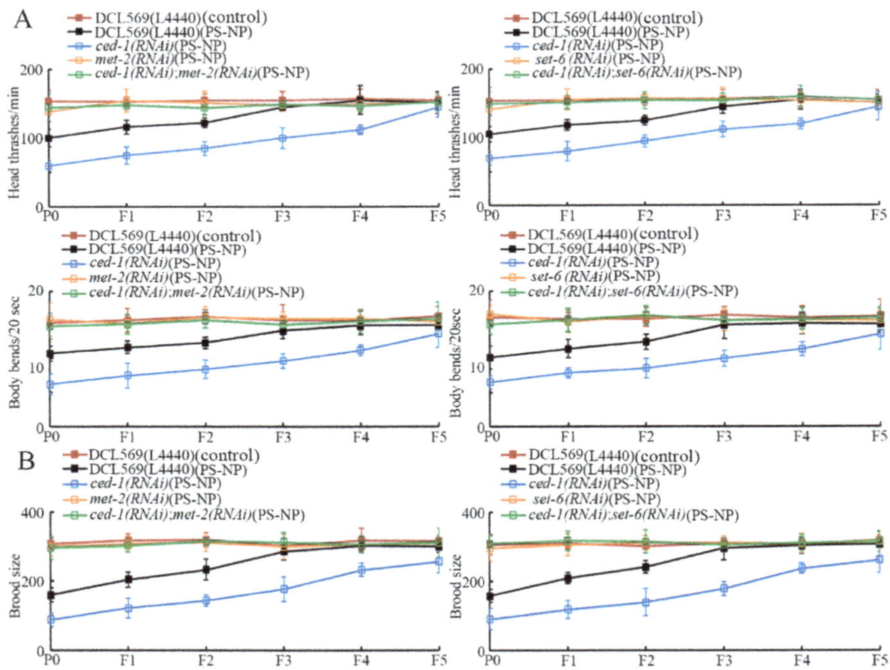

Fig. 15.4 Genetic interaction between CED-1 and MET-2 or SET-6 in regulating transgenerational PS-NP toxicity in decreasing locomotion behavior (**a**) and in reducing brood size [30]. PS-NP exposure concentration was 1 μg/L. The curve of DCL569(L4440)(PS-NP) was significantly ($P < 0.01$) different from that of DCL569(L4440) (control). The curves of *ced-1(RNAi)*(PS-NP), *met-2(RNAi)*(PS-NP), and *set-6(RNAi)*(PS-NP) were significantly ($P < 0.01$) different from that of DCL569(L4440)(PS-NP). The curves of *ced-1(RNAi);met-2(RNAi)*(PS-NP) and *ced-1(RNAi);set-6(RNAi)*(PS-NP) were significantly ($P < 0.01$) different from that of *ced-1(RNAi)*(PS-NP)

neurotoxicity and transgenerational reproductive toxicity of PS-NPs induced by germline RNAi of *ced-1* could be inhibited by germline RNAi of *met-2* and *set-6*, and the *ced-1(RNAi);met-2(RNAi)* and *ced-1(RNAi);set-6(RNAi)* showed resistance to transgenerational neurotoxicity and transgenerational reproductive toxicity of

PS-NPs (Fig. 15.4) [30]. That is, in the germline, transgenerational inhibition in CED-1 mediates transgenerational pollutant toxicity by causing transgenerational activation of MET-2 and SET-6.

In *C. elegans*, some ligands functioned together with their receptors to regulate the transgenerational pollutant toxicity [32–38]. Among these ligands, after PS-NP exposure, germline RNAi of *met-2* caused inhibition in INS-39 and WRT-3, and germline RNAi of *set-6* resulted in inhibition in INS-39, EFN-3, and WRT-3 [30]. Additionally, after PS-NP exposure, germline RNAi of *met-2* caused inhibition in DAF-2, receptor of INS-39, and germline RNAi of *set-6* resulted in inhibition in DAF-2 and VAB-1 (receptor of EFN-3) [30]. Therefore, inhibition in CED-1 causes transgenerational activation of MET-2 and SET-6 to mediate transgenerational pollutant toxicity via resulting in certain transgenerational response of secreted ligands, such as transgenerational activation of insulin, Ephrin, and Hedgehog ligands.

15.3 Role of Germline NHRs

15.3.1 Transgenerational Response of Germline NHRs

Exposure to pollutants can induce transgenerational response of certain germline NHRs. PS-NP was used as the example of pollutants. Among 33 germline NHRs, PS-NP exposure caused the inhibition in germline NHR-12 and activation of germline NHR-14, NHR-47, and DAF-12 (Fig. 15.5) [39]. Moreover, exposure to 1 and 10 μg/L PS-NPs induced transgenerational inhibition in germline NHR-12 and transgenerational activation of germline NHR-14, NHR-47, and DAF-12 (Fig. 15.6) [39]. After 10 μg/L PS-NPs exposure, transgenerational inhibition in germline NHR-12 and transgenerational activation of germline NHR-14, NHR-47, and DAF-12 could be detected till to the F4-G (Fig. 15.6) [39].

15.3.2 Function of Germline NHRs in Regulating Transgenerational Toxicity of Pollutants

Functional analysis indicated that germline RNAi of *nhr-12* caused susceptibility to transgenerational pollutants toxicity and germline RNAi of *nhr-14*, *nhr-47*, and *daf-12* induced resistance to transgenerational pollutant toxicity. The transgenerational neurotoxicity and transgenerational reproductive toxicity of PS-NPs could be strengthened by germline RNAi of *nhr-12* and inhibited by germline RNAi of *nhr-14*, *nhr-47*, and *daf-12* (Figs. 15.7 and 15.8) [39]. Therefore, transgenerational inhibition in germline NHR-12 and transgenerational activation of germline NHR-14, NHR-47, and DAF-12 mediate the induction of transgenerational pollutant toxicity.

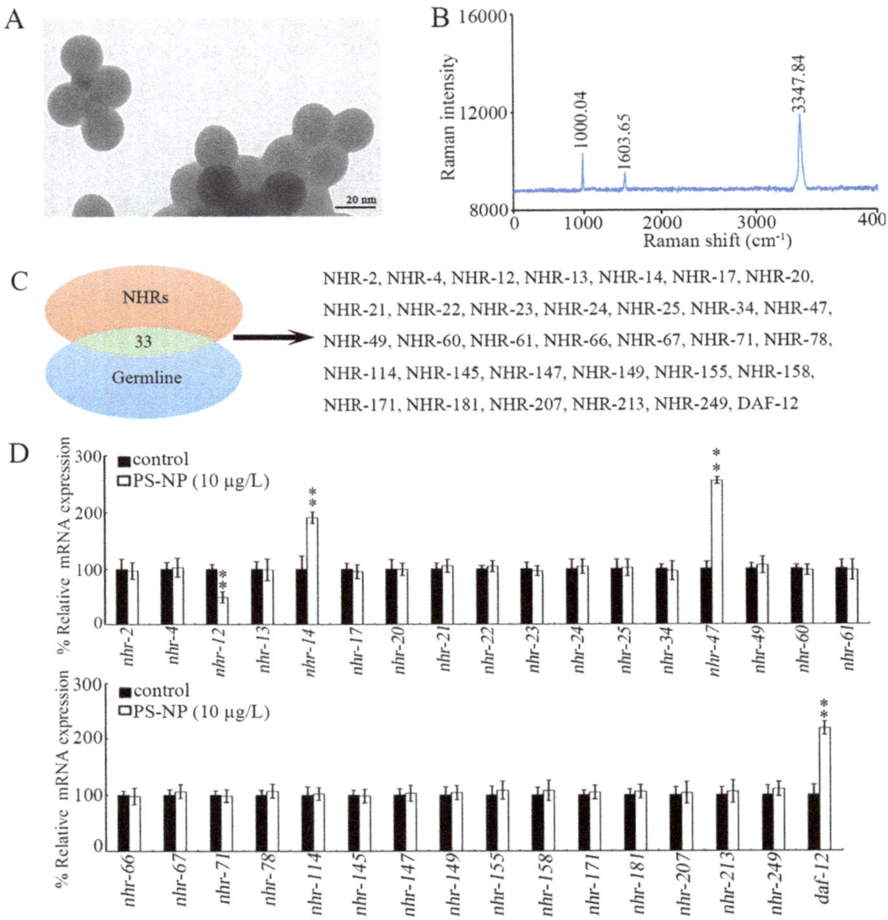

Fig. 15.5 Effect of PS-NP exposure on expression of germline NHR genes [39]. (**a**) TEM image of PS-NPs before sonication. (**b**) Raman spectrum of PS-NP. (**c**) NHR genes expressed in the germline. (**d**) Effect of PS-NP (10 μg/L) exposure on expressions of germline NHR genes. In total, 30 intact gonads were used for the qRT-PCR assay for each treatment. $^{**}P < 0.01$ vs. control

15.3.3 Downstream Targets of Germline NHR-12, NHR-14, NHR-47, and DAF-12 in Regulating Transgenerational Toxicity of Pollutants

In the germline, certain secreted ligands were identified as downstream targets of NHR-12, NHR-14, NHR-47, and DAF-12 to regulate the transgenerational pollutant toxicity. After PS-NP exposure, germline RNAi of *nhr-12* caused activation of germline INS-3, INS-39, DAF-28, LIN-44, and EFN-3; germline RNAi of *nhr-14* induced inhibition in germline INS-39 and EFN-3; germline RNAi of *nhr-47* caused

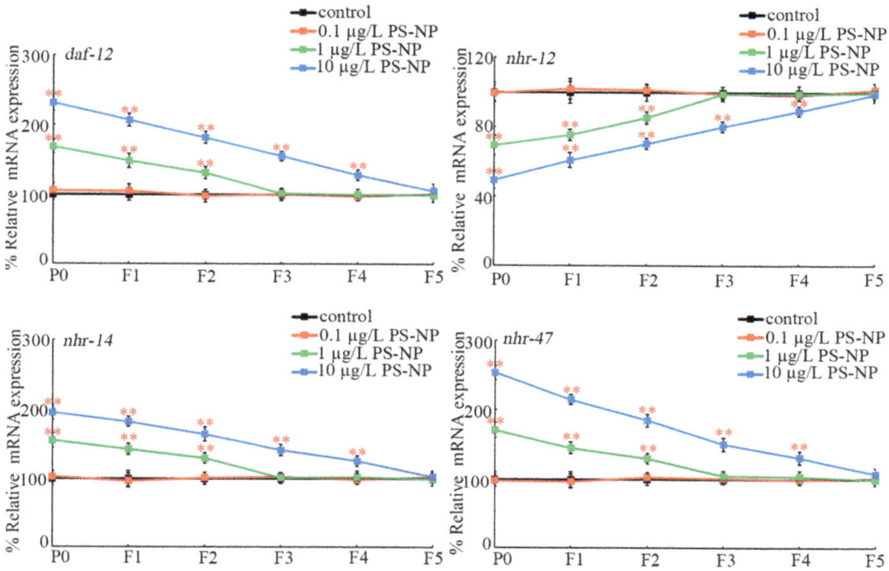

Fig. 15.6 Transgenerational expressions of germline *daf-12*, *nhr-12*, *nhr-14*, and *nhr-47* after exposure to PS-NPs at P0-G [39]. In total, 30 intact gonads were used for the qRT-PCR assay for each treatment. $^{**}P < 0.01$ vs. control

inhibition in germline INS-3, INS-39, DAF-28, and EFN-3; and germline RNAi of *daf-12* resulted in inhibition in germline INS-39 and EFN-3 (Fig. 15.9) [39]. After PS-NP exposure, germline RNAi of *nhr-12* at the P0-G caused activation of DAF-2 (receptor of INS-3, INS-39, and DAF-28), MIG-1 (receptor of LIN-44), and VAB-1 (receptor of EFN-3) at the F1-G, and germline RNAi of *nhr-14*, *nhr-47*, and *daf-12* at the P0-G induced inhibition in DAF-2 and VAB-1 at the F1-G (Fig. 15.10) [39], which further supported the above observation.

15.4 Perspectives

The identified germline GPCRs and NHRs are helpful to answer the question on the role of their activation or suppression as the forces to drive the formation of transgenerational toxicity of stresses and pollutants. Firstly, in *C. elegans*, CED-1 was identified as the germline GPCR, and its inhibition mediated induction of transgenerational pollutant toxicity. GPCRs are a superfamily, and there are still other GPCRs expressed in the germline. Nevertheless, it remains unclear whether other germline GPCRs can participate in the control of transgenerational pollutant toxicity.

During the control of transgenerational pollutant toxicity, MET-2 and SET-6 were identified as the downstream targets of germline CED-1. MET-2 and SET-6

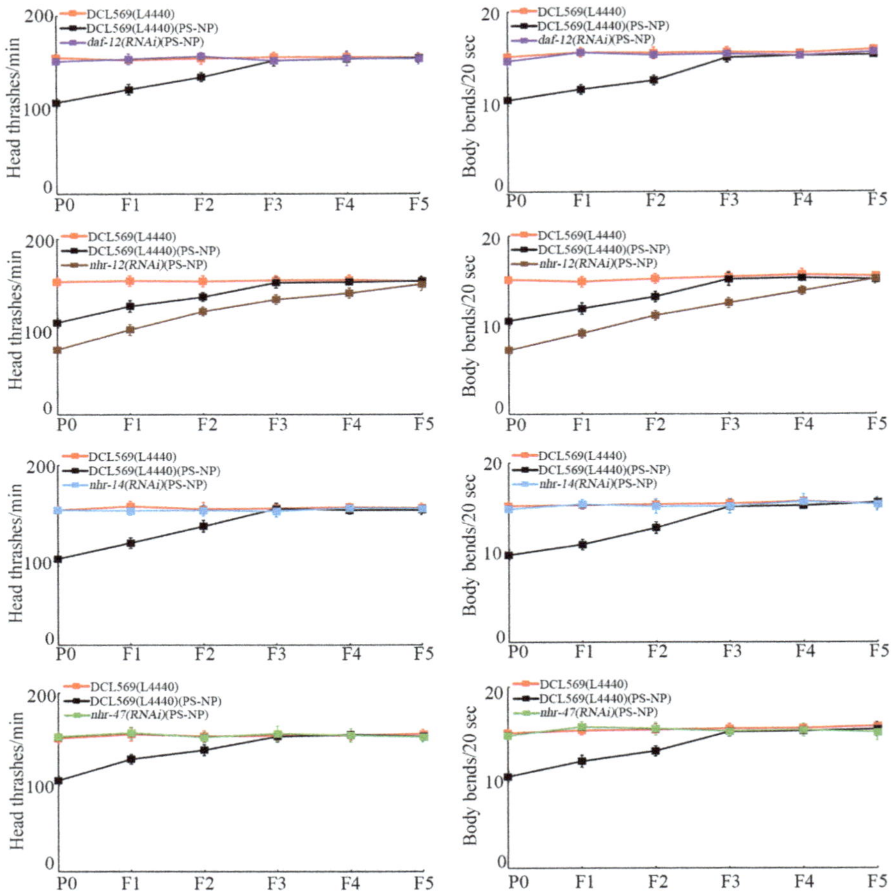

Fig. 15.7 Effect of RNAi of *daf-12*, *nhr-12*, *nhr-14*, and *nhr-47* on transgenerational PS-NP toxicity in decreasing locomotion behavior [39]. Exposure concentration of PS-NPs was 1 μg/L. The curves of DCL569(L4440)(PS-NP) were significantly ($P < 0.01$) different from those of DCL569(L4440). The curves of *daf-12(RNAi)*(PS-NP), *nhr-12(RNAi)*(PS-NP), *nhr-14(RNAi)*(PS-NP), and *nhr-47(RNAi)*(PS-NP) were significantly ($P < 0.01$) different from those of DCL569(L4440)(PS-NP). Significance between curves was tested by Kaplan-Meier analysis

functions reflect a form of epigenetic regulation. In cells, GPCRs activate or inhibit the downstream signaling cascade to regulate pollutant toxicity though affecting the binding Gα subunits [40, 41]. On the one hand, in the germline, how the Gα subunits mediate the role of CED-1 in affecting MET-2 and SET-6 activities and functions needs to be further examined. On the other hand, besides the epigenetic regulation signals, other molecular signals in the cytoplasm (such as MAPK signals) may also act downstream of germline GPCRs to mediate their function in regulating transgenerational pollutant toxicity.

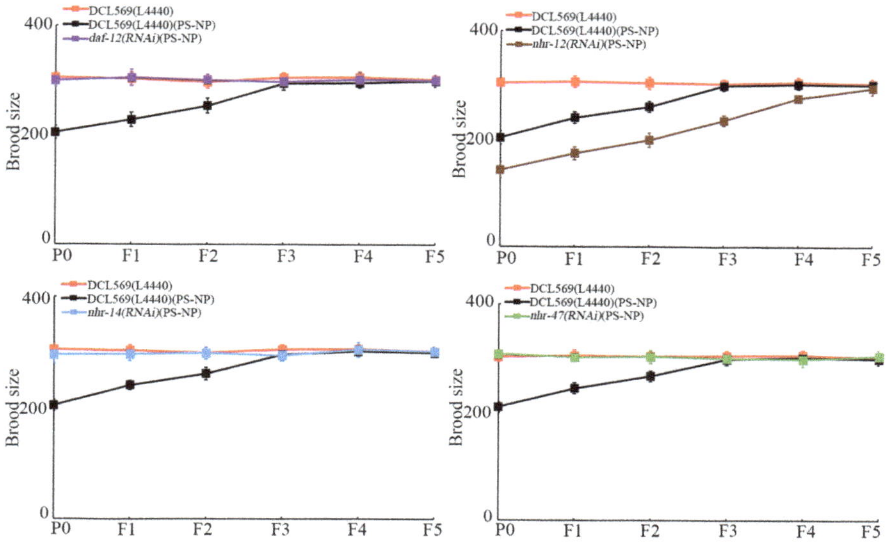

Fig. 15.8 Effect of RNAi of *daf-12*, *nhr-12*, *nhr-14*, and *nhr-47* on transgenerational PS-NP toxicity in reducing brood size [39]. Exposure concentration of PS-NPs was 1 μg/L. The curves of DCL569(L4440)(PS-NP) were significantly ($P < 0.01$) different from those of DCL569(L4440). The curves of *daf-12(RNAi)*(PS-NP), *nhr-12(RNAi)*(PS-NP), *nhr-14(RNAi)*(PS-NP), or *nhr-47(RNAi)*(PS-NP) were significantly ($P < 0.01$) different from those of DCL569(L4440)(PS-NP). Significance between curves was tested by Kaplan-Meier analysis

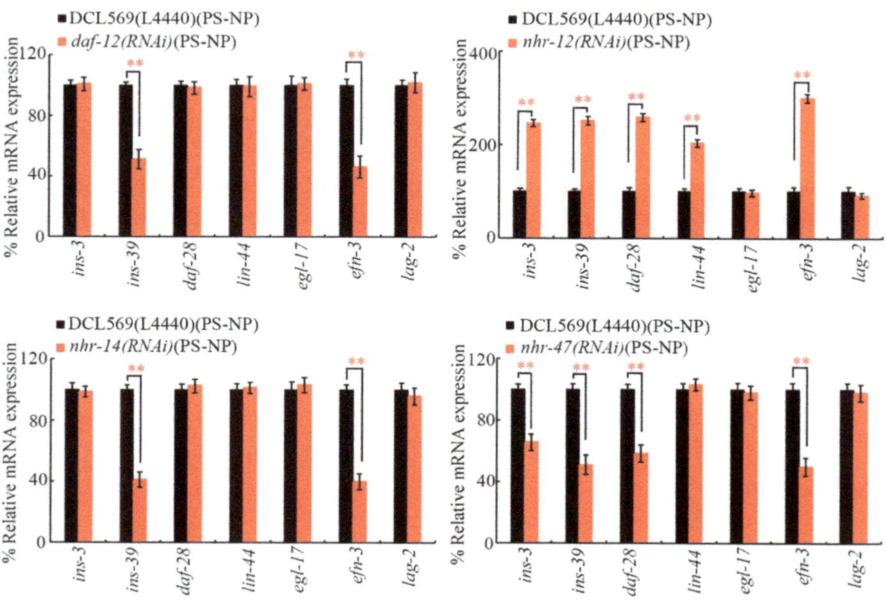

Fig. 15.9 Effect of germline RNAi of *daf-12*, *nhr-12*, *nhr-14*, and *nhr-47* on expressions of secreted ligand genes in PS-NP-exposed nematodes [39]. Exposure concentration of PS-NPs was 1 μg/L. In total, 30 intact gonads were used for the qRT-PCR assay for each treatment. $^{**}P < 0.01$

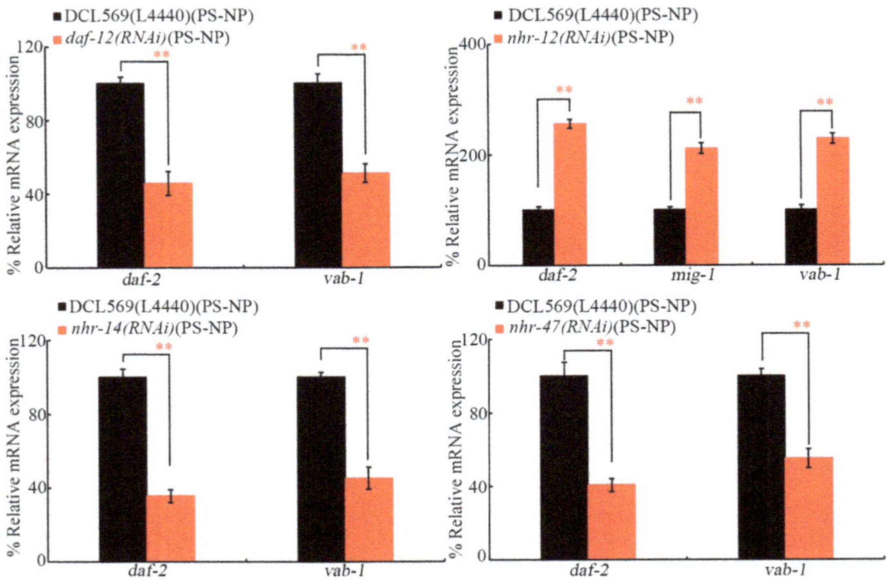

Fig. 15.10 Effect of germline RNAi of *daf-12*, *nhr-12*, *nhr-14*, and *nhr-47* on expressions of *daf-2*, *vab-1*, and/or *mig-1* at F1-G of PS-NP-exposed nematodes [39]. Exposure concentration of PS-NPs was 1 μg/L. $^{**}P < 0.01$

With the nanoplastic as the example of pollutants, 4 germline NHRs (NHR-12, NHR-14, NHR-47, and DAF-12) were found to exhibit the transgenerational response to nanoplastic exposure. In *C. elegans*, there are 33 NHRs expressed in the germline. This implies that exposure to other pollutants with the more severe toxicity will induce transgenerational response of more germline NHRs.

In the germline, the identified four germline NHRs regulate transgenerational pollutant toxicity by causing transgenerational response of certain secreted ligands, such as insulin, Ephrin, and Hedgehog ligands. As the transcriptional factors, normally, NHR-12, NHR-14, NHR-47, and DAF-12 have a large number of downstream targets. That is, even in the germline, other response genes could also be activated or inhibited by these germline NHRs to mediate their function in regulating transgenerational pollutant toxicity.

References

1. Wang D-Y (2018) Nanotoxicology in *Caenorhabditis elegans*. Springer Nature Singapore Pte Ltd., Singapore
2. Wang D-Y (2020) Exposure toxicology in *Caenorhabditis elegans*. Springer Nature Singapore Pte Ltd., Singapore
3. Wang D-Y (2022) Toxicology at environmentally relevant concentrations in *Caenorhabditis elegans*. Springer Nature Singapore Pte Ltd., Singapore

4. Wang D-Y (2019) Molecular toxicology in *Caenorhabditis elegans*. Springer Nature Singapore Pte Ltd., Singapore
5. Wang D-Y (2019) Target organ toxicology in *Caenorhabditis elegans*. Springer Nature Singapore Pte Ltd., Singapore
6. Liu H-L, Tian L-J, Wang D-Y (2021) Notch receptor GLP-1 regulates toxicity of simulated microgravity stress by activating germline-intestine communication of insulin signaling in *C. elegans*. Biochem Biophys Res Commun 534:248–253
7. Wang S-T, Liu H-L, Qu M, Wang D-Y (2021) Response of tyramine and glutamate related signals to nanoplastic exposure in *Caenorhabditis elegans*. Ecotoxicol Environ Saf 217:112239
8. Yang Y-H, Wu Q-L, Wang D-Y (2021) Dysregulation of G protein-coupled receptors in the intestine by nanoplastic exposure in *Caenorhabditis elegans*. Environ Sci Nano 8:1019–1028
9. Yang Y-H, Dong W-T, Wu Q-L, Wang D-Y (2021) Induction of protective response associated with expressional alterations in neuronal G protein-coupled receptors in polystyrene nanoparticle exposed *Caenorhabditis elegans*. Chem Res Toxicol 34:1308–1318
10. Liu Z-Y, Hua X, Zhao Y, Bian Q, Wang D-Y (2024) Polyethylene nanoplastics cause reproductive toxicity associated with activation of both estrogenic hormone receptor NHR-14 and DNA damage checkpoints in *C. elegans*. Sci Total Environ 906:167471
11. Liu Z-Y, Bian Q, Wang D-Y (2025) 6-PPD quinone induces response of nuclear hormone receptors in the germline associated with formation of reproductive toxicity in *Caenorhabditis elegans*. J Hazard Mater 495:138815
12. Hua X, Wang D-Y (2022) Cellular uptake, transport, and organelle response after exposure to microplastics and nanoplastics: current knowledge and perspectives for environmental and health risks. Rev Environ Contam Toxicol 260:12
13. Qu M, Kong Y, Yuan Y-J, Wang D-Y (2019) Neuronal damage induced by nanopolystyrene particles in nematode *Caenorhabditis elegans*. Environ Sci Nano 6:2591–2601
14. Qu M, Qiu Y-X, Kong Y, Wang D-Y (2019) Amino modification enhances reproductive toxicity of nanopolystyrene on gonad development and reproductive capacity in nematode *Caenorhabditis elegans*. Environ Pollut 254:112978
15. Qu M, Nida A, Kong Y, Du H-H, Xiao G-S, Wang D-Y (2019) Nanopolystyrene at predicted environmental concentration enhances microcystin-LR toxicity by inducing intestinal damage in *Caenorhabditis elegans*. Ecotoxicol Environ Saf 183:109568
16. Qu M, Xu K-N, Li Y-H, Wong G, Wang D-Y (2018) Using *acs-22* mutant *Caenorhabditis elegans* to detect the toxicity of nanopolystyrene particles. Sci Total Environ 643:119–126
17. Qiu Y-X, Liu Y-Q, Li Y-H, Li G-J, Wang D-Y (2020) Effect of chronic exposure to nanopolystyrene on nematode *Caenorhabditis elegans*. Chemosphere 256:127172
18. Qiu Y-X, Luo L-B, Yang Y-H, Kong Y, Li Y-H, Wang D-Y (2020) Potential toxicity of nanopolystyrene on lifespan and aging process of nematode *Caenorhabditis elegans*. Sci Total Environ 705:135918
19. Li D, Deng Y-J, Wang S-T, Du H-H, Xiao G-S, Wang D-Y (2020) Assessment of nanopolystyrene toxicity under fungal infection condition in *Caenorhabditis elegans*. Ecotoxicol Environ Saf 197:110625
20. Liu H-L, Shao H-M, Guo Z-J, Wang D-Y (2020) Nanopolystyrene exposure activates a fat metabolism related signaling-mediated protective response in *Caenorhabditis elegans*. NanoImpact 17:100204
21. Dong S-S, Qu M, Rui Q, Wang D-Y (2018) Combinational effect of titanium dioxide nanoparticles and nanopolystyrene particles at environmentally relevant concentrations on nematodes *Caenorhabditis elegans*. Ecotoxicol Environ Saf 161:444–450
22. Zhao Y-Y, Li D, Rui Q, Wang D-Y (2020) Toxicity induction of nanopolystyrene under microgravity stress condition in *Caenorhabditis elegans*. Sci Total Environ 703:135623
23. Liu H-L, Wang D-Y (2021) Intestinal mitochondrial unfolded protein response induced by nanoplastic particles in *Caenorhabditis elegans*. Chemosphere 267:128917

24. Tang M, Ding G, Lu X, Huang Q, Du H, Xiao G, Wang D (2022) Exposure to nanoplastic particles enhances *Acinetobacter* survival, biofilm formation, and serum resistance. Nanomaterials 12:4222
25. Zhao L, Qu M, Wong G, Wang D-Y (2017) Transgenerational toxicity of nanopolystyrene particles in the range of μg/L in nematode *Caenorhabditis elegans*. Environ Sci Nano 4:2356–2366
26. Sun L-M, Liao K, Wang D-Y (2021) Comparison of transgenerational reproductive toxicity induced by pristine and amino modified nanoplastics in *Caenorhabditis elegans*. Sci Total Environ 768:144362
27. Liu H-L, Tian L-J, Wang S-T, Wang D-Y (2021) Size-dependent transgenerational toxicity induced by nanoplastics in nematode *Caenorhabditis elegans*. Sci Total Environ 790:148217
28. Zhao Y, Chen J, Wang R, Pu X, Wang D (2023) A review of transgenerational and multigenerational toxicology in the *in vivo* model animal *Caenorhabditis elegans*. J Appl Toxicol 43:122–145
29. Yang Y-H, Dong W-T, Wu Q-L, Wang D-Y (2021) Response of G protein-coupled receptor CED-1 in germline to polystyrene nanoparticles in *Caenorhabditis elegans*. Nanoscale Adv 3:1997–2006
30. Wu J-W, Shao Y-T, Hua X, Wang Y-X, Wang D-Y (2024) Nanoplastic at environmentally relevant concentrations induces toxicity across multiple generations associated with inhibition in germline G protein-coupled receptor CED-1 in *Caenorhabditis elegans*. Chemosphere 364:143011
31. Zhang L, Wang S-T, Zhao Y, Bi K, Wang D-Y (2022) Increase in germline methyltransferases governing methylation of histone H3K9 is associated with transgenerational nanoplastic toxicity in *Caenorhabditis elegans*. Environ Sci Nano 9:265–274
32. Wu J-W, Shao Y-T, Hua X, Wang D-Y (2024) Activated hedgehog and insulin ligands by decreased transcriptional factor DAF-16 mediate transgenerational nanoplastic toxicity in *Caenorhabditis elegans*. J Hazard Mater 480:135909
33. Hua X, Cao C, Zhang L, Wang D-Y (2023) Activation of FGF signal in germline mediates transgenerational toxicity of polystyrene nanoparticles at predicted environmental concentrations in *Caenorhabditis elegans*. J Hazard Mater 451:131174
34. He W-M, Gu A-H, Wang D-Y (2023) Sulfonate-modified polystyrene nanoparticle at precited environmental concentrations induces transgenerational toxicity associated with increase in germline Notch signal of *Caenorhabditis elegans*. Toxics 11:511
35. Liu H-L, Zhao Y-L, Hua X, Wang D-Y (2022) Induction of transgenerational toxicity is associated with the activated germline insulin signals in nematodes exposed to nanoplastic at predicted environmental concentrations. Ecotoxicol Environ Saf 243:114022
36. Xu R-R, Hua X, Rui Q, Wang D-Y (2022) Alteration in Wnt signaling mediates induction of transgenerational toxicity of polystyrene nanoplastics in *C. elegans*. NanoImpact 28:100425
37. Zhao Y, Hua X, Bian Q, Wang D-Y (2022) Nanoplastic exposure at predicted environmental concentrations induces activation of germline Ephrin signal associated with toxicity formation in the *Caenorhabditis elegans* offspring. Toxics 10:699
38. Hua X, Zhao Y, Yuan Y-J, Zhang L, Bian Q, Wang D-Y (2022) Nanoplastics cause transgenerational toxicity through inhibiting germline microRNA *mir-38* in *C. elegans*. J Hazard Mater 437:129302
39. Liu Z-Y, Wang Y-X, Bian Q, Wang D-Y (2024) Transgenerational response of germline nuclear hormone receptor genes to nanoplastics at predicted environmental doses in *Caenorhabditis elegans*. Toxics 12:420
40. Yang Y-H, Wu Q-L, Wang D-Y (2021) Neuronal Gα subunits required for the control of response to polystyrene nanoparticles in the range of μg/L in *C. elegans*. Ecotoxicol Environ Saf 225:112732
41. Yang Y-H, Du H-H, Xiao G-S, Wu Q-L, Wang D-Y (2020) Response of intestinal Gα subunits to nanopolystyrene in nematode *Caenorhabditis elegans*. Environ Sci Nano 7:2351–2359

Chapter 16
Protective Responses Against Transgenerational Toxicity

Abstract Some protective responses can be activated to be against toxicity of stresses and pollutants. We here discussed the role of protective responses–related signals as the "effectors" for formation of transgenerational toxicity induced by stresses and pollutants. We first introduced the transgenerational oxidation response, mitochondrial unfolded protein response (mt UPR), endoplasmic reticulum unfolded protein response (ER UPR), and innate immune response to pollutants. Moreover, we introduced and discussed the molecular basis for induction of transgenerational oxidation response, mt UPR response, and innate immune response to pollutants.

Keywords Transgenerational protective response · Effector · *Caenorhabditis elegans*

16.1 Introduction

For the study of toxicology, the animal model of *Caenorhabditis elegans* initially receives the attention due to its high sensitivity to stresses and pollutants [1–3]. More importantly, as a classic model animal, knowledge systems of molecular toxicology and target organ toxicology have been raised in *C. elegans* [4, 5]. Besides the introduced molecular signaling pathways, metabolic signals, and epigenetic regulation–related signals from Chaps. 5, 6, 7, 8, 9, 10, 11, 12, 13, 14, and 15, some protective responses–related signals also showed response to stresses and pollutants and regulated their toxicity induction.

In *C. elegans*, antioxidation system refers to the network including enzymatic antioxidants (superoxide dismutases (SODs), catalases (CTLs), glutathione peroxidases (GPXs), glutathione S-transferases (GSTs), and peroxiredoxins (PRDXs)) and non-enzymatic antioxidants (glutathione (GSH), thioredoxin (TRX) system, and small-molecule antioxidants) to maintain cellular redox homeostasis by neutralizing reactive oxygen species (ROS) and preventing oxidative damage. Among SOD family required for the antioxidation system, Mn-SODs (SOD-2 and SOD-3)

D. Wang, *Transgenerational Toxicology in Caenorhabditis elegans*,
https://doi.org/10.1007/978-981-92-1389-4_16

acting as targets of transcriptional factor DAF-16 were involved in controlling toxicity of stresses and pollutants [6–9]. Mitochondrial unfolded protein response (mt UPR) is a cytoprotective response to detect proteotoxic stress within the mitochondrial matrix, including accumulation of misfolded or unfolded proteins by activating nuclear transcriptional program to restore mitochondrial homeostasis. Exposure to stresses and pollutants all could induce the response of mt UPR in nematodes [10–13]. Endoplasmic reticulum unfolded protein response (ER UPR) refers to intracellular signaling network to sense accumulation of misfolded or unfolded proteins in ER lumen by using tripartite stress-response signaling pathway (IRE-1/XBP-1, PEK-1/ATF-4, and ATF-6) to restore ER protein-folding homeostasis. Certain ER UPR could be activated by pollutants to mediate the protective strategy against pollutant toxicity [14, 15]. In *C. elegans*, innate immune response is a non-adaptive defense system to provide rapid protection against a broad spectrum of microbial pathogens and xenotoxic threats by relying on cell-autonomous and systemic signaling pathways within somatic cells to activate the protective transcriptional response and to restore the homeostasis. The immune effectors regulated by p38 MAPK, insulin, and stress-response pathways include antimicrobial peptides (secreted peptides that disrupt microbial membranes (e.g., CNC, NLP, ABF families)), lysozymes that degrade bacterial cell walls, C-type lectins (e.g., CLEC family) that agglutinate microbes, detoxification enzymes (cytochrome P450s, UDP-glucuronosyltransferases (UGTs), glutathione S-transferases (GSTs)), mucins, and barrier proteins. Exposure to pollutants induced innate immune response, and pollutants with more severe toxicity could cause the innate immunosuppression [16–19].

Considering that the protective response–related signals usually reflect the final and directly produced functional change or output, they can be considered as the "effectors" for formation of transgenerational toxicity of stresses and pollutants. In this chapter, we introduced transgenerational antioxidation response, mt UPR response, ER UPR response, and innate immune response and discussed the role of related signaling in regulating the transgenerational pollutants toxicity.

16.2 Transgenerational Antioxidation Response

16.2.1 Transgenerational Antioxidation Response to Pollutants

With SOD-3 expression to reflect the antioxidation response, exposure to pollutants can induce different forms of transgenerational antioxidation response. The 6-PPD quinone (6-PPDQ) could cause several aspects of toxicity in nematodes, including organ toxicity, lifespan reduction, and metabolic disruptions [20–32]. With the 6-PPDQ as the example of pollutants, transgenerational activation of SOD-3 was induced by 6-PPDQ exposure (Fig. 16.1) [33]. After exposure to 10 μg/L 6-PPDQ

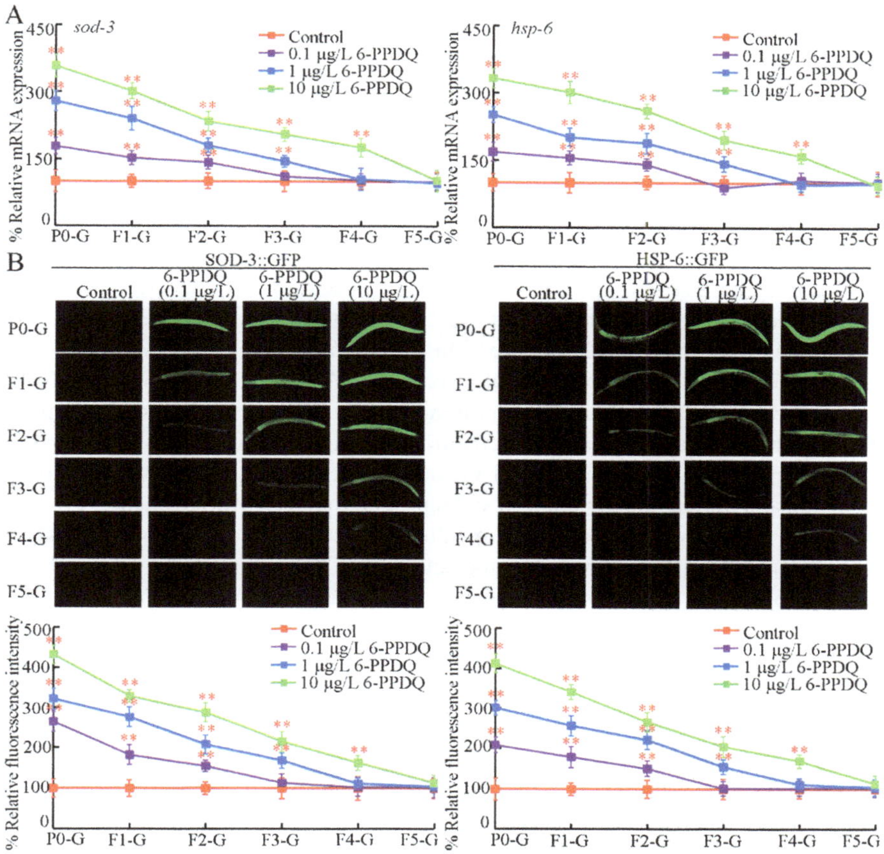

Fig. 16.1 Transgenerational response of SOD-3 and HSP-6 after 6-PPDQ exposure at P0-G [33]. (**a**) Transgenerational change of *sod-3* and *hsp-6* expressions after 6-PPDQ exposure at P0-G. (**b**) Transgenerational change of SOD-3::GFP and HSP-6::GFP expressions after 6-PPDQ exposure at P0-G. $^{**}P < 0.01$ vs. control

at parental generation (P0-G), transgenerational activation of SOD-3 could be observed till to the F4-G (Fig. 16.1) [33]. Similarly, exposure to polystyrene microplastics (PS-MPs) induced transgenerational activation of SOD-3, SOD-4, and SOD-5 till to the F3-G [34]. Nanoplastic exposure could cause multiple toxicities on *C. elegans*, including the induction of transgenerational toxicity [10, 35–49]. PS nanoparticle (PS-NP) was further selected as the example of pollutants. Moreover, exposure to 100 nm PS-NP (1 μg/L) induced transgenerational activation of SOD-3 till to the F1-G, whereas the exposure to 20 nm PS-NP (1 μg/L) caused transgenerational inhibition in SOD-3 till to the F2-G [50].

16.2.2 Function of MEV-1, AAK-2, DAF-2, and DAF-16 in Regulating Transgenerational Antioxidation Response to Pollutants

MEV-1, a component of mitochondrial complex II, participated in the control of transgenerational antioxidation response induced by pollutants. Exposure to 1 μg/L PS-NPs induced transgenerational intestinal oxidative stress and transgenerational inhibition in MEV-1 till to the F2-G [51]. After PS-NP exposure, RNAi of *mev-1* at F3-G could cause severe intestinal oxidative stress and toxicity on locomotion and reproduction [51]. Meanwhile, after PS-NP exposure, RNAi of *mev-1* at F3-G caused the inhibition in SOD-3 expression (Fig. 16.2) [51].

In insulin signaling pathway, insulin receptor DAF-2 and DAF-16 all regulated transgenerational antioxidation response induced by pollutants. Exposure to 1 μg/L PS-NPs induced transgenerational activation of DAF-2 till to the F2-G [51]. After PS-NP exposure, RNAi of *daf-2* at F1-G could result in inhibition in oxidative stress induction and toxicity on locomotion and reproduction [51]. Moreover, after PS-NPs exposure, RNAi of *daf-2* at F1-G resulted in activation of SOD-3 expression (Fig. 16.2) [51]. Different from this, the activated SOD-3 by 6-PPDQ was inhibited by RNAi of *daf-16*, as well as by RNAi of *aak-2* encoding the AMPK (Fig. 16.3) [33].

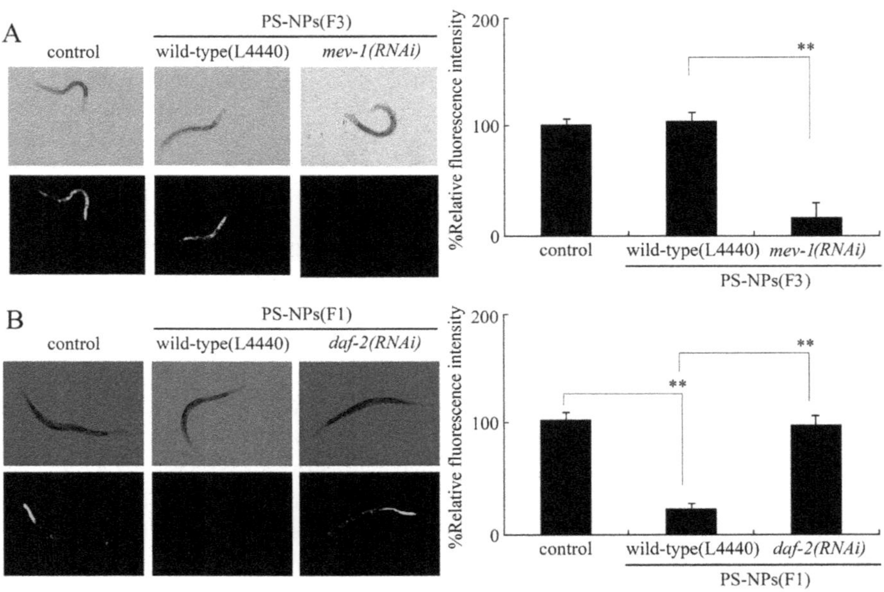

Fig. 16.2 Effect of RNAi knockdown of *mev-1* or *daf-2* on expression of SOD-3::GFP in offspring of PS-NP-exposed nematodes [51]. (**a**) Effect of RNAi knockdown of *mev-1* on expression of SOD-3::GFP in offspring of PS-NP-exposed nematodes. RNAi knockdown of *mev-1* was performed at F3-G. (**b**) Effect of RNAi knockdown of *daf-2* on expression of SOD-3::GFP in offspring of PS-NP-exposed nematodes. RNAi knockdown of *daf-2* was performed at F1-G. L4440, empty vector. Exposure concentration of PS-NP was 1 μg/L. Bars represent means ± SD. $^{**}P < 0.01$

16.2.3 Function of Glucose Metabolism–Related Signals in Regulating Transgenerational Antioxidation Response to Pollutants

Glucose metabolism–related signals were also involved in controlling transgenerational antioxidation response induced by pollutants. Accompanied with transgenerational activation of SOD-3, transgenerational inhibition in HXK-1, HXK-3, PYK-1, and PYK-2, enzymes governing glycolysis, was induced by 6-PPDQ exposure [33]. Moreover, the activated SOD-3 by 6-PPDQ was inhibited by RNAi of *hxk-1*, *hxk-3*, *pyk-1*, and *pyk-2* (Fig. 16.3) [33].

16.3 Transgenerational mt UPR Response

16.3.1 Transgenerational mt UPR Response to Pollutants

Exposure to pollutants could also induce different forms of transgenerational mt UPR response. With 6-PPDQ as the example of pollutants, transgenerational activation of mt UPR was induced (Fig. 16.1) [33]. After exposure to 10 μg/L 6-PPDQ, transgenerational activation of mt UPR was observed till to the F4-G (Fig. 16.1) [33]. Moreover, exposure to 100 nm PS-NP induced transgenerational activation of mt UPR; however, exposure to 20 nm PS-NP caused transgenerational suppression in mt UPR [50, 52, 53]. After exposure 20 nm PS-NPs (100 μg/L), transgenerational suppression in mt UPR could be detected till to the F5-G (Fig. 16.4) [53].

16.3.2 Transcriptional Factors (ATFS-1, DVE-1, and UBL-5) Mediate Transgenerational mt UPR Response to Pollutants

ATFS-1, DVE-1, and UBL-5 are transcriptional factors to activate the mt UPR response [54–56]. Their transgenerational activation mediated transgenerational mt UPR response to pollutants. Accompanied with the transgenerational suppression in mt UPR induced by PS-NPs, transgenerational inhibition in ATFS-1, DVE-1, and UBL-5 was caused (Fig. 16.5) [53]. After exposure to 20 nm PS-NPs (1 μg/L), transgenerational inhibition in ATFS-1, DVE-1, and UBL-5 was detected till to the F2-G (Fig. 16.5) [53]. Moreover, the transgenerational suppression in mt UPR was further strengthened by RNAi of *atfs-1*, *dve-1*, and *ubl-5*, and more severe transgenerational suppression in mt UPR was observed in *atfs-1(RNAi)*, *dve-1(RNAi)*, and *ubl-5(RNAi)* (Fig. 16.6) [53]. RNAi of *atfs-1*, *dve-1*, and *ubl-5* also caused the susceptibility to transgenerational PS-NP toxicity on locomotion and reproduction

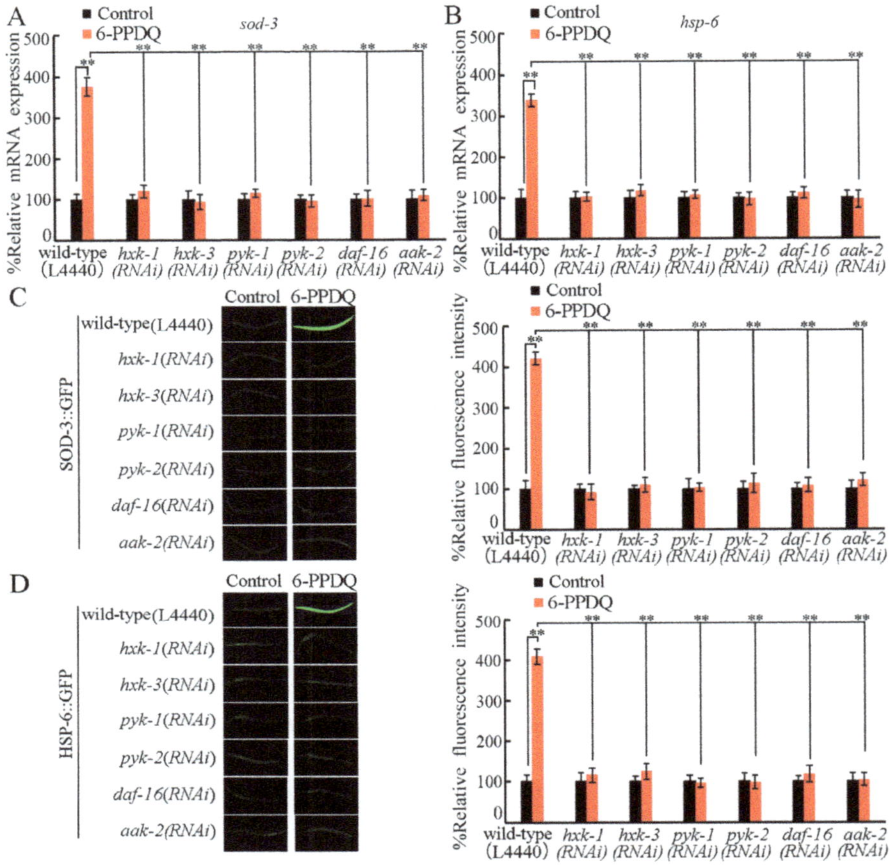

Fig. 16.3 Effect of RNAi of *hxk-1*, *hxk-3*, *pyk-1*, *pyk-2*, *daf-16*, and *aak-2* on activation of SOD-3 and HSP-6 induced by 6-PPDQ [33]. (**a**) Effect of RNAi of *hxk-1*, *hxk-3*, *pyk-1*, *pyk-2*, *daf-16*, and *aak-2* on *sod-3* expression in 6-PPDQ-exposed nematodes. (**b**) Effect of RNAi of *hxk-1*, *hxk-3*, *pyk-1*, *pyk-2*, *daf-16*, and *aak-2* on *hsp-6* expression in 6-PPDQ-exposed nematodes. (**c**) RNAi of *hxk-1*, *hxk-3*, *pyk-1*, *pyk-2*, *daf-16*, and *aak-2* on SOD-3::GFP expression in 6-PPDQ-exposed nematodes. (**d**) Effect of RNAi of *hxk-1*, *hxk-3*, *pyk-1*, *pyk-2*, *daf-16*, and *aak-2* on HSP-6::GFP expression in 6-PPDQ-exposed nematodes. The 6-PPDQ exposure concentration was 10 μg/L. $^{**}P < 0.01$

[53]. The 20 nm PS-NPs (1 μg/L) induced transgenerational suppression in mt UPR could be recovered at the F3-G (Fig. 16.4) [53]. However, after RNAi of *atfs-1*, *dve-1*, and *ubl-5* at the F3-G, the recovered mt UPR could be further inhibited (Fig. 16.7) [53].

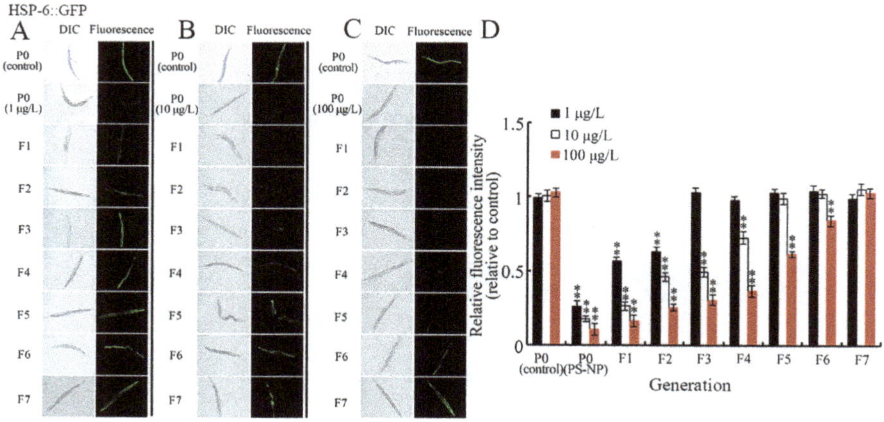

Fig. 16.4 Transgenerational mt UPR response after PS-NP exposure [53]. (**a**) Transgenerational mt UPR response after 1 μg/L PS-NP exposure. (**b**) Transgenerational mt UPR response after 10 μg/L PS-NP exposure. (**c**) Transgenerational mt UPR response after 100 μg/L PS-NP exposure. (**d**) Comparison of relative fluorescence intensity of HSP-6:GFP. $^{**}P < 0.01$ vs. P0(control). Control, without PS-NP exposure

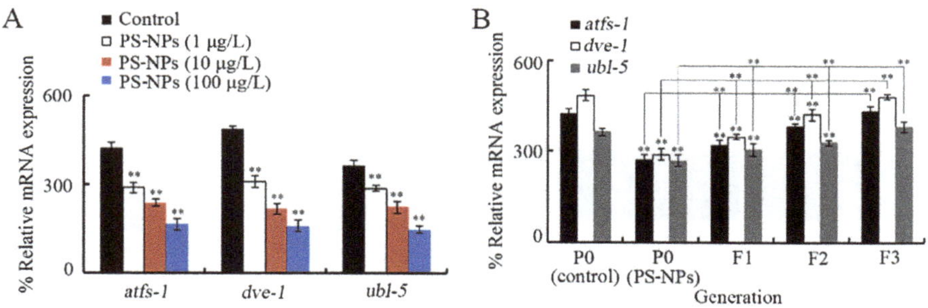

Fig. 16.5 Effect of PS-NP exposure on transgenerational expression of *atfs-1*, *dve-1*, and *ubl-5* [53]. (**a**) Effect of PS-NP exposure on expressions of *atfs-1*, *dve-1*, and *ubl-5* at P0-G. $^{**}P < 0.01$ vs. control. (**b**) Transgenerational expression of *atfs-1*, *dve-1*, and *ubl-5* after PS-NP (1 μg/L) exposure at P0-G. $^{**}P < 0.01$ vs. P0(control) (if not specially indicated). Control, without PS-NP exposure

16.3.3 Function of DAF-2, DAF-16, and AAK-2 in Regulating Transgenerational mt UPR Response to Pollutants

Both DAF-16 and AAK-2 were required for induction of transgenerational mt UPR response to pollutants. Exposure to 6-PPDQ caused the transgenerational inhibition in DAF-16 and AAK2 [33]. Accompanied with this, the activated mt UPR response by 6-PPDQ could be inhibited by RNAi of *daf-16* and *aak-2* (Fig. 16.3) [33]. Different from this, the PS-NP-induced transgenerational suppression in mt UPR could be inhibited by RNAi of *daf-2* at the F1-G (Fig. 16.7) [53].

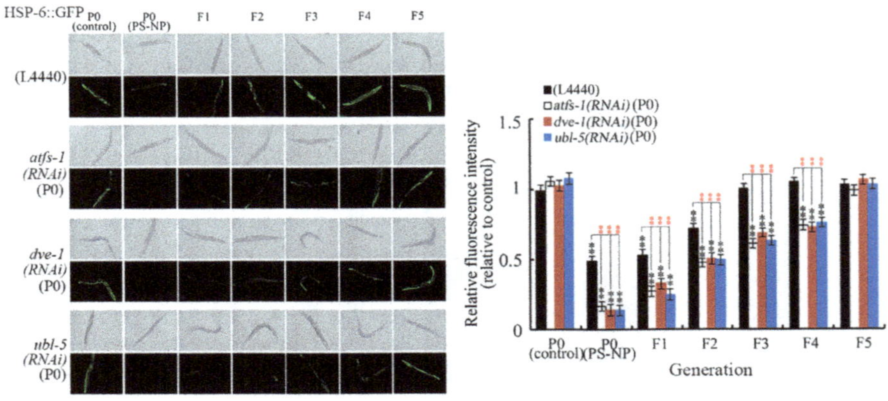

Fig. 16.6 Transgenerational mt UPR response in PS-NP (1 μg/L)-exposed nematodes with RNAi of *atfs-1*, *dve-1*, and *ubl-5* at P0-G [53]. $^{**}P < 0.01$ vs. P0(control) (if not specially indicated). Control, without PS-NP exposure

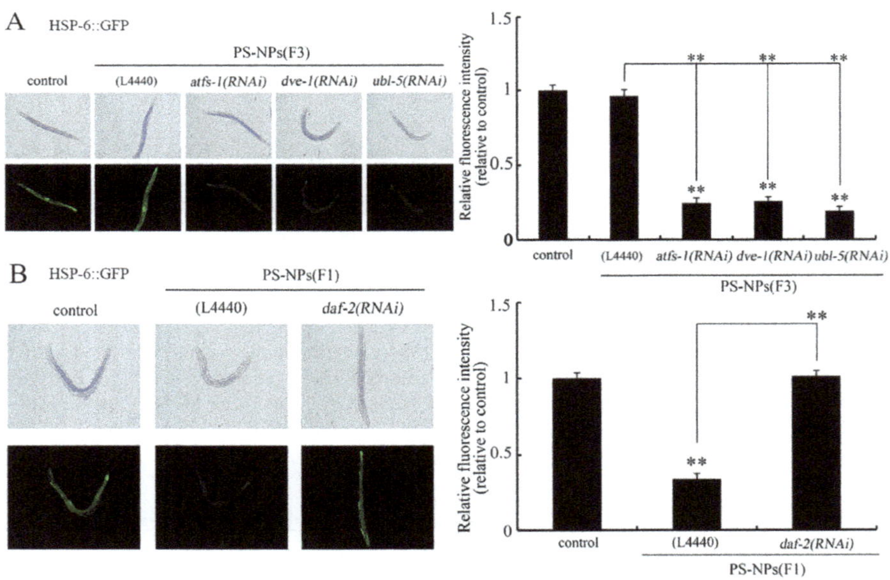

Fig. 16.7 Effect of *atfs-1*, *dve-1*, and *ubl-5* RNAi at F3-G (**a**) or *daf-2* RNAi at F1-G (**b**) on mt UPR response in PS-NP exposed nematodes [53]. Exposure concentration of PS-NP was 1 μg/L. $^{**}P < 0.01$ vs. control (if not specially indicated). Control, without PS-NP exposure

16.3.4 Function of Wnt Signaling in Regulating Transgenerational mt UPR Response to Pollutants

BAR-1, the β-catenin in Wnt signaling pathway, played an important function in regulating transgenerational mt UPR response to pollutants. The transgenerational suppression in mt UPR was strengthened by RNAi of *bar-1* [57]. After PS-NP exposure, compared to transgenerational suppression in mt UPR till to F2-G in wild type, transgenerational suppression in mt UPR was observed till to the F4-G in *bar-1(RNAi)* [57]. Moreover, after PS-NP exposure, *bar-1* RNAi could cause inhibition in DVE-1 [57], suggesting that inhibition in DVE-1 mediated function of *bar-1* RNAi in strengthening PS-NP-induced transgenerational suppression in mt UPR.

16.3.5 Function of Glucose Metabolism–Related Signals in Regulating Transgenerational mt UPR Response to Pollutants

Glucose metabolism–related signals also participated in the control of transgenerational mt-UPR response to pollutants. 6-PPDQ was selected as the example of pollutants. 6-PPDQ-induced activation of mt UPR response could be inhibited by RNAi of *hxk-1*, *hxk-3*, *pyk-1*, and *pyk-2*, encoding enzymes required for controlling glycolysis (Fig. 16.3) [33].

16.4 Transgenerational ER UPR Response

The PS-NP was used as the example of pollutants. Accompanied with transgenerational induction of oxidative stress from P0-G to F1-G, exposure to 100 nm PS-NP (1 μg/L) induced transgenerational activation of ER-UPR till to the F1-G [50]. In contrast, accompanied with transgenerational induction of oxidative stress from P0-G to F2-G, exposure to 20 nm PS-NP (1 μg/L) caused transgenerational suppression in ER UPR till to the F2-G [50].

16.5 Transgenerational Innate Immune Response

16.5.1 Transgenerational Innate Immune Response to Pollutants

Exposure to pollutants can cause the transgenerational innate immune response. The 6-PPDQ was selected as the example of pollutants. After the exposure, 0.1 μg/L 6-PPDQ activated innate immune response reflected by antimicrobial gene

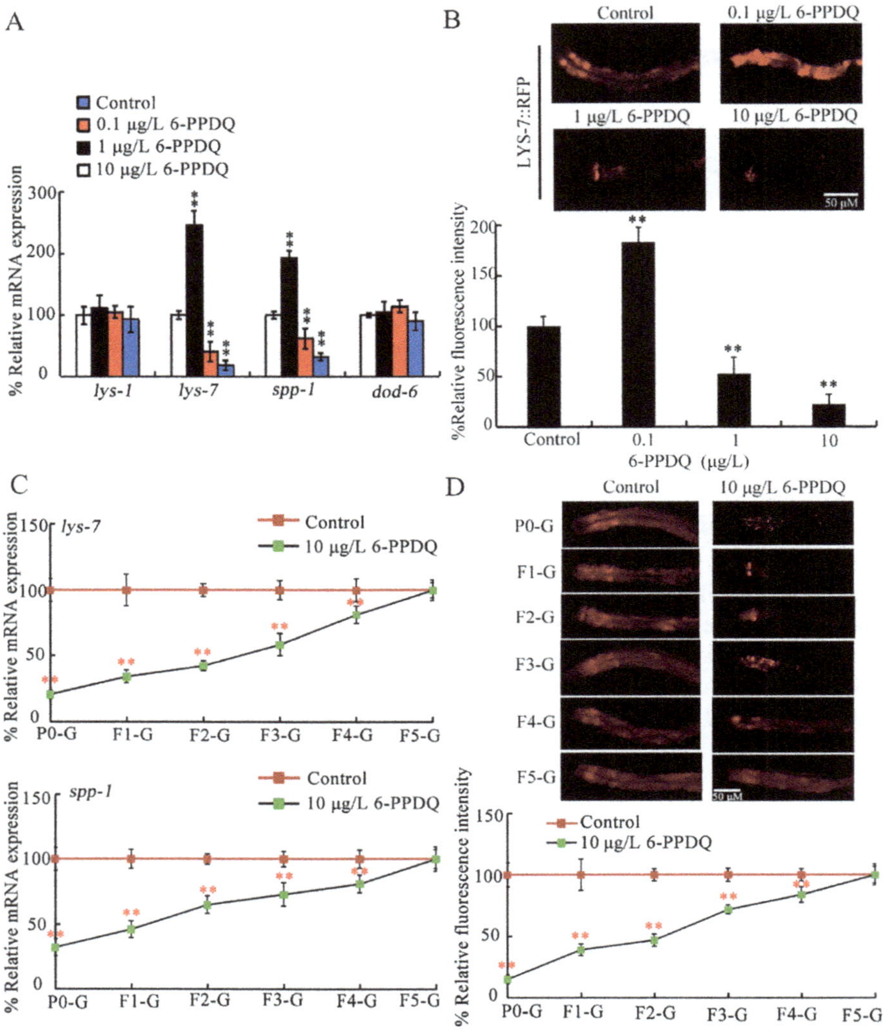

Fig. 16.8 Transgenerational innate immune response after 6-PPDQ exposure [58]. (**a**) Effect of 6-PPDQ exposure on expressions of *lys-1*, *lys-7*, *spp-1*, and *dod-6* at P0-G. (**b**) Effect of 6-PPDQ exposure on expression of LYS-7::RFP at P0-G. (**c**) Transgenerational expressions of *lys-7* and *spp-1* after exposure to 10 μg/L 6-PPDQ at P0-G. (**d**) Transgenerational expression of LYS-7::RFP after exposure to 10 μg/L 6-PPDQ at P0-G. $^{**}P < 0.01$ vs. control

expressions, whereas 1–10 μg/L 6-PPDQ inhibited innate immune response at the P0-G (Fig. 16.8) [58]. This implied that different concentrations of 6-PPDQ could induce different forms of transgenerational innate immune response. After exposure to 10 μg/L 6-PPDQ, transgenerational inhibition in innate immune response could be observed till to the F4-G (Fig. 16.8) [58].

16.5.2 Function of Intestinal Barrier–Related Signals in Regulating Transgenerational Innate Immune Response to Pollutants

Certain intestinal barrier–related signals regulated transgenerational innate immune response induced by pollutants. Accompanied with transgenerational intestinal oxidative stress and enhanced damage on intestinal barrier induced by 6-PPDQ, transgenerational inhibition in intestinal barrier–related signals (ACS-22, ERM-1, HMP-2, and PKC-3) was detected [58]. Moreover, the activated innate immune response by 0.1 μg/L 6-PPDQ could be suppressed by RNAi of *acs-22*, *erm-1*, *hmp-2*, and *pkc-3* (Fig. 16.9) [58], suggesting the involvement of these genes in modulating transgenerational innate immune response induced by pollutants.

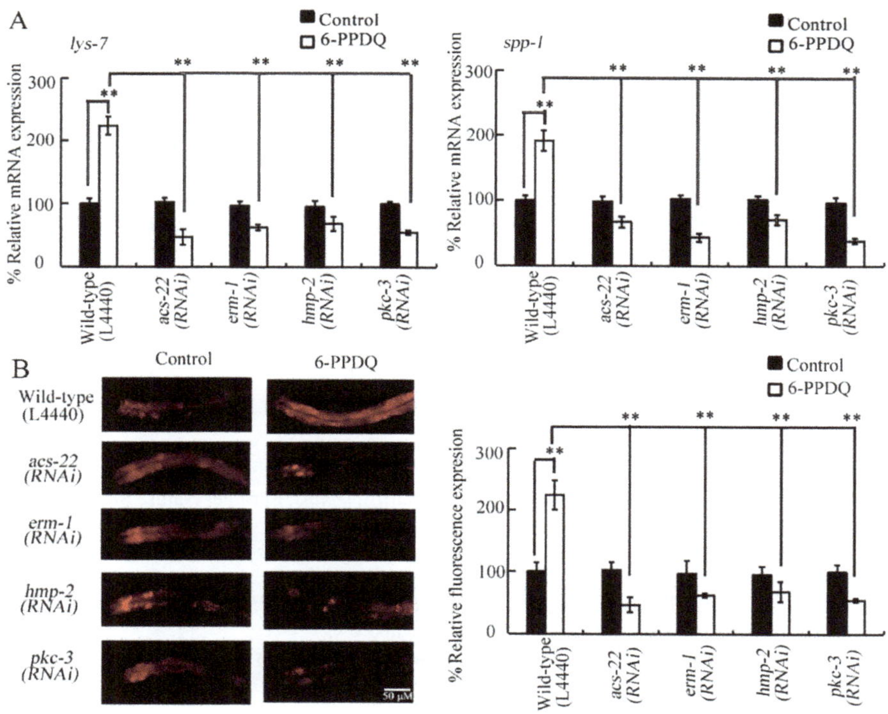

Fig. 16.9 Effect of RNAi of *acs-22*, *erm-1*, *hmp-2*, and *pkc-3* on innate immune response in 6-PPDQ-exposed nematodes [58]. (**a**) Effect of RNAi of *acs-22*, *erm-1*, *hmp-2*, and *pkc-3* on expressions of *lys-7* and *spp-1* in 6-PPDQ-exposed nematodes. (**b**) Effect of RNAi of *acs-22*, *erm-1*, *hmp-2*, and *pkc-3* on LYS-7::RFP expression in 6-PPDQ-exposed nematodes. Exposure concentration of 6-PPDQ was 0.1 μg/L. $^{**}P < 0.01$

16.6 Perspectives

First of all, besides the introduced transgenerational antioxidation response, mt UPR response, ER UPR response, and innate immune response to pollutants, exposure to stresses and pollutants potentially induce the other forms of protective responses, such as autophagy and mitophagy, protective response of different organelle, and certain biochemical metabolism responses. For example, exposure to 6-PPDQ could cause dynamic response of mitophagy [59].

In this chapter, we discussed the role of protective responses related signals as the "effectors" for induction of transgenerational toxicity of stresses and pollutants. Nevertheless, so far, the information for transgenerational response of protective responses–related signals is very limited. With transgenerational oxidation response as the example, only transgenerational SOD-3 response has been well described after pollutant exposure. The functions of these protective responses–related signals in regulating transgenerational pollutant toxicity need to be further thoroughly examined.

Considering the role of protective responses–related signals as the "effectors" for induction of transgenerational toxicity of stresses and pollutants, at least the following three questions need to be further answered:

1. How does signaling communication between germline ligands and their receptors regulate transgenerational response of protective responses–related signals to stresses and pollutants?
2. How do metabolic molecules and metabolic signaling regulate transgenerational response of protective responses related signals to stresses and pollutants?
3. How do germline epigenetic regulation–related signals regulate transgenerational response of protective responses–related signals to stresses and pollutants?

References

1. Wang D-Y (2018) Nanotoxicology in *Caenorhabditis elegans*. Springer Nature Singapore Pte Ltd., Singapore
2. Wang D-Y (2020) Exposure toxicology in *Caenorhabditis elegans*. Springer Nature Singapore Pte Ltd., Singapore
3. Wang D-Y (2022) Toxicology at environmentally relevant concentrations in *Caenorhabditis elegans*. Springer Nature Singapore Pte Ltd., Singapore
4. Wang D-Y (2019) Molecular toxicology in *Caenorhabditis elegans*. Springer Nature Singapore Pte Ltd., Singapore
5. Wang D-Y (2019) Target organ toxicology in *Caenorhabditis elegans*. Springer Nature Singapore Pte Ltd., Singapore
6. Hua X, Wang D-Y (2023) Exposure to 6-PPD quinone at environmentally relevant concentrations inhibits both lifespan and healthspan in *C. elegans*. Environ Sci Technol 57:19295–19303
7. Shao H-M, Han Z-Y, Krasteva N, Wang D-Y (2019) Identification of signaling cascade in the insulin signaling pathway in response to nanopolystyrene particles. Nanotoxicology 13:174–188

8. Kong Y, Liu H-L, Li W-J, Wang D-Y (2019) Intestine-specific activity of insulin signaling pathway in response to microgravity stress in *Caenorhabditis elegans*. Biochem Biophys Res Commun 517:278–284
9. Zhao Y-L, Yang R-L, Rui Q, Wang D-Y (2016) Intestinal insulin signaling encodes two different molecular mechanisms for the shortened longevity induced by graphene oxide in *Caenorhabditis elegans*. Sci Rep 6:24024
10. Liu H-L, Wang D-Y (2021) Intestinal mitochondrial unfolded protein response induced by nanoplastic particles in *Caenorhabditis elegans*. Chemosphere 267:128917
11. Hua X, Wang D-Y (2024) 6-PPD quinone at environmentally relevant concentrations induced damage on longevity in *C. elegans*: mechanistic insight from inhibition in mitochondrial UPR response. Sci Total Environ 954:176275
12. Wang Y-X, Hu G-C, Wang D-Y (2025) Increased S-adenosyl methionine strengthens the suppression in mitochondrial unfolded protein response induced by 6-PPD quinone at environmentally relevant concentrations in *Caenorhabditis elegans*. Environ Pollut 386:127231
13. Liu P-D, Li D, Li W-J, Wang D-Y (2019) Mitochondrial unfolded protein response to microgravity stress in nematode *Caenorhabditis elegans*. Sci Rep 9:16474
14. Qu M, Liu Y-Q, Xu K-N, Wang D-Y (2019) Activation of p38 MAPK signaling-mediated endoplasmic reticulum unfolded protein response by nanopolystyrene particles. Adv Biosyst 3:1800325
15. Yang Y, Shao H, Wu Q, Wang D (2020) Lipid metabolic response to polystyrene particles in nematode *Caenorhabditis elegans*. Environ Pollut 256:113439
16. Wu J-W, Shen S-H, Wang D-Y (2024) 6-PPD quinone at environmentally relevant concentrations induces immunosenescence by causing immunosuppression during the aging process. Chemosphere 368:143719
17. Ren M-X, Zhao L, Lv X, Wang D-Y (2017) Antimicrobial proteins in the response to graphene oxide in *Caenorhabditis elegans*. Nanotoxicology 11:578–590
18. Shakoor S, Sun L-M, Wang D-Y (2016) Multi-walled carbon nanotubes enhanced fungal colonization and suppressed innate immune response to fungal infection in nematodes. Toxicol Res 5:492–499
19. Wu Q, Zhao Y, Fang J, Wang D (2014) Immune response is required for the control of in vivo translocation and chronic toxicity of graphene oxide. Nanoscale 6:5894–5906
20. Hua X, Feng X, Liang G-Y, Chao J, Wang D-Y (2023) Long-term exposure to tire-derived 6-PPD quinone causes intestinal toxicity by affecting functional state of intestinal barrier in *Caenorhabditis elegans*. Sci Total Environ 861:160591
21. Hua X, Feng X, Liang G-Y, Chao J, Wang D-Y (2023) Exposure to 6-PPD quinone at environmentally relevant concentrations causes abnormal locomotion behaviors and neurodegeneration in *Caenorhabditis elegans*. Environ Sci Technol 57:4940–4950
22. Hua X, Feng X, Liang G-Y, Chao J, Wang D-Y (2023) Long-term exposure to 6-PPD quinone reduces reproductive capacity by enhancing germline apoptosis associated with activation of both DNA damage and cell corpse engulfment in *Caenorhabditis elegans*. J Hazard Mater 454:131495
23. Liu Z-Y, Bian Q, Wang D-Y (2024) Exposure to 6-PPD quinone causes ferroptosis activation associated with induction of reproductive toxicity in *Caenorhabditis elegans*. J Hazard Mater 471:134356
24. Hu D-Y, Wang Y-X, Hu G-C, Liu R, Wang D-Y (2025) 6-PPD quinone inhibited retinoic acid synthesis mediates toxicity through feedback loop between ALH-3/DHS-19-SEX-1 axis and intestinal signals in *Caenorhabditis elegans*. J Environ Exposure Assess 4:40
25. Wang W, Li Y-H, Wang D-Y (2025) Long-term exposure to 6-PPD quinone inhibits glutamate synthesis and glutamate receptor function associated with its toxicity induction in *Caenorhabditis elegans*. Toxics 13:434
26. Wang Y-X, Liang G-Y, Chao J, Wang D-Y (2024) Comparison of intestinal toxicity in enhancing intestinal permeability and in causing ROS production of six PPD quinones in *Caenorhabditis elegans*. Sci Total Environ 927:172306

27. Liu Z-J, Li Y-H, Wang D-Y (2024) Exposure to 6-PPD quinone disrupts glucose metabolism associated with lifespan reduction by affecting insulin and AMPK signals in *Caenorhabditis elegans*. Chemosphere 363:142975
28. Wang Y-X, Wang D-Y (2024) Exposure to 6-PPD quinone enhances glycogen accumulation in *Caenorhabditis elegans*. Environ Pollut 359:124600
29. Wan X, Liang G-Y, Wang D-Y (2025) 6-PPD quinone at environmentally relevant concentrations disrupts citric acid cycle in *Caenorhabditis elegans*: Role of reduction in acetyl CoA and pyruvate contents. Environ Chem Ecotoxicol 7:1119–1129
30. Hua X, Liang G-Y, Chao J, Wang D-Y (2024) Exposure to 6-PPD quinone causes damage on mitochondrial complex I/II associated with lifespan reduction in *Caenorhabditis elegans*. J Hazard Mater 472:134598
31. Hua X, Wang D-Y (2025) 6-PPD quinone causes alteration in ubiquinone-mediated complex III associated with toxicity on mitochondrial function and longevity in *Caenorhabditis elegans*. J Environ Chem Eng 13:116571
32. Hua X, Wang D-Y (2025) 6-PPD quinone at environmentally relevant concentrations activates feedback response of electron transport chain to mediate damage on mitochondrial function and longevity in *Caenorhabditis elegans*. Environ Chem Ecotoxicol 7:2356–2365
33. Liu Z-J, Li Y-H, Wang D-Y (2024) Transgenerational response of glucose metabolism in *Caenorhabditis elegans* exposed to 6-PPD quinone. Chemosphere 367:143653
34. Chen H, Hua X, Li H, Wang C, Dang Y, Ding P, Yu Y (2021) Transgenerational neurotoxicity of polystyrene microplastics induced by oxidative stress in *Caenorhabditis elegans*. Chemosphere 272:129642
35. Hua X, Wang D-Y (2022) Cellular uptake, transport, and organelle response after exposure to microplastics and nanoplastics: current knowledge and perspectives for environmental and health risks. Rev Environ Contam Toxicol 260:12
36. Qu M, Kong Y, Yuan Y-J, Wang D-Y (2019) Neuronal damage induced by nanopolystyrene particles in nematode *Caenorhabditis elegans*. Environ Sci Nano 6:2591–2601
37. Qu M, Qiu Y-X, Kong Y, Wang D-Y (2019) Amino modification enhances reproductive toxicity of nanopolystyrene on gonad development and reproductive capacity in nematode *Caenorhabditis elegans*. Environ Pollut 254:112978
38. Qu M, Nida A, Kong Y, Du H-H, Xiao G-S, Wang D-Y (2019) Nanopolystyrene at predicted environmental concentration enhances microcystin-LR toxicity by inducing intestinal damage in *Caenorhabditis elegans*. Ecotoxicol Environ Saf 183:109568
39. Qu M, Xu K-N, Li Y-H, Wong G, Wang D-Y (2018) Using *acs-22* mutant *Caenorhabditis elegans* to detect the toxicity of nanopolystyrene particles. Sci Total Environ 643:119–126
40. Qiu Y-X, Liu Y-Q, Li Y-H, Li G-J, Wang D-Y (2020) Effect of chronic exposure to nanopolystyrene on nematode *Caenorhabditis elegans*. Chemosphere 256:127172
41. Qiu Y-X, Luo L-B, Yang Y-H, Kong Y, Li Y-H, Wang D-Y (2020) Potential toxicity of nanopolystyrene on lifespan and aging process of nematode *Caenorhabditis elegans*. Sci Total Environ 705:135918
42. Li D, Deng Y-J, Wang S-T, Du H-H, Xiao G-S, Wang D-Y (2020) Assessment of nanopolystyrene toxicity under fungal infection condition in *Caenorhabditis elegans*. Ecotoxicol Environ Saf 197:110625
43. Liu H-L, Shao H-M, Guo Z-J, Wang D-Y (2020) Nanopolystyrene exposure activates a fat metabolism related signaling-mediated protective response in *Caenorhabditis elegans*. NanoImpact 17:100204
44. Dong S-S, Qu M, Rui Q, Wang D-Y (2018) Combinational effect of titanium dioxide nanoparticles and nanopolystyrene particles at environmentally relevant concentrations on nematodes *Caenorhabditis elegans*. Ecotoxicol Environ Saf 161:444–450
45. Zhao Y-Y, Li D, Rui Q, Wang D-Y (2020) Toxicity induction of nanopolystyrene under microgravity stress condition in *Caenorhabditis elegans*. Sci Total Environ 703:135623

46. Tang M, Ding G, Lu X, Huang Q, Du H, Xiao G, Wang D (2022) Exposure to nanoplastic particles enhances *Acinetobacter* survival, biofilm formation, and serum resistance. Nanomaterials 12:4222
47. Zhao L, Qu M, Wong G, Wang D-Y (2017) Transgenerational toxicity of nanopolystyrene particles in the range of μg/L in nematode *Caenorhabditis elegans*. Environ Sci Nano 4:2356–2366
48. Sun L-M, Liao K, Wang D-Y (2021) Comparison of transgenerational reproductive toxicity induced by pristine and amino modified nanoplastics in *Caenorhabditis elegans*. Sci Total Environ 768:144362
49. Zhao Y, Chen J, Wang R, Pu X, Wang D (2023) A review of transgenerational and multigenerational toxicology in the *in vivo* model animal *Caenorhabditis elegans*. J Appl Toxicol 43:122–145
50. Liu H-L, Tian L-J, Wang S-T, Wang D-Y (2021) Size-dependent transgenerational toxicity induced by nanoplastics in nematode *Caenorhabditis elegans*. Sci Total Environ 790:148217
51. Zhang L, Wang S-T, Zhao Y-L, Nurdebek B, Bu Y-Q, Wang D-Y (2022) Long-term exposure to polystyrene nanoparticles causes transgenerational toxicity by affecting the function and expression of MEV-1 and DAF-2 signals in *Caenorhabditis elegans*. NanoImpact 26:100403
52. Liu H-L, Zhao Y-L, Hua X, Wang D-Y (2022) Induction of transgenerational toxicity is associated with the activated germline insulin signals in nematodes exposed to nanoplastic at predicted environmental concentrations. Ecotoxicol Environ Saf 243:114022
53. Xu R-R, Hua X, Rui Q, Wang D-Y (2022) Polystyrene nanoparticles caused dynamic alteration in mitochondrial unfolded protein response from parents to the offspring in *C. elegans*. Chemosphere 308:136154
54. Nargund AM, Pellegrino MW, Fiorese CJ, Baker BM, Haynes CM (2012) Mitochondrial import efficiency of ATFS-1 regulates mitochondrial UPR activation. Science 337:587–590
55. Tian Y, Garcia G, Bian Q, Steffen KK, Joe L, Wolff S, Meyer BJ, Dillin A (2016) Mitochondrial stress induces chromatin reorganization to promote longevity and UPR(mt). Cell 165:1197–1208
56. Haynes CM, Petrova K, Benedetti C, Yang Y, Ron D (2007) ClpP mediates activation of a mitochondrial unfolded protein response in *C. elegans*. Dev Cell 13:467–480
57. Xu R-R, Hua X, Rui Q, Wang D-Y (2022) Alteration in Wnt signaling mediates induction of transgenerational toxicity of polystyrene nanoplastics in *C. elegans*. NanoImpact 28:100425
58. Wang Y-X, Wang D-Y (2024) Transgenerational intestinal toxicity of 6-PPD quinone in causing ROS production, enhancement in intestinal permeability and suppression in innate immunity in *C. elegans*. Environ Pollut 363:125208
59. Hua X, Wang D-Y (2025) Environmentally relevant concentration of 6-PPD quinone inhibits two types of mitophagy to cause mitochondrial dysfunction and lifespan reduction in *Caenorhabditis elegans*. Environ Sci Process Impacts 27:1928–1940

Chapter 17
Basic Conclusions for Transgenerational Toxicology in *Caenorhabditis elegans*

Abstract Based on the obtained data and clues in *Caenorhabditis elegans*, we raised eight basic conclusions for the transgenerational toxicology. Moreover, with the focus on different exposure scenarios, the future research on at least pollutant toxicology across multiple generations was further discussed.

Keywords Transgenerational toxicology · Basic conclusions · *Caenorhabditis elegans*

Based on the obtained data and clues from *Caenorhabditis elegans*, we raise the following basic conclusions for the transgenerational toxicology:

1. *Transgenerational molecular alterations are associated with transgenerational toxicity of pollutants at multiple aspects*

The transgenerational molecular alterations associated with transgenerational toxicity of pollutants include transgenerational gene expression plasticity, alteration in epigenetic-related signals and germline ligand genes, and alteration in protective molecular responses. Exposure to pollutants at high concentrations or with more severe toxicity may also induce transgenerational genome instability, which is possibly linked to induction of transgenerational toxicity.

2. *Transgenerational translocation usually provides limit contribution to transgenerational toxicity of pollutants*

Usually, transgenerational translocation of both particle pollutants and organic pollutants provides limit contribution to their transgenerational toxicity induction. Transgenerational translocation of pollutants is partially associated with transgenerational damage on intestinal barrier function. However, the effect of transgenerational damage on intestinal barrier function is not limited to the influence on transgenerational translocation of pollutants.

3. *Germline secreted ligands function together with their receptors in the offspring to regulate the induction of transgenerational toxicity of pollutants*

D. Wang, *Transgenerational Toxicology in Caenorhabditis elegans*,
https://doi.org/10.1007/978-981-92-1389-4_17

Different germline secreted ligands can act together with their receptors in the offspring to regulate the transgenerational toxicity of pollutants. At least two forms of transgenerational communication between germline secreted ligands and their receptors to regulate transgenerational toxicity of pollutants are suggested. One form is the sequential transgenerational communication between the same germline secreted ligands and their receptors. Another form is the sequential transgenerational communication between different germline secreted ligands. The molecular network for transgenerational communication between germline secreted ligands and their receptors in regulating transgenerational toxicity of pollutants is shown in Fig. 17.1.

4. *Metabolic molecules and metabolic signaling participate in the control of transgenerational toxicity of pollutants*

On the one hand, exposure to pollutants can induce transgenerational disruption in certain biochemical metabolisms. On the other hand, certain metabolic molecules and metabolic signaling are involved in controlling transgenerational toxicity of pollutants. During this regulation process, some metabolic molecules may be considered as signal molecules.

5. *Germline epigenetic regulation–related signals are required for transgenerational toxicity induction of pollutants*

The first group of germline epigenetic regulation–related signals are small RNAs (such as microRNAs and long non-coding RNAs) and RNAi-associated signaling. The second group of germline epigenetic regulation–related signals are histone methylation, histone acetylation, and DNA methylation–related signals. In the germline, these two groups of epigenetic regulation–related signals regulate

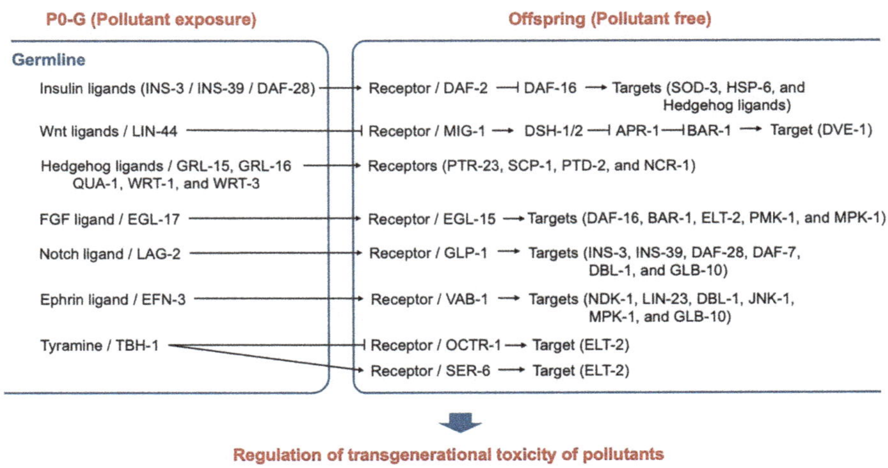

Fig. 17.1 Molecular network for transgenerational communication between germline secreted ligands and their receptors in regulating transgenerational toxicity of pollutants

transgenerational toxicity of pollutants by controlling expression of their targeted genes. The molecular network for the role of germline epigenetic regulation–related signals in regulating transgenerational toxicity of pollutants is shown in Fig. 17.2.

6. *Germline secreted ligands and their receptors, metabolic molecules and metabolic signaling, and germline epigenetic regulation–related signals may organize a complex network to govern transgenerational toxicity of pollutants*

The first possibility is that, in the offspring, the receptors of germline secreted ligands and/or their downstream transcriptional factors regulate transgenerational toxicity of pollutants by controlling metabolic molecules and metabolic signaling and germline epigenetic regulation–related signals. The second possibility is that metabolic molecules and metabolic signaling regulate transgenerational toxicity of pollutants by affecting germline secreted ligands and germline epigenetic regulation–related signals. The third possibility is that germline epigenetic regulation–related signals regulate transgenerational toxicity of pollutants by regulating expression of germline secreted ligand genes and metabolic signaling genes. Are there any other possible mechanisms that may regulate the transgenerational toxicity of pollutants? The answer is very likely YES.

7. *Germline G protein–coupled receptors and nuclear hormone receptors can be considered as the "drivers" for transgenerational toxicity induction of pollutants*

Germline G protein–coupled receptors (GPCRs) can be considered as the "drivers" for transgenerational toxicity induction of water-soluble pollutants and particle pollutants by regulating the downstream germline secreted ligands, metabolic signaling, and germline epigenetic regulation–related signals. Germline nuclear hormone receptors (NHRs) can be considered as the "drivers" for transgenerational toxicity induction of fat-soluble pollutants and particle pollutants with very small sizes by regulating the downstream germline secreted ligands, metabolic signaling, and germline epigenetic regulation–related signals.

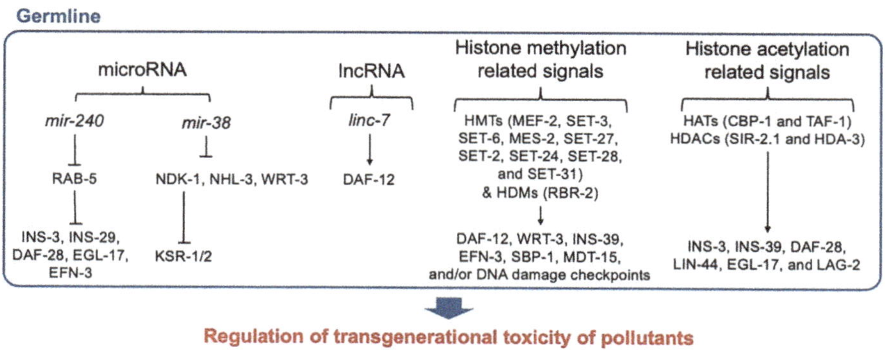

Fig. 17.2 Molecular network for the role of germline epigenetic regulation–related signals in regulating transgenerational toxicity of pollutants

8. *Protective responses–related signals can be considered as the "effectors" for transgenerational toxicity induction of pollutants*

Accompanied with transgenerational responses, some protective responses–related signals (such as related signals for antioxidation, mitochondrial unfolded protein response, and innate immune response) participate in controlling transgenerational toxicity of pollutants. The protective responses–related signals can be considered as the "effectors" for transgenerational toxicity induction of pollutants under the control of germline secreted ligands and their receptors, metabolic molecules and metabolic signaling, and germline epigenetic regulation–related signals.

What we described exposure scenarios here is that the parental hermaphrodite nematodes are exposed to stresses and pollutants, while their offspring (F1-Fn) will be cultured under the normal condition. Besides this exposure scenario, transgenerational toxicity of pollutants can also be analyzed with the aid of genetic crossing [1, 2], which will be helpful to confirm the existence of transgenerational communication and to reflect transgenerational toxicity occurred in heterosexual animals. In addition, continuous multi-generation exposure to pollutants can also be perform to mimic the possible real scenario of pollutants exposure for some environmental organisms [3].

References

1. Zhang Q, Wang Z, Zhang W, Wen Q, Li X, Zhou J, Wu X, Guo Y, Liu Y, Wei C, Qian W, Tian Y (2021) The memory of neuronal mitochondrial stress is inherited transgenerationally via elevated mitochondrial DNA levels. Nat Cell Biol 23:870–880
2. Wang S, Meyer DH, Schumacher B (2023) Inheritance of paternal DNA damage by histone-mediated repair restriction. Nature 613:365–274
3. Zhao Y, Chen J, Wang R, Pu X, Wang D (2023) A review of transgenerational and multigenerational toxicology in the in vivo model animal *Caenorhabditis elegans*. J Appl Toxicol 43:122–145

www.ingramcontent.com/pod-product-compliance
Ingram Content Group UK Ltd.
Pitfield, Milton Keynes, MK11 3LW, UK
UKHW020632260926
475829UK00003BA/41
* 9 7 8 9 8 1 9 2 1 3 8 8 7 *